abelian groups

LECTURE NOTES IN PURE AND APPLIED MATHEMATICS

Additional Volumes in Preparation

abelian groups

proceedings of the 1991 Curaçao conference

edited by

Laszlo Fuchs
Tulane University
New Orleans, Louisiana

Rüdiger Göbel
University of Essen
Essen, Germany

Marcel Dekker, Inc.   **New York • Basel • Hong Kong**

Library of Congress Cataloging-in-Publication Data

Abelian groups : proceedings of the 1991 Curaçao conference / edited
by Laszlo Fuchs, Rüdiger Göbel.
    p. cm. -- (Lecture notes in pure and applied mathematics : v.
146)
  Includes bibliographical references and index.
  ISBN 0-8247-8901-6 (alk. paper)
  1. Abelian groups--Congresses.   I. Fuchs, László.   II. Göbel R.
(Rüdiger).   III. Series.
QA180.A25   1993
512'.2--dc20
                                93-18116
                                   CIP

The publisher offers discounts on this book when ordered in bulk quantities. For more information, write to Special Sales/Professional Marketing at the address below.

This book is printed on acid-free paper.

Marcel Dekker, Inc.
270 Madison Avenue, New York, New York 10016

Current printing (last digit):
10 9 8 7 6 5 4 3 2 1

PRINTED IN THE UNITED STATES OF AMERICA

Preface

The Caribbean Mathematics Foundation, located at the University of Florida, Gainesville, Florida, and the University of the Netherlands Antilles, Willemstad, Curacao, generously consented to sponsor an International Conference on Torsion-free Abelian Groups in Curacao. Abelian group theorists enthusiastically welcomed this unique opportunity, because in view of the rapid developments in abelian groups, and, in particular, in torsion-free abelian groups, regular conferences are most stimulating. Experts from all over the world were invited to this meeting on the Caribbean island and the participants reported on the most recent developments in their own research.

The conference on abelian groups was preceded by a "mini conference" on Set-Theoretic Methods in Abelian Group Theory providing an opportunity to study interactions of set theory and almost-free groups (Menachem Magidor), abelianized automorphism groups (Manfred Droste), as well as interactions between set theory and radicals (Laszlo Fuchs), and Butler groups (Rüdiger Göbel).

A welcome feature of the main conference was a kind of Oberwolfach atmosphere—carried over from the mountains to the seashore: picturesque surroundings with a sandy beach by the magnificent ocean replacing Oberwolfach's gorgeous fir trees, and uncountably many corals. Additional incentive for discussion and steps into new areas of research was provided by the isolated site of the hotel complex, from which there was essentially no chance to escape.

The papers presented in this Proceedings reflect a portion of the activity on the island during the conference. They include an article about Friedrich Levi, one of the forefathers of abelian group theory, surveys by D. Arnold and C. Vinsonhaler (on finite rank Butler groups), by P. Eklof (on set-theoretical methods), by R. Göbel (on endomorphism algebras), by P. Hill (on infinite rank torsion-free groups), and research papers carefully refereed according to the standards of major international journals. It is our hope that this Proceedings will stimulate interest and research in abelian groups, connecting this branch of algebra in a unique way with recent progress in logic, model theory, infinite combinatorics, and axiomatic set theory. All the articles are in final form and no version of them will be published elsewhere.

It is our privilege to extend our sincere thanks to all who contributed to the success of the conference and to the publication of this Proceedings. We gratefully acknowledge generous financial help from the University of the Netherlands Antilles, the Government of Curaçao, and the University of Florida—their support made the Conference possible. We also thank the Banco di Caribe and Maduro & Curiel's Bank for their financial support. Above all, thanks are due to the director of the Caribbean Mathematics Foundation, Professor Jorge Martinez of University of Florida, for his initiative and for his continuing help in running a very successful conference.

We wish to thank Barbara Levi for providing us a photograph of her late husband, Friedrich Levi, taken at the time of his retirement from Freie Universität Berlin.

Laszlo Fuchs
Rüdiger Göbel

Contents

Contributors

D. ARNOLD Baylor University, Waco, Texas

CLAUDIA BÖTTINGER University of Essen, Essen, Germany

H. H. BRUNGS University of Alberta, Edmonton, Alberta, Canada

MANFRED DUGAS Baylor University, Waco, Texas

PAUL C. EKLOF University of California, Irvine, California

THEODORE G. FATICONI Fordham University, Bronx, New York

L. FUCHS Tulane University, New Orleans, Louisiana

ANTHONY J. GIOVANNITTI University of Southern Mississippi, Hattiesburg, Mississippi

RÜDIGER GÖBEL University of Essen, Essen, Germany

H. PAT GOETERS Auburn University, Auburn University, Alabama

JUTTA HAUSEN University of Houston, Houston, Texas

PAUL HILL Auburn University, Auburn University, Alabama

ADOLF MADER University of Hawaii, Honolulu, Hawaii

WARREN MAY University of Arizona, Tucson, Arizona

CHARLES MEGIBBEN Vanderbilt University, Nashville, Tennessee

C. METELLI University of Naples, Naples, Italy

OTTO MUTZBAUER University of Würzburg, Würzburg, Germany

ED OXFORD Baylor University, Waco, Texas

R. S. PIERCE University of Arizona, Tucson, Arizona

K. M. RANGASWAMY University of Colorado, Colorado Springs, Colorado

J. D. REID Wesleyan University, Middletown, Connecticut

LUIGI SALCE University of Padova, Padova, Italy

WILLIAM ULLERY Auburn University, Auburn University, Alabama

C. VINSONHALER · University of Connecticut, Storrs, Connecticut

W. WICKLESS University of Connecticut, Storrs, Connecticut

Conference Participants

H. R. Antonius, Surinam
David M. Arnold, Baylor University, Waco,TX
Khalid Benabdallah, Université de Montréal, Montréal, Canada
Claudia Böttinger, Universität Essen, Essen, Germany
Hans Brungs, University of Alberta, Edmonton, Canada
Manfred Droste, Universität Essen, Essen, Germany
Manfred Dugas, Baylor University, Waco, TX
Paul Eklof, University of California, Irvine, CA
Theodore G. Faticoni, Fordham University, Bronx, NY
Temple H. Fay, University of Southern Mississippi, Hattiesburg, MS
Laszlo Fuchs, Tulane University, New Orleans, LA
Anthony J. Giovannitti, University of Southern Mississippi, Hattiesburg, MS
Brendan Goldsmith, Dublin Institute of Technology, Dublin, Ireland
C. W. Gorisson, Surinam
Rüdiger Göbel, Universität Essen, Essen, Germany
Alfred Hales, University of California, Los Angeles, CA
Jutta Hausen, University of Houston, Houston, TX
Paul Hill, Auburn University, Auburn,AL
Kin-ya Honda, St. Paul's University, Tokyo, Japan
John Irwin, Wayne State University, Detroit, MI
Patrick Keef, Whitman College, Walla Walla, WA
Toshiko Koyama, Ochanomizu University, Japan
Wolfgang Liebert, Universität München, München, Germany
Adolf Mader, University of Hawaii, Honolulu, HI
Menachem Magidor, Hebrew University, Jerusalem, Israel
Jorge Martinez, University of Florida, Gainesville, FL
Warren May, University of Arizona, Tucson, AZ
Claudia Metelli, Università di Napoli, Napoli, Italy
Otto Mutzbauer, Universität Würzburg, Würzburg, Germany
John O'Neill, University of Detroit, Detroit, MI
Takashi Okuyama, Toba National College of Maritime Technology, Ikegami-cho, Japan
K.M. Rangaswamy, University of Colorado, Colorado Springs, CO

Fred Richman, Florida Atlantic University, Boca Raton, FL
Luigi Salce, Università di Padova, Padova, Italy
Gregory Schlitt, McMaster University, Hamilton, Canada
Noriko Tone, Tokyo Denki University, Tokyo, Japan
William Ullery, Auburn University, Auburn, AL
Charles I. Vinsonhaler, University of Connecticut, Storrs, CT
William Wickless, University of Connecticut, Storrs, CT
Birge Zimmermann-Huisgen, University of Utah, Salt Lake City, UT

abelian groups

Friedrich Wilhelm Levi
1888–1966

Friedrich Wilhelm Levi, 1888–1966

L. Fuchs

Department of Mathematics, Tulane University, New Orleans, LA 70118

R. Göbel

Fachbereich Mathematik und informatik, Universität Essen
Essen, Germany

Friedrich Wilhelm Daniel Levi, son of Georg Levi and Emma Blum, was born on February 6, 1888 in Mülhausen (Elsass). His father was a prominent lawyer, who later became chairman of the board of judges.

Levi went to elementary school in Strassburg 1894-96, continued his schooling in a humanistic gymnasium in Strassburg 1896-1904, and in the lyceum in Colmar (Elsass) 1904-1906. After graduating from high-school, he served in the military for a year. Simultaneously, he was enrolled as mathematics student at the University of Würzburg, 1906-1907. After completing his military service, he returned to his home town to study mathematics and physics at the University of Strassburg (1907-1911). His professors were F. Schur, R. von Mises, A. Speiser, Timerding and H. Weber.

He received his Ph.D. in 1911 in Strassburg. His thesis adviser was Heinrich Weber, the author of the well-known two-volume "Lehrbuch der Algebra", who introduced him to number theory and algebra, and had a great deal of influence on him. Even after his Ph.D. he continued to attend lectures by D. Hilbert, E. Landau, E. Madelung, O. Toeplitz, H. Weyl, and O. Hölder, P. Koebe at the universities of Göttingen and Leipzig, respectively, until 1914.

In July 1914 (just when World War I broke out), he submitted his Habilitationsschrift (for venia legendi) at University of Leipzig, but was granted the Privatdozent status only in 1919, after serving in the front 1914-1918. He was decorated several times (including Iron Cross II), and was discharged as second lieutenant. From 1920 on, he taught at University of Leipzig as Privatdozent; in 1923 he became extraordinarius without regular salary, and a year later he was promoted to professorship. In April 1935 the nazi government fired him.

1

In September 1935, he was offered the Hardinge chair of higher mathematics at the University of Calcutta, India. Accepting this offer, in January 1936 he moved with his family to Calcutta where he was appointed as the head of the Mathematics Department. He modernized the mathematics curricula and was very active in promoting mathematics at Indian universities. In 1948 he became professor of mathematics at the prestigious Tata Institute of Fundamental Research in Bombay. He held this position until 1952 when he returned to his native country.

Between 1952 and 1956 he was professor of mathematics at the Free University of Berlin. In 1956 he became emeritus. Two years later he moved to Freiburg where he lectured at the University of Freiburg until 1961, first as a visiting professor, later as an honorary professor.

Levi's mathematical interest was manyfold. Undoubtedly, his main interest lay in algebra, and it is this field where he made the most lasting contributions. But he also published several papers of interest on a variety of problems in geometry, topology, analysis and set theory. He promoted mathematics also through his lectures and his organizational work.

He gave talks on his research at a number of universities in India, England, Holland, and Germany. His talks for mathematicians as well as for students of mathematics were very inspiring. He is generally viewed as the father of algebra in India.

In 1917 he married Barbara Fitting; they had three children (Paul, Charlotte and Susanne) and eight (or more) grandchildren. He passed away January 1, 1966, shortly before his 78th birthday, in Freiburg. His widow lives in London.

Levi will be remembered not only for his mathematical contributions, but also as a very kind and charming person.

LIST OF PUBLICATIONS

1. *Körper und Integritätsbereiche 3. Grades*, Ph. D. Thesis (Strassburg,1911).
2. Arithmetische Gesetze im Gebiete diskreter Gruppen, Rend. Circ. Mat. Palermo 35 (1912), 225-236.
3. Kubische Zahlkörper und binäre kubische Formenklassen, Berichte Sächs. Akad. Wiss. 66 (1914).
4. *Abelsche Gruppen mit abzählbaren Elementen*, Habilitationsschrift (Leipzig, 1917).
5. Über die gegenseitige Lage von 5 und 6 Punkten in der projektiven Ebene, Berichte Sächs. Akad. Wiss. 74 (1922).
6. Beiträge zu einer Analysis der stetigen periodischen Kurven, Berichte Sächs. Akad. Wiss. 75 (1923).
7. Über stetige periodische Kurven, Berichte Sächs. Akad. Wiss. 75 (1923).
8. Einige topologische Anzahlbestimmungen, Christiaan Huyghens 8 (1923).
9. Streckenkomplexe auf Flächen, Math. Z. 16 (1923), 148-158.
10. Streckenkomplexe auf Flächen. II, Math. Z. 22 (1925), 45-61.
11. Die Singularitäten der Kurven in beliebigen affin zusammenhängenden Räumen, Berichte Sächs. Akad. Wiss. 77 (1925).
12. Die Teilung der projektiven Ebene durch Gerade oder Pseudogerade, Berichte Sächs. Akad. Wiss. 78 (1926).
13. Über repartitive Mengeneigenschaften, J. reine ang. Math. 161 (1929), 101-106.

and prime ideals. There are lots of special results depending on the number theoretic nature of the relevant constants.

In paper [3] he investigates the relation between cubic number fields Ω and classes of binary cubic forms $f(x, y)$ in order to overcome the difficulty in obtaining information about the factorization of rational primes in cubic fields where the discriminants have common unessential divisors. It is proved that the factorization of the principal ideal (p) (p a rational prime number) corresponds to the decomposition of the related binary cubic form into irreducible factors mod p. Using this correspondence, Levi derives necessary and sufficient conditions for the existence of common unessential divisors of the discriminants. This allows him to give, for the cubic case, elementary proofs of theorems by Dedekind and Minkowski which were proved by sophisticated ideal-theoretic and analytic methods, respectively. It is also shown that (p) is divisible by the square of a prime ideal if and only if the discriminant of the field is divisible by p.

2. Abelian Groups

Levi has made a major contribution to the theory of Abelian groups in his Habilitationsschrift [4]. This was submitted to the University of Leipzig in 1914, but was published only in 1917 when he was serving on the front in World War I.

It is most regrettable that this pivotal work was printed as a separate piece of publication, and not published in a major journal. Its influence would have been considerably greater and the development of Abelian group theory would have been probably faster if it were accessible to those interested in Abelian group theory. As a matter of fact, he discovered several theorems which we erroneously attribute to others, due to our ignorance of the content of this paper. Hopefully, in the future credit will be given where it belongs.

The paper contains important results on torsion groups, on torsion-free groups, on mixed groups, as well as on automorphism groups. In most theorems he restricts himself to countable groups, but points out that in several cases transfinite induction can be used to extend the results to groups of higher cardinalities. We must admit that the proofs are not always as rigorous and clear as we would demand nowadays, but they definitely display an amazingly deep insight into the crux of the problems.

For countable <u>torsion</u> groups, he proves the important results:

(a) the union of divisible subgroups is a divisible subgroup;

(b) a divisible group is a direct sum of groups of type p - these are nothing else than groups what we now call groups of type p^∞.

Satz 3 states that divisible subgroups are always direct summands; unfortunately, the proof is wrong.

In Satz 4 Levi goes too far when he claims that every countable torsion group is the direct sum of cyclic groups and groups of type p (in the review of this paper in *Fortschritte der Mathematik*, the predecessor of *Zentralblatt für Mathematik*, this mistake was already pointed out, with reference to Prüfer's work). What he should have stated is that a countable torsion group has this structure whenever every element of infinite height is contained in a divisible subgroup. The proof for the reduced part is, however, unconvincing.

He discovers several important features of <u>torsion-free</u> groups. Concentrating on the pure subgroups generated by one or a finite set of elements, he represents the elements in these subgroups geometrically by points with rational coordinates on the line or in a finite dimensional euclidean space.

(a) The 'characteristic' of an element is introduced, pointing out the distinction between divisibility by finitely or infinitely many powers of a prime p. After defining the equivalence of characteristics, he notices that the equivalence classes of characteristics are in a bijective correspondence with the isomorphy classes of rank 1 groups. (This result is usually attributed to R. Baer [B3].)

(b) He discovers the connection between torsion-free groups and p-adic integers. By making use of this connection, he introduces 'multidimensional characteristics'. Though these are rather complicated objects which have been ignored so far, by making use of non-periodic p-adic integers, he manages to construct an indecomposable torsion-free group of rank 2, moreover, indecomposable groups of any finite rank. And this happened 20 years before Pontryagin [P] !

(c) What we call today the *inner type* of a finite rank torsion-free group A was actually introduced by Levi. He calls it 'Hauptcharakteristik' and proves that the intersection of types of elements in a maximal independent system of A is an invariant of A.

(d) The notion of quasi-isomorphism is actually due to him. He calls two finite rank groups A and B 'isolog' if A is a finite extension of a group isomorphic to B. He proves that if $A = B_1 \oplus C_1$ and $A = B_2 \oplus C_2$ where $B_1 \cap C_2 = 0$ and $B_2 \cap C_1 = 0$, then B_1 and B_2 as well as C_1 and C_2 are isolog.

(e) He notices the failure of the Krull-Schmidt property for finite rank torsion-free groups (40 years before B. Jónsson [J]). The following example is given. Let $\{a, b, c\}$ be a basis of a $\mathbf{Q}$-vectorspace V. The group A is the subgroup of V generated by $a5^{-n}, b5^{-n}, c7^{-n}$ for all integers $n > 0$, and by $(b+c)/3$. Putting $a' = a+b$, $b' = 3a+2b$, we have

$$A = \langle a5^{-\infty} \rangle \oplus \langle b5^{-\infty}, c7^{-\infty}, (b+c)/3 \rangle = \langle a'5^{-\infty} \rangle \oplus \langle b'5^{-\infty}, c7^{-\infty}, (b'+2c)/3 \rangle.$$

In these two direct decompositions of A, the first summands are isomorphic, but the second ones are not.

For <u>mixed</u> groups, his main concern is the splitting property. In this paper a splitting mixed group is called a *Séguierian group* ; de Séguier erroneously claims in [Se] on p. 97 that "un groupe abélien G qui contient des éléments d'ordre infini est évidemment le produit direct du groupe formé de ses éléments d'ordre fini et du groupe formé de ses éléments d'ordre infini."

His main results on mixed groups are:

(a) In a mixed group G with torsion part T, there is a neat subgroup N such that $T \oplus N$ is essential in G.

(b) A criterion for splitting is given and applied to conclude that mixed groups with finite torsion subgroup are splitting.

(c) This criterion is used to verify the existence of non-splitting mixed groups. The counterexample is an extension of a direct sum of cyclic p-groups by the group of rationals with denominators powers of p. (There is a misprint which invalidates the example; on p. 30, the relation $px_i = 0$ should read $p^i x_i = 0$.)

The second part of paper [4] is devoted to automorphisms as well as to isomorphisms between subgroups.

Levi's earlier paper [2] is devoted to an extension of Dedekind's theory of integrity to group theory. Although a non-commutative extension is included, the main results are concerned with torsion-free abelian groups. Levi gives the definition of (torsion-free) rank and shows that this in an invariant. Moreover he establishes the existence of $2^{\aleph_0}$ non-isomorphic torsion-free abelian groups of rank 1. "Primitive systems" are used for calculations in finite rank torsion-free abelian groups.

In a later paper, [53], the ring of those endomorphisms of an abelian group A is investigated which carry every subgroup of A into itself. It is shown that this ring is $\cong \mathbf{Z}$ if A contains elements of infinite order; otherwise, it is an epimorphic image of the direct product of the p-adic integers (for different primes p).

3. Non-Commutative Group Theory

Levi's papers on non-commutative groups are devoted to the foundation of group theory, combinatorial methods with special emphasis on free constructions, and interactions between group theory, graph theory and ordered structures. Levi is one of the pioneers of a flourishing branch of group theory, known as combinatorial group theory, [MKS], [LS].

His papers [19] and [67] deal with the foundation of group theory. The question about irreducible systems of group axioms was investigated jointly with Baer [19]. Its results are discussed at length in Kurosh's text book [Ku, Vol.1 pp. 16–25]. In [67] the author considers uniform systems of axioms for groups and for alternative systems, based on a relational language.

A series of papers [16, 21, 23, 29, 30, 32, 36, 39] deal with free products of groups and their subgroups. The first paper [16] presents a simple proof of the Nielsen-Schreier theorem, showing that subgroups of free groups are free. Later Neumann-Wiegold [NW] classified all varieties of groups having the subgroup property (i.e. subgroups of free groups of the variety are again such). There are only very few such varieties, so called Schreier varieties, namely the class of all groups, all abelian groups and all elementary abelian p-groups for any fixed prime p.

The strategy of Levi's proof [16], which is now included in text books [MKS, p. 105 ff.] or [Sp, p. 157 ff.] is similar to the proof of Pontryagin's criterion for freeness in abelian groups, see [F, Vol.1, p. 93]. From the present point of view, it might be interesting to generalize both Pontryagin's criterion and Nielsen-Schreier's theorem to Schreier varieties using Neumann-Wiegold's result and Levi's strategy. Examples of (abelian and non-abelian) groups discussed in Higman [H1], [H2], and in the results quoted there and in Richkov [R] showing that such an extension will only be possible consequences for representations of elements in free groups, see [Ku, Vol. 2, p. 133].

Levi [21] continues the investigation of subgroups of free groups. He shows that the intersection of any infinite set of characteristic subgroups is trivial see [MKS, p.109, Cor. 2.12]. Moreover, an interesting result is proved for fully invariant subgroups of free groups. It is shown that the lattice of fully invariant subgroups has the finite intersection property [21, Satz 4].

Levi's joint paper [23] with Baer contains celebrated results and proofs related to Kurosh's extension of the Nielsen-Schreier theorem to free products [Ni], [S], [Ku1]: A subgroup of a free product P is a free product whose factors are a free group and subgroups conjugate to subgroups of the factors of P. The elegant proof is based on an idea of Cayley's (1878), which transfers the abstract group theoretic setting to fundamental groups of topological complexes. The free product becomes a fundamental group of a coloured complex and the subgroup in question becomes the fundamental group of a suitable covering complex which allows to identify its topological and group-theoretical factors.

As a byproduct a refinement theorem (hence a uniqueness theorem) of Krull-Remak-Schmidt type follows as well [23, Verfeinerungssatz]. This is of particular interest, because it is no longer true for direct sums of abelian groups, not even under severe restrictions, see Krull [Kr], Baer [B1] or Göbel [G]. The decompositions in a Krull-Remak-Schmidt theorem are replaced by decompositions into conjugate subgroups. It is this extra freedom to choose factors up to inner automorphisms that makes a uniqueness theorem possible.

Levi [29] gives an explicit construction of the commutator subgroup G' of a free product $G = A * B$ in terms of A', B' and A/A', B/B' . If $\bar{A}$, $\bar{B}$ are subsets of A and B, respectively, representing all cosets of A mod A' and B mod B', then the main theorem of [29] gives the following description of G' by conjugate subgroups and commutators:

$$G' = \prod_{b \in \bar{B}}{}^* b^{-1}A'b * \prod_{a \in \bar{A}}{}^* a^{-1}B'a * \prod_{a \in \bar{A}, b \in \bar{B}}{}^* \langle a^{-1}b^{-1}ab \rangle.$$

This very useful representation was generalized in [48]. This paper deals with generalized products $G = A \circ B$ and describes the commutator subgroup G' in terms of $A', B', A/A'$ and B/B'. This notion is used in subsequent papers of several authors, see [MKS, p.410 ff.] or [Ku, Vol.2, pp.146 ff.].

If $m(A)$ denotes the minimal cardinality of generating sets of A (which may be infinite), then it follows from Gruschko's theorem (from 1940, see [Ku, Vol.2, p.260]) that

$$m(A * B) = m(A) + m(B).$$

Levi [32] settled the case $m(A * B) = 2$ simultaneously and independently, and showed that $A \neq 1 \neq B$ must be cyclic in this case. It is still an open problem to find a finite epimorphic image $E = E(A, B)$ of $A * B$ with $m(E) = m(A) + m(B)$.

The first of the two papers [36, 39] in "The mathematics student" is a survey on automorphism groups, free groups and combinatorial methods discussed by Levi as a main speaker of a group theory conference. The second paper gives a proof of the following theorem on elements $a \neq 1$ in free groups: Powers a^m and a^n are conjugate if and only if $m = n$. This result is applied to settle a question on paper folding:

"A slip of paper (with ends) A, A_m may be folded at A_1, $A_2, \ldots, A_{m-1}$, into m congruent parts; so one obtains $2m$ "fields" on the two sides of the paper." The ends A and A_m are joined without twisting the slip. Given n pairs of — say "complementary" — colours a, a', b, b', ... , d, d', the $2m$ fields $A_i A_{i+1}$ are to be painted of one colour each, but fields on opposite sites of the paper have complementary colours. For slips coloured according to this rule it will be proved:

"If the cyclic order of the colours on one side of the slip taken in the clockwise orientation tallies with the anti-clockwise cyclic order of the colours on the other side, then the slip can be folded, so that the fields on the innerside are touching two by two, contacting fields having complementary colours."

Levi also mentions the problem whether free groups are orderable and reduces the question to the case of rank 2. This problem was answered a few years later by Iwasawa and independently by B.H. Neumann, see [Fu, p.78].

In [30] the uniqueness of elements in a free group as reduced words is derived from a result on partially ordered sets. The paper [21] is particular interesting, however it is unfortunately not often quoted in recent text books of group theory. It deals with the holomorph of a group G. Using Baer [B2], the set $\mathrm{Hol}\,G$ of all holomorphic permutations σ of G [taking products $ab^{-1}c$ to products $(a\sigma)(b\sigma)^{-1}(c\sigma)$] is a semidirect product of G as right regular representation and Aut G, see [Ro2, pp.36, 37]. This splitting, known for classical groups as the Poincaré group, suggests a geometric interpretation of $\mathrm{Hol}\,G$. Levi uses what we now call "model theoretic arguments" and replaces the group axioms in terms of a relation with three entries by a relational structure with four entries. Such a relation R on a set G is called a "pencil" if it satisfies the following axioms.

(1) If a, b, $c \in G$, then $R(a, b, c, x)$ has a unique solution $x \in G$,

(2) $R(a, a, b, b)$, $R(a, b, b, a)$ hold for a, $b \in G$.

(3) (a) $R(a, b, c, d) \wedge R(b, f, g, c) \models R(f, a, d, g)$

(b) $R(a, b, c, d) \wedge R(d, c, f, g) \models R(g, f, b, a)$.

It follows that Klein's 4-group acts naturally on (G, R) as a group of automorphisms. We select an element $e \in G$ and let $a \cdot b$ be the unique element $x \in G$ with $R(a, e, b, x)$. This makes $(G, \cdot)$ into a group. Conversely, a group $(G, \cdot)$ can be made into a pencil (G, R) by setting $R(a, b, c, d)$ if and only if $ab^{-1}cd^{-1} = 1$ in $(G, \cdot)$.

The pencil (G, R) can be extended to a combinatorial topological complex $C(G, R)$ viewing $R(a, b, c, d)$ as a quadrangular topological cell with vertices a, b, c, d and sides $\bar{ab}$, $\bar{dc}$ painted blue, $\bar{ad}$, $\bar{bc}$ painted red. Among other interesting results about $C(G, R)$ it is shown that the group Aut $C(G, R)$ of automorphisms of $C(G, R)$ is canonically isomorphic to the holomorph $\mathrm{Hol}(G, \cdot)$, hence $C(G, R)$ is the desired geometry!

Two papers [46, 49] deal with the maps $m^* : G \to G$ $(a \to a^m)$ for $m \in \mathbf{Z}$. It is not hard to show that there exists a "Levi number" $L = L(G) \in \mathbf{N} \cup \{0\}$ such that the set EG of all $m \in \mathbf{Z}$ with m^* an endomorphism, is a union of certain cosets of $\mathbf{Z}$ modulo L. Moreover if $m \in EG$, then $1 - m \in EG$ and $m(1 - m) \equiv 0 \bmod L$. All non-negative integers are represented by some $L(G)$ and G is abelian if and only if $L(G) = 2$.

The theory of Levi numbers becomes very interesting and non-trivial if it is applied to classifications of non-abelian groups. The classification of groups G with $3 \in EG$ is carried out in detail [33] using earlier results on commutator identities, cf.[Ro2, p.358]. The main result of [33] (in modern terminology) says, that any Engel group of class 2 must be nilpotent of class ≤ 3, see also [Ro1, Vol.2, pp. 43 ff.]. Levi [49] continues this investigation and derives informations about abelian groups G from EG.

The joint paper [20] with van der Waerden is devoted to the Burnside problem. Burnside groups B_{nm} are the factor groups of a free group F_n with n generators modulo the normal subgroup generated by the set of all $x^m (x \in F_n)$. Van der Waerden and Levi give the precise order of the groups B_{n3} which is 3^{e_n} with $e_n = n + \binom{n}{2} + \binom{n}{3}$, see also [V, p. 85, Theorem 5.2.1] and [MKS, section 5.12]. The finiteness of B_{n3} was already shown by Burnside. Moreover, all possible characteristic subgroups of B_{n3}-group are determined. Details are now discussed in text books, see [Ro2, p. 408 ff.].

The Burside problem was finally solved by Novikov and Adjan in 1968 by giving examples of infinite Burnside groups B_{nm}. This is the case for $n > 1$ and odd $m \geq 665$, see [A] for this improved bound. Recently, the bound has been improved (replacing 665 by 115, see [L]).

The restricted Burnside problem about the finiteness of

$$R_{nm} = B_{nm} / \bigcap \{N : N \text{ normal in } B_{nm}\} \ ,$$

was answered positively by Kostrikin (1959) for primes m (and all n). This was extended recently (1990) by Zelmanov replacing primes m by prime powers, see [V] and [Ko] for details. A very nice argument, mentioned by Zelmanov at the ICM in Kyoto (1990), leads to the final solution of the restricted Burnside problem. The classification of all finite simple groups and the solution to the restricted Burnside problem for prime powers show that the hypothesis of the Hall-Higman reduction theorem (1956) is satisfied. Hence the positive solution of the restricted Burnside problem (for all natural numbers) follows from [HH].

4. Ordered Groups

Levi contributed three papers to the theory of ordered groups.

From the title (Arithmetic laws in discrete groups) of the first of these papers [2], published in 1912, one does not get the impression that it is about lattice-ordered groups. He sets himself the goal of investigating groups in which every pair of elements has a G.C.D. Dedekind's axioms on G.C.D. [D] serve as a point of departure; he observes that in the non-commutative case additional properties have to be postulated. After defining in terms of the meet $a \wedge b$ of two elements a, b (which is an idempotent, commutative and associative operation over which group operation distributes) what we would call today a lattice-ordered group, basic properties are established. He also proves that an Abelian group admits a lattice-order if and only if it is torsion-free.

His second paper [35] on ordered groups was published 30 years later, in 1942. It is no longer concealed that the paper deals with ordered groups, though at the time of publication the theory of ordered groups was still in its infancy. Levi observes that if N is a convex subgroup of a totally ordered group G, then G/N carries an induced order. More importantly, he proves that if N is a normal subgroup of a group G, and if both N and G/N are ordered, then there is a group order on G inducing those of N and G/N exactly if the set of positive elements in N is carried into itself by all the inner automorphisms of G.

In addition, he shows that an Abelian group admits a total order if and only if it is torsion-free, and rediscovers Hölder's theorem that an archimedean totally ordered group is order-isomorphic to a subgroup of the real numbers.

In [44] an extension of $\mathbf{Z}$ by itself is constructed which contains non-trivial elements conjugate to their own inverses. This is a torsion-free group admitting no total order. However, torsion-free groups in which the commutator subgroup lies in the center do admit a total group order.

5. Semigroups

His papers [45] and [51] on semigroups deal with group images of semigroups and with free semigroups.

Let S be a monoid and $a, b \in S$. The element a is called a 'left factor' of b if $b = ac$ for some $c \in S$. In [45] it is shown that a semigroup S is a free monoid if and only if it satisfies the following conditions:

(i) if $ab = a'b'$ in S, then either a is a left factor of a' or a' is a left factor of a;

(ii) S has the left cancellation law;

(iii) every $a \in S$ has but a finite number of left factors;

(iv) S has a right identity e;

(v) $ab = e\ (a, b \in S)$ implies $b = e$.

A similar theorem was proved independently by Dubreil-Jacotin [DJ] a few years later.

Several other results are also proved in [45] which are concerned with groups of the form F/N where F is free monoid and N is a 'normal' subsemigroup in the sense that if any two of abc, ac and b belong to N, then so does the third one.

In [51], homomorphisms η of a semigroup S onto groups are examined. The complete inverse image of the group identity under η is a normal subsemigroup N which is 'complete' in the sense that every $a \in S$ is a left factor of some element in N. The main result states that homomorphisms of semigroups S onto groups are uniquely determined by the complete normal subsemigroups of elements mapped upon the group identity.

Paper [31] deals with an application of semigroups. He considers bounded subsets M of the euclidean plane, and shows that the rigid motions of the plane which map M upon a proper subset of M is a commutative semigroup S_M without identity whose cardinality does not exceed continuum (S consists of rotations around a fixed point). Conversely, if S is a semigroup of this kind, then it is isomorphic to S_M for a suitable bounded set M of the plane.

6. Rings and Skewfields

In [22] he concentrates on the irreducibility of the cyclotomic polynomials over prime fields of characteristic $p > 0$. Recall that the mth cyclotomic polynomial over the rational number field is always irreducible, but over the prime field of characteristic p it need not be irreducible; it is if and only if p is a primitive root of the congruence $x^{\phi(m)} \equiv 1 \bmod m$ (ϕ denotes Euler's function). Generalizing Dedekind's idea from 0 characteristic to prime characteristic, he bases the proof on the following result: Let α be a primitive mth root of unity, m prime to p; if $f(x)$ is an irreducible polynomial with integer coefficients, then $f(\alpha) = 0$ implies $f(\alpha^p) = 0$.

In [24] a short proof is given to the fundamental theorem on symmetric polynomials. The proof relies on the fact that the elementary symmetric polynomials are algebraically independent over any field.

Paper [26] discusses inverse modules in fields; this problem is motivated by a problem in plane geometry. Two additive subgroups, A and A', of a field F are said to form an *inverse pair* if the inverses of all the nonzero elements of A are in A', and vice-versa. For instance, the group of homogeneous rational functions of degree m and the group of those of degree $-m$ form an inverse pair in a function field in finitely many indeterminates. It is shown that if the characteristic of F is $\neq 2$, then all the non-trivial inverse pairs are obtained by taking $A = aK$ and $A' = a^{-1}K$, where K is a subfield of F and $a \in F$.

Years later, he raises the question of inverse pairs in skewfields [54]. The non-commutative situation is quite different, e.g. in the skewfield of quaternions the subspace spanned by $1, j, k$ is not a coset of a subfield, but has an inverse pair, viz. itself. Levi also shows that a subgroup A of a skewfield has an inverse pair exactly if $a \in A$ and $0 \neq b \in A$ imply $ab^{-1}a \in A$.

Starting from the observation that a ring R is commutative if and only if $ab - ba = 0$ for all $a, b \in R$, in [56] he introduces "higher commutators" in rings, following standard group theoretical pattern. He forms the 'determinant'

$$[a_1, \ldots, a_n] = \sum \pm a_{i_1} \ldots a_{i_n}$$

where the sum is to be taken over all possible permutations of indices, with positive or negative signs according as the permutation is even or odd. Now the *commutativity rank* of a ring R is n if it contains elements $a_1, \ldots, a_n$ with $[a_1, \ldots, a_n] \neq 0$, but there are no $n+1$ elements in R with nonzero determinant. Particular attention is given to semisimple Artinian rings, which case is easily reduced to simple rings, i.e. to complete matrix rings. Surprisingly, the commutativity rank of a simple ring is connected with a problem of paths in finite graphs. The special case of skewfields is discussed in a subsequent paper [55]. In spite of the excellent idea, the commutativity rank has not been used by many mathematicians; an application may be found in Kaplansky's paper [Ka].

7. Other Algebraic Subjects

His book Algebra I [41], which was first published as lecture notes, is an introduction to abstract algebra for graduate students. It discusses the basic algebraic structures and some number theory (like continued fractions). A few topics which are usually not included in graduate algebra texts are also studied, e.g. approximation of roots, Sturm's and Budan-Fourier's theorems etc. Important features are the numerous examples which make the book especially useful for students. Levi completed the manuscript of the second volume in which abelian groups, finite groups, Galois theory, commutative and non-commutative rings were discussed at length, but this volume has never been printed in book form.

With certain geometric applications in mind, in [57] his attention is focussed on absolute values of linear forms, and he proves the following theorem:

Let $\lambda_1, \ldots, \lambda_m$ be real linear forms in indeterminates $x_1, \ldots, x_n$, and let $k_1, \ldots, k_m$ be real numbers. If

$$k_1 |\lambda_1| + \cdots + k_m |\lambda_m| \geq 0$$

holds for all real values of $x_1, \ldots, x_n$, then the same inequality holds for any vectors $x_1, \ldots, x_n$ of a finite dimensional euclidean space. As a special case one can derive the vector-inequality of Hornich-Hlawka:

$$|x + y + z| + |x| + |y| + |z| - |x + y| - |y + z| - |z + x| \geq 0.$$

This idea led him to a problem in a more general setting [61]. Suppose M is a vectorspace over the reals $\mathbf{R}$ such that M carries a real seminorm $N(a) \geq 0$ $(a \in M)$ subject to the usual requirements: $N(\rho a) = |\rho| N(a)$ for $\rho \in \mathbf{R}$, and the triangle inequality. If $l_1(x), l_2(x), \ldots, r_1(x), r_2(x), \ldots$ with $(x) = (x_1, \ldots, x_n)$ are real linear forms, then define

$$L = N(l_1(x)) + \cdots + N(l_s(x)) \,,$$
$$R = N(r_1(x)) + \cdots + N(r_t(x)) \,,$$

and set $A = L - R$. The problem consists in finding the positive semidefinite functionals A.

If M and N are fixed, then the set $S(N, M)$ of positive semidefinite functionals A form a semimodule over the semiring of positive real numbers which is at the same time a complete lattice. Two seminorms, N and N', of M are regarded as equivalent if $S(N, M) = S(N', M)$.

If $\{e_i \mid i \in I\}$ is a basis of M, then a seminorm N_i is defined on M by assigning to an element $a \in M$ the absolute value of its ith coordinate α_i $(i \in I)$ in the given basis. Norms can be defined by the Minkowski norm of index p in the given basis and their limits:

$$N[p](a) = \left(\sum |\alpha_i|p\right) \quad (p \geq 1)$$

and

$$N[\omega](a) = \lim N[p](a) = \max |\alpha_i|.$$

The maximum element S_1 of the lattice $S(N, M)$ (for non-trivial N) is shown to be $S(N[1], M) = S(N[2], M)$, while the minimum element for a countably infinite dimensional M is equal to $S_0 = S(N[\omega], M)$. For 2-dimensional vector spaces all non-trivial seminorms are equivalent. Levi shows that the classes $S(N, M)$ can be characterized by suitable real numbers.

REFERENCES

[A] S.I. Adjan, *The Burnside Problem and identities in groups*, translated from Russian by J.C. Lennox and J. Wiegold, Springer, Berlin, 1978.

[B1] R. Baer, The decomposition of enumerable, primary, abelian groups into direct summands, Quart. J. Math. Oxford 6 (1935), 217–221.

[B2] R. Baer, Zur Einführung des Scharbegriffs, Journ. für die reine und angew. Math., **160** (1929), 199– 207.

[B3] R. Baer, Abelian groups without elements of finite order, Duke Math. J. 3 (1937), 68-122.

[D] R. Dedekind, Über Zerlegungen von Zahlen durch ihre grössten gemeinsamen Teiler, Braunschweiger Festschrift, 1897.

[DJ] M. L. Dubreil-Jacotin, Sur l'immersion d'un semi-groupe dans un groupe, C. R. Acad. Sci. Paris 225 (1947), 787-788.

[Fu] L. Fuchs, *Teilweise geordnete algebraische Strukturen*, Vandenhoeck und Ruprecht, Göttingen, 1966.

[F] L. Fuchs, *Infinite abelian groups*, Vol. 1,2, Academic Press, New York,1970, 1973.

[G] R. Göbel, Wie weit sind Moduln vom Satz von Krull-Remak-Schmidt entfernt? Jahresber. Deutsch. Math. Ver. **88** (1986), 11–49.

[HH] P. Hall, G. Higman, The p-length of a p-soluble group, and reduction theorems for Burnside's problem, Proc. Lond. Math. Soc. (3) **6** (1956), 1–42.

[H1] G. Higman, Almost free groups, Proc. Lond. Math. Soc. (3) **1** (1951), 284–290.

[H2] G. Higman, Some countably free groups, in *Proceedings of the Singapore conference on group theory*, Walter de Gruyter, Berlin, 1990, pp. 129–150.

[J] B. Jónsson, On direct decomposition of torsion-free abelian groups, Math. Scand. 5 (1957), 230–235; 7 (1959), 361–371.

[Ka] I. Kaplansky, Groups with representations of bounded degree, Can. J. Math. 1 (1949), 105–112.

[Ko] A. J. Kostrikin, *Around Burnside - problem*, Springer Ergebnisberichte Vol. **20**, Berlin, 1990.

[Kr] W. Krull, Matrizen, Moduln und verallgemeinerte Abelsche Gruppen im Bereich der ganzen algebraischen Zahlen II, Sitzungsber. Heidelberg Akad. (1932), Beiträge zur Algebra **19**, 13–38.

[Ku1] A. Kurosch, Die Untergruppen der freien Produkte von beliebigen Gruppen, Math. Ann. **109** (1934), 647–660.

[Ku] A. G. Kurosch, *Gruppentheorie I, II*, Akademie-Verlag, Berlin, 1972.

[LS] R.O. Lyndon, P.E. Schupp, *Combinatorial group theory*, Springer, Berlin, 1977.

[L] I. Lysenok, On the Burnside problem for odd exponents $n \geq 115$, (in Russian), Abstract from the Algebra Conference in Novosibirsk, 1989, p. 75.

[MKS] W. Magnus, A. Karras, D. Solitar, *Combinatorial group theory*, Wiley-Interscience, New York, 1966.

[N] H. Neumann, *Varieties of groups*, Springer, Berlin, 1967.

[NW] P. Neumann, J. Wiegold, Linked products and linked embeddings of groups, Math. Zeitschr. **73** (1960), 1–19.

[Ni] I. Nielsen, Om Regning med ikke-kommutative Faktorer og dens Anvendelse i Gruppeteorien, Matematisk Tidsskrift (1921), 77–94.

[P] L. Pontryagin, The theory of topological commutative groups, Ann. Math. 35 (1934), 361–388.

[R] C. Richkov, Almost free groups in varieties, (in Russian) to appear 1991.

[Ro1] D. J. S. Robinson, *Finiteness conditions and generalized soluble groups* I, II, Springer, Berlin, 1972.

[Ro2] D. J. S. Robinson, *A course in the theory of groups*, Springer Graduate Texts 80, New York, Berlin, 1982.

[S] O. Schreier, Die Untergruppen der freien Gruppen, Abh. Math. Sem. Univ. Hamburg **5** (1927), 161–183.

[Se] J. de Séguier, Éléments de la Théorie des Groupes Abstraits, 1904.

[Sp] W. Specht, *Gruppentheorie*, Springer, Berlin, 1956.

[V] M. Vaughan-Lee, *The restricted Burnside problem*, Clarendon Press, Oxford, 1990.

[W] H. Weber, Lehrbuch der Algebra, vol. I-II, F. Vieweg, Braunschweig, 1895–96.

SURVEY ARTICLES

Finite Rank Butler Groups:
A Survey of Recent Results

D. ARNOLD[1] and C. VINSONHALER[2]

Baylor University, Waco, Texas 76798
University of Connecticut, Storrs, CT 06269

The study of finite rank Butler groups is currently one of the most active areas in abelian group theory. The fact that strongly indecomposable groups of the form $G[A]$ or $G(A)$ can be classified up to quasi–isomorphism and isomorphism by invariants, a rarity for torsionfree groups of finite rank, has helped stimulate research in this area. Demonstrated links between Butler groups and representations of finite posets provide new perspectives on the complexity and structure of these groups. In particular, there is a renewed interest in almost completely decomposable groups.

This paper is devoted to a survey of selected recent results relating to Butler groups, including almost completely decomposable groups. A few new theorems appear in Section 2. Some open questions and research projects are listed at the end of Sections 2 through 5.

The vigor of the subject is evidenced by applications to such diverse areas as representations of finite posets, finite valued p–groups, finitely generated modules over multiple pullback rings and generalizations to additive categories. These applications are not discussed explicitly in this paper; the reader is referred to [AD1], [AD2] and [AV2] for representations; [ARV] and [AV7] for valued p–groups; [A3] for modules over multiple pullback rings; and [AV7] for generalizations to additive categories.

---

[1]Research supported, in part, by NSF grant #DMS 9101000

[2]Research supported, in part, by NSF grant #DMS 9022730

This survey is by no means comprehensive; significant results are omitted without prejudice. The literature on finite rank Butler groups is now so extensive that a comprehensive survey is not feasible in a paper of this length.

Groups of the form G[A] and their duals G(A) are studied in [AD1], [AD2], [FM], [GM], [GU1], [GU2], [HM], [Le], [Ri1], [Ri2], [Y], and a series of papers by the authors. Properties of these groups have been developed in a variety of contexts, including representing and co–representing graphs, representations of finite posets, {0,1} matrices, and equivalence theorems. This, together with the vagaries of publication and the number of researchers working in the area, has led to multiple proofs of many of these properties. Consequently, as a guide to the literature, we begin Section 1 with an annotated list of properties, tools, and techniques for the G[A]'s. As the G(A)'s are dual to the G[A]'s via a quasi–homomorphism duality given in [AV4], analogs for groups of the form G(A) are not stated explicitly. Notation varies from paper to paper, so that we have, in the interest of consistency, made notational choices. Copious references are given, along with examples demonstrating limitations of known techniques.

One of the themes of this paper is an examination of the extent to which known classifications of strongly indecomposable G[A]'s can be generalized to more general classes of Butler groups. An example of a complicated class ripe for investigation is C, the class of Butler groups quasi–isomorphic to a finite direct sum of strongly indecomposable G[A]'s. Included in this class are almost completely decomposable groups. Results along this line (Sections 2 and 3) are fragmentary, but do provide a rich source of problems.

Classification of Butler groups up to quasi–isomorphism, near–isomorphism, or isomorphism is linked to classification of representations of finite posets over fields, respectively factor rings of Z, or Z, as reviewed in Section 4.

Within the past year, there has been a flurry of work on almost completely decomposable groups, as surveyed in Section 5. The study of groups in the class C may be viewed as a simultaneous generalization of the theories of almost completely decomposable groups and strongly indecomposable G[A]'s — in both cases the underlying strongly indecomposables have endomorphism ring a subring of Q.

Two sets of invariants, introduced in [DO], serve as a unifying theme in this paper: $r(\tau,E,G) = (G(\tau)+G(E))/G(E)$ and $\mu(\tau,E,G)$ defined by the exact sequence $0 \rightarrow r(\tau,E,G) \rightarrow (G/G(E))(\tau) \rightarrow \mu(\tau,E,G) \rightarrow 0$ for a type τ and a finite set E of types. In our applications, the $r(\tau,E,G)$'s are quasi–isomorphism invariants while the $\mu(\tau,E,G)$'s are isomorphism and near–isomorphism, but not necessarily quasi–isomorphism, invariants. These invariants are complete sets of invariants for only special classes of Butler groups; in most of these cases the $r(\tau,E,G)$'s are classified by their ranks and the $\mu(\tau,E,G)$'s are subgroups of Q/Z.

1. GROUPS OF THE FORM G[A].

Let $A = (A_1,...,A_n)$ be an n–tuple of subgroups of Q each containing 1 and define $G[A]$ to be $(A_1 \oplus .. \oplus A_n)/K_A$, where K_A is the pure rank–1 subgroup generated by $(1,1,...,1)$, so that $K_A \simeq \cap\{A_i : 1 \le i \le n\}$. Call A (or $G[A]$) cotrimmed if the image of A_i is pure in $G[A]$, equivalently, $A_i \supseteq \cap \{A_j : j \ne i\}$ for each i [AV1, or FM wherein cotrimmed is called regular]. In this case, $G[A] = A_1x_1 + ... + A_nx_n$, a subgroup of $Qx_1 \oplus .... \oplus Qx_{n-1}$ with $x_1 + ... + x_n = 0$ and each A_ix_i pure in $G[A]$. If a given A is not cotrimmed, then $G[A] = G[B] = B_1x_1 + ... + B_nx_n$ with B cotrimmed, where $B = (B_1,...B_n)$ and each B_ix_i is the purification of A_ix_i in $G[A]$. Specifically, $B_i = A_i + \cap\{A_j : j \ne i\}$

Note that if $x = a_1x_1 + ... + a_nx_n \in G[A]$ with each $a_i \in A_i$, then in QG, then $x = x - 0 = x - a_n(x_1+...+x_n) = (a_1 - a_n)x_1 + ... + (a_{n-1} - a_n)x_{n-1}$. Thus, for each i, $G[A]$ is a subgroup of torsion index in $\oplus\{(A_j + A_i)x_j : j \ne i\}$, since any $n - 1$ element subset of the x_i's is independent and rank $G[A] = n - 1$. For a non–empty subset E of $n^+ = \{1,2,...,n\}$, define $\tau(E) = \text{type}\{\cap\{A_j : j \in E\} + \cap\{A_j : j \notin E\})$. More generally, given a partition $P = \{E(1),...,E(r)\}$ of n^+, define $\tau(P) = \cap \{\tau(E(i)) : 1 \le i \le r\}$, noting that $\tau(E) = \tau(P) = \tau(E^c)$ for $P = \{E, E^c\}$ and $E^c = n^+\backslash E$.

LEMMA 1.1: [Le; FM, Theorem 2.3; AV2, Lemma 2.1 (representations); GU1, Proposition 1.1] Suppose that A is cotrimmed. Then typeset $G[A] = \{\tau(P) : P$ a partition of $n^+\}$. In particular, if $0 \ne x = a_1x_1 + ... + a_nx_n \in G[A]$, then type $x = \text{type}(\cap\{(A_i + A_j) : a_i \ne a_j\})$

Given $0 \ne x = a_1x_1 + ... + a_nx_n \in G[A]$, there is an equivalence relation on n^+ defined by i x–equivalent to j if and only if $a_i = a_j$. The x–equivalence classes form a partition $P = \{E(1),...,E(r)\}$ of n^+. Group together the x_i's with common coefficients to write $x = a_{E(1)}x_{E(1)} + ... + a_{E(r)}x_{E(r)}$, where $x_{E(j)} = \Sigma\{x_i : i \in E(j)\}$. This equivalence relation is exploited in [FM], for example. Lemma 1.1 demonstrates its relevance.

Given a type τ, there is also an equivalence relation defined on n^+ by i τ–equivalent to j if and only if either $i = j$ or else there is a sequence $i(1) = i, i(2), ... , i(m) = j$ with $\tau \nleq \text{type}(A_{i(k)} + A_{i(k+1)})$ for $1 \le k \le m - 1$ [AV3, Section 2]. It will be convenient at times to regard τ–equivalence as a relation on the n–tuple A rather than on the index set. This equivalence relation, developed in [AV8], is used extensively in a series of papers by the authors. The connection between these two equivalence relations is given by:

LEMMA 1.2: [AV2, Lemma 2.3 (representations)] Suppose that $G[A]$ is cotrimmed, $\tau \in$ typeset $G[A]$, and $0 \ne x = a_1x_1 + ... + a_nx_n \in G[A]$. Then type $x \ge \tau$ if and only if τ–equivalence implies x–equivalence.

The following lemma is dual to [AV8, Theorem 1.10.b and Corollary 1.11.c]; the representation version is [AV2, Lemma 2.3]; and a mild variation is given in [GM, Theorem 1.8].

LEMMA 1.3: Suppose that $G = G[A]$ where A is cotrimmed, and that $\tau \in$ typeset G. Then $G(\tau) \simeq G[B]$, where $B = (B_1,...,B_r)$, type $B_i = \tau(E(i))$, and $E(1),...,E(r)$ are the τ–equivalence classes of n^+.

Theorem 1.4 is a consequence of [AV3, Theorem 3.1] and the duality of [AV4] and is also proved in [GM, Lemma 4.1] (also, see [AV10]). An alternate version of Theorem 1.4 is given in [FM, Proposition 3.2].

THEOREM 1.4: Suppose that $H = G[A]$ with $A = (A_1,...A_n)$ cotrimmed. Then H is quasi–isomorphic to $G[B(1)] \oplus ... \oplus G[B(m)]$, where each $B(i)$ is a subtuple of A with $G[B(i)]$ strongly indecomposable; if $i \neq j$, then $B(i)$ and $B(j)$ have at most one A_k in common; and each A_k is in some $B(i)$.

The proof of Theorem 1.4 is based on the following fact: Let $\tau_i =$ type A_i and let $E_0 = \{A_i\}$, $E_1, ... , E_r$ be the τ_i–equivalence classes in A. Then $G[A]$ is quasi–isomorphic to $G[D(1)] \oplus ... \oplus G[D(r)]$, where $D(j) = E_j \cup \{A_i\}$ [see proof of AV2, Theorem 2.4]. Moreover, r is the rank of $G[A](\tau_i)$ by Lemma 1.3.

Define $\Delta(A) = \{\tau : G[A](\tau)$ has rank 1 and type $\tau\}$; equivalently, $\Delta(A) = \{\tau \in$ typeset $G[A]$: rank $G[A](\tau) = 1\}$. Note that if $\tau \in$ (typeset $G[A])\backslash\Delta(A)$, then there are at least 3 τ–equivalence classes in n^+ (Lemma 1.3).

COROLLARY 1.5: Let $G = G[A]$ with $A = (A_1,...A_n)$ cotrimmed. Then G is strongly indecomposable if and only if each $\tau_i =$ type A_i is in $\Delta(A)$. In this case, $G[A](\tau_i) = A_i x_i$ for each i and the endomorphism ring of $G[A]$ is isomorphic to a subring of Q.

There are a number of published proofs of Corollary 1.5 from different perspectives: the duality of [AV4] applied to [AV1, Theorem 3], the proof of which uses the theory of quasi–representing graphs; the representation version [AV2, Theorem 2.4] which translates immediately via a category equivalence to a group–theoretic proof; a description of the dual of a strongly indecomposable G[A], called a **Richman–Butler** group, in [HM] (this description, a geometric version of τ–equivalence, avoids the theory of representing graphs); [FM, Theorem 3.3]; [FM, Corollary 3.5] a group theoretic proof via Theorem 1.4; and [GU1, Theorem 1.2] which is a variation of the proof in [FM].

If G[A] is strongly indecomposable, then the types of the A_i's are pairwise incomparable by Corollary 1.5. The following example shows that the converse is, in general, not true. Given primes p_i and p_j, denote $\mathbf{R}^i = Z[1/p_i]$, the subring of Q generated by $1/p_i$, and $\mathbf{R}^{ij} = R^i + R^j$.

EXAMPLE 1.6: [FM, Example 2.0] Let $p_1,...,p_4$ be 4 distinct primes and $A = (A_1,...,A_5) = (R^{12},R^{23},R^{34},R^{14},R^{13})$. Then $G[A]$ is not strongly indecomposable, but the types of the A_i's are pairwise incomparable.

Clearly, each element of $\Delta(A)$ is a maximal element of the typeset of $G[A]$. The converse is not true, in general, even if $G[A]$ is strongly indecomposable:

EXAMPLE 1.7: Let $\mathbf{R}_i$ denote the localization of Z at the ith prime and $\mathbf{R}_S = \cap \{R_i; i \in S\}$ for S a subset of Z^+. Let $A = (A_1,...A_8) = (R_{56},R_{125},R_{124},R_{127},R_{78},R_{138},R_{134},R_{136})$ Then A is cotrimmed, $G[A]$ is strongly indecomposable, and type R_{123} is a maximal element of typeset $G[A]$ that is not in $\Delta(A)$.

Proof: Note that $R_S + R_T = R_{S \cap T}$ and that $R_S \cap R_T = R_{S \cup T}$. A routine computation using these formulas shows that A is cotrimmed, i.e. for each i, $A_i \supset \cap \{A_j; j \neq i\} = R_S$, where $S = \{1,2,...,8\}$. Moreover, it can be seen that for each i there are exactly two τ_i-equivalence classes, $\{A_i\}$ and $\{A_j : j \neq i\}$. Thus, $G[A]$ is strongly indecomposable by Lemma 1.3 and Corollary 1.5. The R_{123}-equivalence classes are $\{A_1,A_2,A_8\}$, $\{A_4,A_5,A_6\}$, $\{A_3,A_7\}$, so that type $R_{123} \notin \Delta(A)$ by Lemma 1.3. Moreover, type $R_{123} = \tau(E)$ for $E = \{3,7\}$), so type $R_{123} \in$ typeset $G[A]$ by Lemma 1.1.

Finally, type R_{123} is maximal in typeset $G[A]$, as can be seen by first observing that type R_{12} and type R_{13} are not in typeset $G[A]$. For example, type $R_{12} >$ type R_{125} and type R_{125} is maximal in typeset $G[A]$, as noted above. Furthermore, type R_{23} is not in typeset $G[A]$ as a consequence of Lemma 1.1 and the fact that each occurence of 2 or 3 as a subscript is accompanied by a 1, whence type $R_{23} \nleq \tau(E)$ for each subset E of 8^+.

The next theorem is a rare instance of $G = G[A]$ strongly indecomposable implying $G(\sigma)$ strongly indecomposable; a tool for an induction argument that has been used in [GM], [HM], and [Y].

THEOREM 1.8: [HM, Theorem 4.7; GM, Corollary 3.3] Assume that $G = G[A]$ is cotrimmed and strongly indecomposable. Let $\tau \in \Delta(A)$ with $\tau \neq$ type A_i for each i, let $E(1)$ and $E(2)$ be the two τ-equivalence classes of n^+, and let X_τ be a subgroup of Q of type τ. Denote $\sigma =$ type $(\cap\{A_i : i \in E(1)\})$. Then $G(\sigma) \simeq G[A']$ is strongly indecomposable, where $A' = (X_\tau, A_i : i \in E(1))$.

2. GROUPS QUASI–ISOMORPHIC TO A STRONGLY INDECOMPOSABLE G[A].

We begin with quasi–isomorphism invariants for strongly indecomposable G[A]'s. For a Butler group G and finite set of types E, define **G(E)** to be the pure subgroup of G generated by $\{G(\tau): \tau \in E\}$ (this terminology is not standard, as some references do not require purity).

Let G be a finite rank Butler group, τ a type, and E a nonempty set of types. Note that if $G(\tau)$ has rank 1, e.g. when $G = G[A]$ and $\tau \in \Delta(A)$, then $r(\tau,E,G) = 0$ or $G(\tau)$, as $G(E)$ is pure in G. More generally, if τ is a maximal element of typeset G, then $G(\tau)$ is τ–homogeneous completely decomposable. Thus, $G(\tau) \cap G(E)$, being a pure subgroup of a τ–homogeneous completely decomposable group, is a summand: $G(\tau) \simeq r(\tau,E,G) \oplus (G(\tau) \cap G(E))$. Moreover, in this case, $r(\tau,E,G)$ is characterized, up to isomorphism, by its rank.

Various special cases of the $r(\tau,E,G)$'s have appeared previously in the literature on finite rank Butler groups. If σ is a type, G is a finite rank Butler group, and $E[\sigma] = \{\delta \in \text{typeset } G: \delta \nleq \sigma\}$, then $G(E[\sigma]) = G[\sigma]$, where **G[σ]** $= \cap \{\ker f: f \in \text{Hom}(G,Q),$ type $f(G) \leq \sigma\}$ is the σ–**radical of** G. It is shown in [AV6, or ARV for the representation version] that $r(\tau,E[\tau],G)$ is isomorphic to a maximal τ–homogeneous completely decomposable quasi–summand of the Butler group G. Thus, the ranks of the $r(\tau,E[\tau],G)$'s are a complete set of (quasi–) isomorphism invariants for completely decomposable groups. Morever, the $r(\tau,E[\sigma],G)$'s, for types τ and σ, are a complete set of quasi–isomorphism invariants for **CT–groups**, groups of the form $G(A)$ having a representing graph with pairwise incomparable labels of edges [AV5].

Groups of the form G[A] having a co–representing graph with pairwise incomparable labels of edges are called **co–CT groups**. A complete set of quasi–isomorphism invariants for co–CT groups (necessarily strongly indecomposable) is $\{$rank $G(M): M \subseteq$ typeset $G\}$ [Le]. This set of invariants also classifies all strongly indecomposable G[A]'s up to quasi–isomorphism, as a consequence of the duality [AV4] applied to [AV1, Theorem 6]. The next theorem is a mild variation of this classification. We have only replaced the rank G(M)'s by the equivalent rank $r(\tau,E,G)$'s. An easy proof, following the representation version in [AV2, Theorem 11], is included.

THEOREM 2.1: [AV4] Suppose that $G = G[A]$ and $H = G[B]$ are strongly indecomposable with $A = (A_1,...A_n)$ and $B = (B_1,...B_n)$ cotrimmed. Then G and H are quasi–isomorphic if and only if

(a) type $A_j \in \Delta(B)$ and type $B_j \in \Delta(A)$ for $1 \leq j \leq n$;

(b) rank $r(\tau,E,G) = $ rank $r(\tau,E,H)$ for each element τ and non–empty subset E of either $\{$type $A_j: 1 \leq j \leq n\}$ or $\{$type $B_j: 1 \leq j \leq n\}$ with $\tau \notin E$.

Proof: It is routine that (a) and (b) hold if G and H are quasi–isomorphic.

For the converse, assume (a) and (b) are true. If $\tau \in \Delta(A)$, then $G(\tau)$ has rank 1. Suppose, in addition, that E is a subset of $\Delta(A)$ with $\tau \notin E$ and $\{G(\sigma): \sigma \in E\}$ is a Z–independent set (of rank–1 subgroups of G). Then $r(\tau,E,G) = 0$, provided that $\{G(\sigma): \sigma \in E \cup \{\tau\}\}$ is Z–dependent, and $r(\tau,E,G) = G(\tau)$ has rank 1, otherwise. Let $M = \{$type $A_i : 1 \le i \le n\}$, a subset of $\Delta(A) \cap \Delta(B)$ by (b) and Corollary 1.5.

In view of the definition of $G = G[A]$, $G(M) = G$ has rank $n-1$ while rank $G(E) =$ card E (i.e. $\{G(\sigma) : \sigma \in E\}$ is Z–independent) for any proper subset E of M. As a consequence of (b) and remarks of the previous paragraph applied to H, rank $H(M) = n-1$ while rank $H(E) =$ card E for any proper subset E of M. Thus, for each $1 \le i \le n$, there is $0 \ne h_i$ in $H(\tau_i)$ such that $h_1 + ... + h_n = 0$ in H. Define $f': A_1 \oplus .. \oplus A_n \to H$ by $f'(1_i) = h_i$ for each i. Then f' induces a well defined quasi–homomorphism $f : G \to H$; in fact, f is a monomorphism since rank(image f) = rank $H(M) = n-1 =$ rank G. A similar argument, interchanging G and H and replacing M by $\{$type $B_j : 1 \le j \le n\}$, gives a monormophism $g : H \to G$. Hence, G and H are quasi–isomorphic, as desired.

Once again, there are a number of proofs of this result: [AV1] using quasi–representing graph techniques and duality; the representation version in [AV2]; [HM] for the duals of $G[A]$'s using a "uniformity condition" on supports of special elements of the dual group; [FM] translating quasi–isomorphism into a $\{0,1\}$ matrix theoretic condition that illuminates the "anisotropic projective Q–vector space" nature of these groups; and [GM, Corollary 3.6]. Quasi–isomorphism for $G[A]$'s that need not be strongly indecomposable is discussed in Section 3.

We briefly discuss the $\{0,1\}$ matrix version of Theorem 2.1. Suppose that $G = G[A] = A_1x_1 + ... + A_nx_n$ and $H = G[B] = B_1y_1 + ... + B_ny_n$ are cotrimmed, strongly indecomposable, and quasi–isomorphic. Then each type $B_j \in \Delta(B) = \Delta(A)$ is maximal in typeset G and rank $G($type $B_j) = 1$. Thus, type $B_j = \tau(E(j))$ for some subset $E(j)$ of n^+ determined by type B_j–equivalence on the entries of A. Define M to be an n by n $\{0,1\}$ matrix with j^{th} column entries given by $m_{ij} = 1$ if and only if $i \in E(j)$. This may be symbolized by the matrix equation $T(A)M = T(B)$, where $T(A) = ($type $A_1,...,$type $A_n)$. Since G and H are quasi–isomorphic, it follows that M is invertible and **admissible**, i.e. det $M_j^{\#} \ne 0$ for each $1 \le j \le n$, where $M_j^{\#}$ is the matrix obtained from M by replacing the jth column by a column of all 1's. Note that admissibility of M is equivalent to (b) of Theorem 2.1 (also see [GM, Corollary 3.6]).

THEOREM 2.2: [FM, Theorem 4.6; GM, Corollary 3.5] Suppose that $G = G[A]$ and $H = G[B]$ are cotrimmed and strongly indecomposable. Then G and H are quasi–isomorphic if and only if

(i) rank G = rank H;

(ii) typeset G = typeset H; and

(iii) there is an admissible $\{0,1\}$ matrix M with $T(A)M = T(B)$.

In this case M is admissible if and only if all row sums of E have the same parity and det $M_n^{\#}$ is an odd integer.

Also included in [FM] is an explicit algorithm for finding the quasi–isomorphism classes of strongly indecomposable $G[A]$'s of a given rank. This algorithm has been implemented on a computer by a colleague of C. Metelli for "small" ranks.

For $G[A]$ cotrimmed strongly indecomposable, Peter Yom [Y] has classified all cotrimmed n–tuples $B = (B_1,...B_n)$ which give $G[B]$ quasi–isomorphic to $G[A]$. The classification involves "two–vertex exchanges" which replace two entries in a given n–tuple $A = (A_1,...,A_n)$. Specifically, $G[A]$ is quasi–isomorphic to $G[B]$ if and only if there is a series of "two–vertex exchanges" transforming $(A_1,...,A_n)$ to $(B_1,...B_n)$.

Next we turn to isomorphism invariants for the strongly indecomposable $G[A]$'s.

THEOREM 2.3: [AV3, HM, FM] If $G[A]$ and $G[B]$ are cotrimmed, strongly indecomposable, and quasi–isomorphic, then they are isomorphic if and only if a vector of canonical subgroups of Q determined by A and $\Delta(A)$ is a non–zero rational multiple of a vector of canonical subgroups of Q determined by B and $\Delta(B)$.

The classification in [FM, Theorem 5.4] uses $\{0,1\}$ matrices as in Theorem 2.2, and will not be spelled out here.

The goal of the next theorem is to replace the above equivalence of vectors with the invariants $\mu(\tau,E,G)$, invariants that do not depend on the group being of the form $G[A]$. This allows a mild generalization of Theorem 2.3 in Theorem 2.5, where only one of the groups needs to be of the form $G[A]$. The invariants first arose in the classification of rigid uniform almost completely decomposable groups [DO], as surveyed in Section 5. The next lemma contains some calculations of $\mu(\tau,E,G)$ for a Butler group G quasi–isomorphic to a strongly indecomposable $G[A]$. We first introduce some additional notation. If $G = G[A] = A_1x_1 + ... + A_nx_n$, is cotrimmed strongly indecomposable, we denote $E_{ij} = \{A_1,...,A_n\}\backslash\{A_i,A_j\}$. Then there is an epimorphism $\theta_{ij}:G \rightarrow A_i + A_j \subsetneq Q$ given by $\sum a_r x_r \rightarrow a_i - a_j$, with ker $\theta_{ij} = G[E_{ij}]$ (see [AV4]). The map θ_{ij} induces a unique map of any full subgroup of QG into Q. All such induced maps will be denoted by the same symbol, θ_{ij}.

LEMMA 2.4:   Assume that $G = G[A] = A_1x_1 + ... + A_nx_n$ is cotrimmed strongly indecomposable.

(a) $\mu(\tau_i, E_{ij}, G) \simeq (A_i + A_j)/A_i = \theta_{ij}(G)/A_i$

(b)   If H is any subgroup of QG which contains G as a subgroup of finite index, then
$$\mu(\tau_i, E_{ij}, H) = \theta_{ij}(H)/B_i, \text{ where } B_ix_i \text{ is the pure subgroup of } H \text{ generated by } x_i.$$

(c)   For each i there is a full embedding $\oplus\{\theta_{ij} : j \neq i\} : G \to \oplus\{A_i + A_j : j \neq i\}$.

Proof: If $\tau = \tau_i$, and $E = E_{ij}$, then $G(\tau) + G(E) = A_ix_i + G(E)$ and $G(E) \subseteq \oplus\{Qx_m : m \neq i,j\}$ has rank $n - 2$.   Hence,   rank $G/G(E) = 1$ and $G/G(E) = (G/G(E))(\tau)$.   Similar calculations hold if G is replaced by H.   Parts (a) and (b) follow from the definition of the μ invariants and the remarks preceding Lemma 2.4.

THEOREM 2.5:   Suppose that $G = G[A]$ is cotrimmed strongly indecomposable with $A = (A_1,...A_n)$ and type $A_i = \tau_i$ and H is a Butler group quasi–isomorphic to G.   Then G and H are isomorphic if and only if $\mu(\tau_i, E_{ij}, G) \simeq \mu(\tau_i, E_{ij}, H)$ for all $i \neq j$ in n^+.

Proof: It is clear that the $\mu(\tau, E, G)$'s   are preserved under isomorphism.   Conversely, since $G = G[A]$ is cotrimmed strongly indecomposable, we can write
$$G = A_1x_1 + ... + A_{n-1}x_{n-1} + A_nx_n$$
as a subgroup of a vector space $Qx_1 \oplus ... \oplus Qx_{n-1}$ with $G(\tau_i) = A_ix_i$ for $1 \leq i \leq n - 1$ and $G(\tau_n) = A_nx_n$, where $x_n = -x_1 - ... - x_{n-1}$.   We may further assume that $G \subseteq H \subset QG$, with the index of G in H minimal.   One immediate consequence of the minimality assumption is that for each prime p,   at least one A_ix_i is p–pure in H.

We will first show that, in fact, each A_ix_i is pure in H.   It is enough to prove this locally, that is, to show that for a fixed prime p,   each A_ix_i is p–pure in H.   Fix p,   and assume that i is an index such that A_ix_i is p–pure in H.   We may also assume that A_i is not p–divisible, since if every such p–pure subgroup were p–divisible,   all would be p–divisible by the minimality of the index of G in H.   In this case, G would be p–divisible, whence p–pure in H.   If $\mu(\tau_i, E_{ij}, G)_p$, the p–component of $\mu(\tau_i, E_{ij}, G)$,   is   0,   then by Lemma 2.4.a, $(A_j)_p \subseteq (A_i)_p$;   and by hypothesis $\mu(\tau_i, E_{ij}, H)_p = 0$.   It follows from Lemma 2.4.b that $\theta_{ij}(H)_p = (A_i)_p$.   Again applying Lemma 2.4, we obtain the isomorphisms $\mu(\tau_j, E_{ij}, G)_p \simeq (A_i + A_j)_p/(A_j)_p = (A_i)_p/(A_j)_p = \mu(\tau_j, E_{ij}, H)_p \simeq \theta_{ij}(H)_p/(B_j)_p = (A_i)_p/(B_j)_p$.   Since A_i is not p–divisible by assumption, we may conclude that $(A_j)_p = (B_j)_p$, and therefore that A_jx_j is p–pure in H.   Suppose, on the other hand, $\mu(\tau_i, E_{ij}, G)_p \simeq \mu(\tau_i, E_{ij}, H)_p$ is not zero. Then $(A_i)_p \subseteq (A_j)_p$ by Lemma 2.4.a, and $\theta_{ij}(H)_p = (A_j)_p$ by Lemma 2.4.b. By 2.4.a again, $\mu(\tau_j, E_{ij}, G)_p = 0 = \mu(\tau_j, E_{ij}, H)_p$.   The latter group is isomorpic to $(A_j)_p/(B_j)_p$ by 2.4.b. Thus, $(A_j)_p = (B_j)_p$ and in this case also, A_jx_j is p–pure in H.   Note that in both cases, $\theta_{ij}(H)_p$ is the larger of $(A_i)_p$ and $(A_j)_p$

We next show that $G = H$. Again it is enough to show this locally, that is, $G_p = H_p$ for each prime p. Choose i so that $(A_j)_p$ is minimal among the subgroups $(A_j)_p$. By the arguments in the previous paragraph $\theta_{ij}(G)_p = \theta_{ij}(H)_p = (A_j)_p$ for all $j \neq i$. From this inequality we may draw two conclusions. First, by Lemma 2.4.c, there is an embedding $\oplus\{\theta_{ij} : j\neq i\}:G_p \to \oplus\{(A_j)_p : j\neq i\}$. This embedding must be an isomorphism, since $G_p = \oplus\{(A_j)_p x_j : j\neq i\}$. Second, there is an induced embedding of H_p into $\oplus\{(A_j)_p : j\neq i\}$. Since $G_p \subseteq H_p$, it follows that $G_p = H_p$ and the proof is complete.

REMARK: Each $\mu(\tau,E,G)$ in the above proof is isomorphic to a subgroup of Q/Z. In this case, $\mu(\tau,E,G) \simeq G(E \cup \{\tau\})/G(E)$, since $G/G(E) = (G/G(E))(\tau)$ has rank 1, where $G = G[A]$.

Two Butler groups G and H are **nearly–isomorphic** if for each prime p, there is a monomorphism $f:G \to H$ such that $H/f(G)$ is finite with order relatively prime to p [La2]. For Butler groups with endomorphism rings principal ideal domains, near–isomorphism and isomorphism coincide [A1]. Thus, Theorem 2.5 could also be stated in terms of near–isomorphism. It is demonstrated in [KM], [MV3] and [DO] that classification of almost completely decomposable groups up to isomorphism leads to complicated number–theoretic problems, while, at least in special cases, classification up to near–isomorphism is more tractable. However, any generalization of Theorem 2.5 to obtain near–isomorphism (or isomorphism) invariants for Butler groups quasi–isomorphic to a strongly indecomposable $G[A]$, is bound to be more complicated. This is demonstrated by the following example and theorem:

EXAMPLE 2.6: Choose 3 distinct primes $p,q,r \notin \{2,3,5,7,11\}$ and let $A_1 = Z[1/p]$, $A_2 = Z[1/q]$, and $A_3 = Z[1/r]$. Define $G = A_1 x_1 + A_2 x_2 + A_3 x_3 + Z(x_1+2x_2+3x_3)/7$ and $H = A_1 x_1 + A_2 x_2 + A_3 x_3 + Z(x_1+5x_2+11x_3)/7$ as subgroups of $Qx_1 \oplus Qx_2$ with $x_1+x_2+x_3 = 0$. Then G and H are both quasi–isomorphic to $G[A_1,A_2,A_3]$ and $\mu(\tau,E,G) \simeq \mu(\tau,E,H)$ for each type τ and finite set E of types, but G and H are not (nearly) isomorphic,.

Proof: A routine argument shows that G and H are strongly indecomposable with $G(\tau_i) = H(\tau_i) = A_i x_i$ for each i, where $\tau_i = $ type A_i. Thus, the endomorphism rings of both G and H are subrings of Q (Corollary 1.5) so that G and H are nearly isomorphic if and only if they are isomorphic. In this case $G = qH$ for some rational q; and $|q| = 1$ since $qA_i = A_i$ for each i.

Since rank $G = 2$ and both G and H are quasi–isomorphic to $G[A]$ for $A = (A_1,A_2,A_3)$, it is sufficient to assume that both $G(\tau)$ and $G(E)$ have rank 1. In this case we can take $E = \{\sigma\}$ with $\sigma \in$ typeset $G =$ typeset $H =$ typeset $G[A]$. But typeset $G[A] = \{\tau(P) : P$ is a partition of $n^+\}$, from which it follows that typeset $G = \{\tau_1,\tau_2,\tau_3,$type

Z}. Hence, it suffices to consider $\tau = \tau_i$ and $\sigma = \tau_j$ for some i,j. If i = j, then $\mu(\tau,E,G) = \mu(\tau,E,H) = 0$, as $(G/G(E))(\tau) = 0 = (H/H(E))(\tau)$. If $i \neq j$, then an application of Lemma 2.4.b shows that $\mu(\tau,E,G) \simeq \mu(\tau,E,H)$. For example, if i = 1 and j = 3, then $\mu(\tau,E,G) = (A_1+A_2+Z(1/7))/A_1 \simeq \mu(\tau,E,H) = (A_1+A_2+Z(4/7))/A_1$.

THEOREM 2.7: Suppose that G and H are Butler groups quasi–equal to a cotrimmed strongly indecomposable G[A] with $A = (A_1,...,A_n)$, $\tau_i =$ type A_i, and let $C_G = G(\tau_1)+...+G(\tau_n)$ and $C_H = H(\tau_1)+...+H(\tau_n)$. Then G and H are isomorphic if and only if

(a) $\mu(\tau,E,C_G) \simeq \mu(\tau,E,C_H)$ for each element τ and n–2 element subset E of $\{\tau_i : 1 \leq i \leq n\}$ with $\tau \notin E$, in which case there is $q \in Q$ with $qC_G = C_H$.

(b) The q from (a) induces an isomorphism $G/C_G \to H/C_H$.

Proof: ($\leftarrow$) Write $G[A] = A_1x_1 + ... + A_nx_n$, with each $A_ix_i = G[A](\tau_i)$ pure in G[A] and $x_1 + ... + x_n = 0$ and assume that both G and H are subgroups of QG[A] quasi–equal to G[A]. Then $G(\tau_i) = A_i'x_i \simeq A_ix_i$ is the purification of $A_ix_i \cap G$ in G and $C_G = G(\tau_1) + ... + G(\tau_n)$ is a subgroup of finite index in G quasi–equal to G[A]. Similarly, $C_H = H(\tau_1) + ... + H(\tau_n)$ is a subgroup of finite index in H quasi–equal to both G[A] and C_G with $H(\tau_i) = A_i''x_i \simeq A_ix_i$. It follows that $C_G \simeq G[A']$ and $C_H \simeq G[A'']$ for $A' = (A_1',...,A_n')$ and $A'' = (A_1'',...,A_n'')$ with each $A_i \simeq A_i' \simeq A_i''$. By (a) and Theorem 2.5, $C_G \simeq C_H$. Since C_G and C_H are quasi–equal, by Corollary 1.5 there is $q \in Q$ with $qC_G = C_H$. In view of (b), $q:G \to H$ is a well defined isomorphism.

($\to$) Since G and H are isomorphic, C_G and C_H are isomorphic, whence (a) follows from Theorem 2.5. In particular, since G and H are quasi–equal, there is $q \in Q$ with $qG = H$, $qC_G = C_H$, and $q:G/C_G \to H/C_H$ an isomorphism.

Note that the invariants in Theorem 2.7.a are invariants that can be computed from G, but that it is not at all clear how to compute the q in Theorem 2.7 from G (also see remark following Theorem 3.1 for a generalization of Theorem 2.7).

OPEN QUESTIONS AND PROBLEMS: Assume that $G = G[A]$ and $H = G[B]$ are cotrimmed and strongly indecomposable with the same rank.

I. Does $\Delta(A) = \Delta(B)$ imply that G and H are quasi–isomorphic?

II. If each type A_i is in $\Delta(B)$ and each type B_j is in $\Delta(A)$, then are G and H quasi–isomorphic?

III. [Fuchs and Metelli] Does typeset G = typeset H imply that G and H are quasi–isomorphic?

IV. What other sets of invariants classify strongly indecomposable groups of the form G[A]?

Clearly, if II is true, then so is I. The connection of either I or II with III is not so clear, as illustrated by Example 1.6. New results of [GM], as discussed following Theorem 4.2, may shed some light on III.

3. DIRECT SUMS OF STRONGLY INDECOMPOSABLE G[A]'s.

In view of Theorem 1.4, each G[A] is quasi–isomorphic to a direct sum of strongly indecomposable groups of the same form. More delicate properties of this decomposition are used in [AV10] to extend Theorem 2.1 to all groups of the form G[A].

THEOREM 3.1: [AV10] Suppose that $G = G[A]$ and $H = G[B]$ for $A = (A_1,...,A_n)$ and $B = (B_1,...,B_n)$ and let L be the lattice of types generated by {type A_i, type B_i : $1 \le i \le n$}. Then G and H are quasi–isomorphic if and only if rank $G(M) =$ rank $H(M)$ for each non–empty subset M of L.

There is an analog of Theorem 2.7 characterizing isomorphism of groups G and H quasi– isomorphic to $G[A]^r$ for $r \ge 1$ given in [AD, Corollary 1.7]. Moreover, there are indecomposable G's of this form of arbitrarily large rank. This indicates some of the difficulties in extending (near–) isomorphism classifications of strongly indecomposable G[A]'s by simple group–theoretic invariants to the class of Butler groups quasi–isomorphic to a finite direct sum of strongly indecomposable G[A]'s.

Direct sum decompositions into strongly indecomposable G[A]'s need not be unique up to isomorphism and order:

EXAMPLE 3.2: Choose 5 distinct primes p,q,p_1,p_2,p_3 and let $A_i = Z[1/p_i]$. Then $G[A_1,A_2,A_3] \oplus G[pA_1,qA_2,A_3] \simeq G[pA_1,A_2,A_3] \oplus G[A_1,qA_2,A_3]$ with all summands strongly indecomposable and pairwise non–isomorphic.

Proof: If $G = G[A_1,A_2,A_3]$, the embedding $p \oplus q:G \to pG \oplus qG \subseteq G[pA_1,A_2,A_3] \oplus G[A_1,qA_2,A_3]$ is split by $r + s$, where $r,s \in Z$ with $rp + sq = 1$. Moreover, the kernel of $r + s$ is easily shown to be isomorphic to $G[pA_1,qA_2,A_3]$. Each of the rank two summands is strongly indecomposable by Corollary 1.5. To show the summands are pairwise non–isomorphic we can repeat the argument used in Example 2.6. Any isomorphism would have to be given by multiplication by some rational. But it is easy to check that no rational works.

Suppose that G and H are finite rank torsion–free abelian groups with endomorphism rings subrings of Q. Then G and H are **isomorphic at a prime** p provided there are homomorphisms f: G → H and g: H → G such that both fg and gf are integers relatively prime to p. Example 3.2 is a special case of:

COROLLARY 3.3:   Suppose   that $G = G_1 \oplus ... \oplus G_m$ and $H = H_1 \oplus ... \oplus H_m$ are quasi–isomorphic, with each G_i and H_j strongly indecomposable of the form G[A]. Then G and H are isomorphic if and only if for each G[A] with $A = (A_1,...,A_n)$ and prime p, the number of i's with $(\mu(\tau,E,G[A]))_p \simeq (\mu(\tau,E,G_i))_p$ is equal to the number of   i's   with $(\mu(\tau,E,G[A]))_p \simeq (\mu(\tau,E,H_j))_p$,   for each element τ and n–2 element subset   E   of {type $A_j : 1 \le j \le n$} with $\tau \notin E$.

Proof: Plainly, we may assume that G_i and H_i are quasi–isomorphic. Since each G_i and H_j have endomorphism rings subrings of Q, G and H are isomorphic if and only if for each G[A] and prime p, there is a bijection from {i : G[A] isomorphic to G_i at p} to {i : G[A] isomorphic to H_i at p} [AHR, Theorem D]. We may restrict to G[A]'s which are quasi–isomorphic to some G_i. However, the proof of Theorem 2.5 shows that G[A] is isomorphic to G_i at p if and only if $(\mu(\tau,E,G[A]))_p \simeq (\mu(\tau,E,G_i))_p$ for each element τ and n–2 element subset E of {type $A_i : 1 \le i \le n$} with $\tau \notin E$.

Finally we note that, elsewhere in these Proceedings, Goeters and Ullery [GU2] show that any quasi–summand of a G[A] is quasi–isomorphic to G[B] for some n–tuple B. This result should provide a useful tool for studying direct sums of G[A]'s.

OPEN QUESTIONS AND PROBLEMS (continued):

V.   Find group–theoretic quasi–isomorphism invariants classifying finite direct sums of strongly indecomposable G[A]'s.

VI. Find necessary and sufficient criteria for a cotrimmed G[A] to be indecomposable.

A necessary condition is a consequence of [FM, Proposition 3.2]: If G[A] is indecomposable, then height $x_i \nleq$ height of x_E for each $i \in n^+$ and proper subset E of n^+ with $E \ne \{i\}$ or $n^+\backslash\{i\}$.

VII. If G[A] and G[B] are quasi–isomorphic, is there an algorithm for obtaining the tuple B from the tuple A. Peter Yom [Y] has solved this problem in the strongly indecomposable case.

The next question is better described as a project.

VIII. Find complete sets of near–isomorphism invariants for (special classes of) indecomposable groups of the form G[A] that are not necessarily strongly indecomposable.

4. REPRESENTATIONS OF FINITE POSETS AND BUTLER GROUPS

Given a commutative ring R and a finite poset S, define **Rep(R,S)** to be the category with objects $U = (U, U_s : s \in S)$, where U is a finitely generated free R–module and each U_s is a submodule of U such that $s \leq t$ in S implies that $U_s \subseteq U_t$. A morphism of representations is an R–module homomorphism $f : U \to U'$ such that $f(U_s) \subseteq U'_s$ for each $s \in S$. Finite direct sums exist in Rep(R,S): $U \oplus V = (U \oplus V, U_s \oplus V_s)$.

The relevance of representations of a finite poset over the rationals Q to the quasi–homomorphism theory of Butler groups is well documented. There is a category equivalence from the quasi–homomorphism category $\mathbf{B_T}$ of Butler groups with typesets in a finite distributive lattice T to Rep(Q,JI(T)op), where $JI(T)^{op}$ is the opposite of the poset of join–irreducible elements of T. Specifically, the equivalence is given by $G \to (QG, QG(\tau))$ [But1,But2]. Consequences of this equivalence are surveyed in [A3]; also see [La3 and references] for a slightly different, but more general, point of view.

Near–isomorphism of classes of Butler groups can also be interpreted in terms of representations of finite posets over finite factor rings of Z [AD]. In contrast to quasi–isomorphism, indecomposability is preserved by near isomorphism [La2 or A1].

We summarize the results of [AD]. To define the classes of Butler groups considered, for $n \geq 2$, let $T_n = \{\tau_0, \tau_1, \tau_2, ..., \tau_n\}$, where $\tau_1, \tau_2, ..., \tau_n$ are pairwise incomparable types with $\tau_i \cap \tau_j = \tau_0$ for each $i \neq j$. Denote by $B(T_n)$ the category of Butler groups with typeset contained in T_n. Included in $B(T_n)$ are the strongly indecomposable G[A]'s characterized by invariants in Theorem 2.5, subject to the condition that the types in A have a pairwise common meet.

A group G in $B(T_n)$ can be written as $H \oplus G_0$ where G_0 is τ_0–homogeneous completely decomposable and H has no rank–1 summands of type τ_0. Then H/C_H is finite, where $C_H = H(\tau_1) + ... + H(\tau_n)$. Up to near isomorphism, it is sufficient to consider the case that H/C_H is a finite p–group [AD, Theorem 2.1].

Given $j \geq 1$ and a prime p such that $(\tau_i)_p$ is finite for each $\tau_i \in T_n$, define $B(T_n, p^j)$ to be the category of groups H in $B(T_n)$ such that H/C_H is a finite group with exponent p^j. In view of the assumptions on the τ_i's, each $H \in B(T_n, p^j)$ is **p–locally free** (H_p is a free Z_p–module). Let $\mathbf{B(T_n, p^j)/p^j}$ be the category with objects the same as $B(T_n, p^j)$ but with morphism sets $\mathrm{Hom}(H, H')/\mathrm{Hom}(C_H, p^j C_{H'})$. The category of $Z/p^j Z$–representations of the poset Λ_{n+1} of $n + 1$ pairwise incomparable elements is denoted by Rep(Z/p^j Z, Λ_{n+1}).

THEOREM 4.1: [AD] There is a full functorial embedding $F : B(T_n, p^j)/p^j \to \mathrm{Rep}(Z/p^j Z, \Lambda_{n+1})$ defined by,

$$F(H) = (C_H/p^j C_H, (H(\tau_1) + p^j C_H)/p^j C_H, ..., (H(\tau_n) + p^j C_H)/p^j C_H, p^j H/p^j C_H).$$

Moreover,

(a) Image F contains the class of representations of the form $U = (U, U_1, ..., U_{n+1})$ such that
     $U, U_1, ..., U_n$ are free Z/p^jZ–modules, $U_i \cap U_r = 0$ for each $1 \le i \ne r \le n + 1$ and
     $U = U_1 + ... + U_n$.
(b) Given $H, H' \in B(T_n, p^j)$, then H is a near summand of H′ if and only if C_H is a near
     summand of $C_{H'}$, and F(H) is a representation summand of F(H′).
(c) If $H \in B(T_n, p^j)$ is almost completely decomposable, then H is indecomposable if and
     only if F(H) is indecomposable.

As a consequence of (b), H is nearly isomorphic to H′ if and only if C_H is nearly
isomorphic to $C_{H'}$, and $F(H) \simeq F(H')$ as representations. The point is that
near–isomorphism of $G = H \oplus G_0 \in B(T_n)$ is determined by a maximal τ_0–homogeneous
completely decomposable summand G_0, near isomorphism of C_H, and representation
isomorphism of the F(H$_j$)'s for H_i/C_H the p$_i$–component of H/C_H. For example, if G is
almost completely decomposable, then C_H is completely decomposable. In this case,
near–isomorphism and indecomposability of G in $B(T_n)$ are determined by G_0 and by
isomorphism and indecomposability of the associated distinguished representations. A detailed
examination of the special classes $B(T_2)$ and $B(T_3)$ is given in [AD], expedited by the
special nature of Rep(Q,Λ_n) and Rep(Z/p^{j}Z,Λ_{n+1}) for $n \le 3$ and by a bound on the rank of
associated representations in Rep(Z,Λ_3).

OPEN QUESTIONS AND PROBLEMS (continued):

IX.   Does $\tau_1 + \tau_2 + \tau_3 =$ type Q   imply that each   G in $B(T_3)$ is a direct sum of
indecomposable groups of rank ≤ 3?
        This is stated as a theorem in [A2], but the proof is flawed. The answer is yes, if G is
almost completely decomposable [AD]. Note that for T_3's of this kind, a representation
interpretation, via Theorem 4.1, is not available as for each prime p, some $(\tau_i)_p$ is infinite.

X. Classify the indecomposable (almost completely decomposable) groups in $B(T_3, p^j)$ for
$j \ge 2$. Given   j,   is there a bound on the ranks of indecomposable almost completely
decomposables?
        This problem is meaningful primarily as a starting point for investigation of groups in
the class C of Butler groups quasi–isomorphic to a direct sum of G[A]'s. In $B(T_3)$, the only
strongly indecomposables are the rank–1's and   G[A$_1$,A$_2$,A$_3$]   (up to quasi–isomorphism).
Nevertheless, these groups remain mysterious; even the structure of C_H for $H \in B(T_3)$ is
unclear (see [AD2]). By contrast, the groups in $B(T_2)$ are well understood ([Lw} and [AD2]).
XI.   There   are   Coxeter   correspondences $C^+, C^- : B_T \to B_T$ preserving   strongly
indecomposability.   Find   complete   sets   of   quasi–isomorphism   invariants   for   strongly

indecomposable groups $C^{+r}G$ or $C^{-r}G$ with G strongly indecomposable of the form $G[A]$ or $G(A)$. Find group theoretic characterizations of these groups.

This problem is discussed in greater detail in [AV9].

Our next result is from current work by Goeters and Megibben. It gives representation criteria for quasi–isomorphism of $G[A]$'s that need not be strongly indecomposable. Recall that for $A = (A_1,...,A_n)$ and a subset E of n^+, $\tau(E) = \text{type}(\cap \{A_i; i \in E\} + \cap \{A_i; i \notin E\})$. We identify $(Z/2Z)^n$ with the set of all subsets of n^+. If $G = G[A]$, denote $n_G(\tau) = \{E \in (Z/2Z)^n : \tau(E) \geq \tau\}$ and define a representation $R(G) = ((Z/2Z)^n, n_G(\tau): \tau \in \text{typeset } G\}$

THEOREM 4.2: [GM] Two groups $G = G[A]$ and $H = G[B]$ are quasi–isomorphic if and only if $\text{rank } G = \text{rank } H$, $\text{typeset } G = \text{typeset } H$, and $R(G) \simeq R(H)$ as representations.

It is now clear that classification of Butler groups up to near–isomorphism is inextricably linked with classification of Z/p^jZ–representations of finite posets, about which little is known for $j > 1$. For $n \geq 4$, virtually nothing is known in the way of classification, up to near–isomorphism in $B(T_n,p^j)$. In particular, $\text{Rep}(Z/pZ,\Lambda_5)$ has wild representation type (see [A3]).

5. ALMOST COMPLETELY DECOMPOSABLE GROUPS.

The purpose of this section is to survey the literature on almost completely decomposable groups, with the emphasis on examples. A **completely decomposable group** is an abelian group which is isomorphic to a direct sum of subgroups of Q, the additive rationals. The set of types of summands is called the **critical typeset**. In 1937, Baer showed that completely decomposable groups are classified up to isomorphism by the number of summands of each type in the critical typeset, a classification result of the sort which Paul Hill has called the "pinnacle" of abelian group theory. Attempts to scale these heights with other classes of torsion–free groups have often foundered below the tree line. However, there have been a number of recent successes within the class of Butler groups, and in particular with **almost completely decomposable** (ACD) groups – those groups which contain a finite rank completely decomposable group as a subgroup of finite index. Some examples will serve to introduce additional concepts. Throughout, p denotes a "suitable" prime.

EXAMPLE 5.1: We define three subgroups of the vector space $Q \oplus Q$:

$$G_0 = Z[1/7] \oplus Z[1/11]$$
$$G_1 = G_0 + Zp^{-1}(1,1)$$
$$G_2 = G_0 + Zp^{-1}(1,2).$$

These groups satisfy the following conditions:

    (1) All of the G_i's are quasi–isomorphic.

    (2) G_1 and G_2 are nearly isomorphic, G_1 and G_0 are not.

    (3) No two of the G_i's are isomorphic (e.g. for $p = 19$).

    (4) G_0 is completely decomposable, G_1 and G_2 are indecomposable.

Condition (1) is immediate, since $pG_i \subseteq G_j$ for all i,j. The groups G_1 and G_2 are nearly isomorphic (condition (2)) because given a prime $q \neq p$, the embedding $pG_1 \subseteq G_2$ has index p which is relatively prime to q. If $q = p$, then the map $(x,y) \to (x,2y)$ embeds G_1 into G_2 with index $2 \neq p$. One checks that there is no embedding of G_0 into G_1 with index prime to p. Condition (3) is far from immediate, but can be obtained by some straightforward calculations (see [A1]). It is worth noting that if $p = 13$, then G_1 and G_2 are isomorphic under the map $(x,y) \to (x,-11y)$. This fact illustrates the number–theoretic problems that are encountered when trying to classify almost completely decomposable groups up to isomorphism. Condition (4) again follows from straightforward calculations.

    How should almost completely decomposable groups be studied? In view of Baer's result, a logical focus would be the various completely decomposable subgroups of finite index. In particular, if we take such a subgroup where the index is minimal, we obtain a **regulating subgroup** as defined by Lady [La1]. He obtained several characterizations of regulating subgroups (see also [MV1]), which we illustrate with an example from [La1].

EXAMPLE 5.2: Let $G_3 = Z[1/15] \oplus Z[1/21] \oplus Z[1/3] \oplus Z[1/11] + Zp^{-2}(1,1,p,p)$.
The subgroup $C = Z[1/15] \oplus Z[1/21] \oplus Z[1/3] \oplus Z[1/11]$ is a regulating subgroup and $G/C \simeq Z(p^2)$. On the other hand, we can replace $Z[1/3](0,0,1,0)$ with $Z[1/3](p^{-1},p^{-1},1,0)$ to obtain a regulating subgroup D with $G/D \simeq Z(p) \oplus Z(p)$ (generators $p^{-2}(1,1,p,p)$ and $(0,0,1,0)$).

    In general, the homogeneous summands G_τ of type τ in a regulating subgroup are obtained by writing $G(\tau) = G_\tau \oplus G^{\#}(\tau)$ via a standard result on Butler groups [But3]. Recall that $G(\tau) = \{x \in G \mid \text{type}(x) \geq \tau\}$; and $G^{\#}(\tau) = G^*(\tau)_* = G(\tau^*)_*$ is the pure subgroup of G generated by the elements of type greater than τ. The three different notations are due to Lady, Arnold and Hill–Megibben, respectively.

    Example 5.2 shows that the finite quotient G/C, C a regulating subgroup, is not an invariant of G, although $|G/C|$ is. Burkhardt circumvented this problem by defining the **regulator** of G as the intersection of all regulating subgroups [Bu1]. The regulator is a completely decomposable fully invariant subgroup of finite index, which Burkhardt characterized in [Bu1]. He then used the regulator to show that if G admitted regulating subgroups C_1 and C_2 such that $G/C_1 \simeq Z(p^n)$ and $G/C_2 \simeq Z(p)^n$, then for any p–group T of cardinality p^n, there is a regulating subgroup D of G such that $G/D \simeq T$. This result can be viewed as a generalization of Lady's example G_3 above. In that example, the regulator is

$$R(G_3) = C \cap D = Z[1/15] \oplus Z[1/21] \oplus pZ[1/3] \oplus Z[1/11].$$

We will return later to the question of which finite p–groups can occur as the quotient G/C of an almost completely decomposable group by a regulating subgroup.

In [KM], Krapf and Mutzbauer used the regulator in obtaining isomorphism characterizations of all groups G containing a fixed completely decomposable R as regulator with the exponent of G/R also fixed. Their results illustrate the complications which arise when isomorphism is studied rather than near isomorphism. One encounters number theoretic problems because automorphisms of a finite quotient of a subgroup A of Q do not always arise as automorphisms of A. Adolf Mader and various coauthors [MV3] and [MM] have probed deeper into this problem by considering the group of extensions of a completely decomposable by a finite group. Some of their results can be found elsewhere in these proceedings.

In a recent paper, Dugas and Oxford have used the regulator concept to great advantage, characterizing, up to near isomorphism, almost completely decomposable groups which are extensions of a rigid completely decomposable C by a torsion group of the form $Z(p^k)^n$. Saying C is rigid means C is isomorphic to a direct sum of subgroups of Q which have incomparable types. In this case, C = R(G) is the regulator of G. The Dugas–Oxford results are illustrated with the following examples.

EXAMPLE 5.3: Let $G_4 = Z[1/3] \oplus Z[1/5] \oplus Z[1/7] \oplus Z[1/11] + Zp^{-2}(1,p,p,3) + Zp^{-2}(p,p,1,2)$.

Dugas and Oxford put (p^2 times) the generators of $G_4/R(G_4)$ into a matrix

$$\begin{bmatrix} 1 & p & p & 3 \\ p & p & 1 & 2 \end{bmatrix}$$

which is then "reduced" (mod p^2) to

$$\begin{bmatrix} 1 & p & 0 & 3-2p \\ 0 & p & 1 & 2-3p \end{bmatrix}$$

The two types represented by the first and third columns are near isomorphism invariants. The canonical form (after a permutation) is then

$$\begin{bmatrix} 1 & 0 & p & 3-2p \\ 0 & 1 & p & 2-3p \end{bmatrix} = [I \ A]$$

EXAMPLE 5.4: Let us examine two canonical forms for ACD groups of rank 6 with "uniform" torsion quotient of exponent p^2. The integers below G and to the right of H are explained below.

$$
\begin{array}{cc}
G & H
\end{array}
$$

$$
\begin{array}{cc}
\begin{bmatrix} 1\,0\,0 & 2\,p & 0 \\ 0\,1\,0 & 0\,1 & -7p \\ 0\,0\,1 & 2\,0 & 2 \end{bmatrix} &
\begin{bmatrix} 1\,0\,0 & 1\,p\,0 \\ 0\,1\,0 & 0\,2\,p \\ 0\,0\,1 & 4\,0\,3 \end{bmatrix}\begin{matrix} 4 \\ 2 \\ 1 \end{matrix} \\
\quad\;\; 2\;4\;\;14 &
\end{array}
$$

In this example, $p = 5$, but we wish to emphasize the places where p occurs as a factor. Indeed, in addition to the type invariants mentioned above, Dugas and Oxford use the "μ–invariants" defined in the introduction. In this case, $\mu(\tau,E,G) = G(\{\tau\} \cup E)/G(E)$, where $E \cup \{\tau\}$ is a subset of the critical typeset of G. These near isomorphism invariants indicate where, in the canonical matrix forms, the prime p occurs as a factor and to what power it

appears. In the above examples p^1 occurs in the same places in each matrix. The third set of near–isomorphism invariants are the "loop invariants" which are defined as follows. A "loop" is a sequence of nonzero entries in the right (non–identity) side of the matrix, whose indices form a loop. In our example the loop is 14,15,25,26,36,34. To obtain a corresponding loop invariant, first factor out all powers of p (the results are unique modulo p). Then form a product using, alternately the resulting entries and their inverses (modulo p^2). In the first matrix this product is $2 \cdot 1^{-1} \cdot 1 \cdot (-7)^{-1} \cdot 2 \cdot 2^{-1} \equiv 2 \cdot 1 \cdot 1 \cdot 2 \cdot 2 \cdot 3 \equiv 4 \pmod 5$; and in the second, $1 \cdot 1^{-1} \cdot 2 \cdot 1^{-1} \cdot 3 \cdot 4^{-1} \equiv 1 \cdot 1 \cdot 2 \cdot 1 \cdot 3 \cdot 4 \equiv 4 \pmod 5$. The equality of these two products allows us to find three integers (2,4,14) to multiply the last three columns of the first matrix, and three integers (4,2,1) to multiply the rows of the second matrix so that the resulting entries on the right (non–identity) side agree modulo p^2. The existence of these integers guarantees near–isomorphism of the corresponding groups. Since $p^2G \subseteq H$ and $p^2H \subseteq G$, to show near–isomorphism, it suffices to produce a monomorphism $\varphi:G \to H$ such that $H/\varphi(G)$ has order prime to p. The map φ called for by the diagram above can be regarded as an automorphism of $Q^6 \supset G$ given by the matrix

$$
\begin{bmatrix} I & & & \\ & 2 & & \\ & & 4 & \\ & & & 14 \end{bmatrix},
$$

where I is a 3×3 identity matrix and the missing entries are all zeroes. The integers to the right of the matrix for H help show that $\varphi(G) \subset H$ and that $H/\varphi(G)$ has order prime to p. Specifically, we can multiply the generator represented by the first row by 4, the second by 2 and the third by 1 and not change the group H. However, the image of the generators of

G under φ is precisely the new set of generators for H (modulo p^2). We leave it to the reader to fill in further details. Having obtained the three sets of near–isomorphism invariants for their special class of ACD groups, Dugas and Oxford place their results in an elegant geometric setting. The reader is referred to [DO] for specifics.

The work of Arnold and Dugas [AD], surveyed in Section 4, has applications to the construction of indecomposable almost completely decomposable groups. Recall that $T_n = \{\tau_0, \tau_1, ..., \tau_n\}$ is a set of types such that for distinct indices $i,j \geq 1$, τ_i, τ_j are incomparable and $\inf\{\tau_i, \tau_j\} = \tau_0$. We also assume that set T_n is a p–locally free set of types: No type τ_i is p–divisible (for a fixed prime p). Theorem 4.1 is used in [AD] to construct indecomposable almost completely decomposables quasi–isomorphic to $A_1{}^m \oplus \cdots \oplus A_n{}^m$, where $\text{type}(A_i) = \tau_i$ and m is any positive integer. In keeping with the theme of this section, we illustrate the concepts with an example in the case $m = n = 3$.

EXAMPLE 5.5: Let $R = Z/p^j Z$ and write the free R–module R^9 as $R^3 e_1 \oplus R^3 e_2 \oplus R^3 e_3$. Define an R–representation of the partially ordered set Λ of four incomparable elements by $U = (R^9, R^3 e_1, R^3 e_2, R^3 e_3, V)$, where $V = V_1 \oplus V_2 \oplus V_3 \oplus V_4$ is an R–module defined as follows:

(i) For $1 \leq i \leq 3$, $V_i = p^{i-1} W_i$, with W_i the image of a diagonal embedding $R \to R^3 e_i$.

(ii) $V_4 = (1 + M)R^3 e_1$, where $M : Re_1 \to Re_2$ has matrix representation

$$M = \begin{bmatrix} 2 & 0 & 0 \\ 1 & 2 & 0 \\ 0 & 1 & 2 \end{bmatrix}$$

The representation U can be shown to be indecomposable, and the machinery in [AD] produces an ACD group quasi–equal to $A_1{}^3 \oplus A_2{}^3 \oplus A_3{}^3$ which is indecomposable.

Next, we return, as promised, to the question of which finite p–groups can occur as the quotient of a given ACD group by a regulating subgroup. The preprint [MV2] contains some partial results on this problem which (it is hoped) will soon become less partial. For the moment, we present an example to illustrate the complexity of any possible solution.

EXAMPLE 5.6: Let $A_1, ..., A_6$ be subgroups of Q whose types form a poset with the following Hasse diagram:

Form $G_5 = A_1 \oplus A_2 \oplus \cdots \oplus A_6 + Zp^{-6}(1,1,p,p,p^3,p^3)$. Then $C = A_1 \oplus A_2 \oplus \cdots \oplus A_6$ is a regulating subgroup of G_5 with $G_5/C \simeq Z/p^6Z = Z(p^6)$. It can be shown that for any $1 \leq k \leq 6$, there is a regulating subgroup C_k with $G_5/C_k \simeq Z(p^k) \oplus Z(p^{6-k})$. There is also a regulating subgroup D such that $G_5/D \simeq Z(p) \oplus Z(p^2) \oplus Z(p^3)$. However, it is impossible to obtain $Z(p^2) \oplus Z(p^2) \oplus Z(p^2)$ as a quotient of G_5 by a regulating subgroup. This anomaly has to do with the powers of p which appear in the extra generator for G_5 as the dedicated reader will be able to check. A complete description of all quotients of an ACD group G by a regulating subgroup, given that there is one quotient which is a cyclic p–group, is to appear in [MV2].

Another way to view a completely decomposable group G is as a subgroup of finite index in a completely decomposable group C. This is the subject of [MV1]. In particular, if the index $|C/G|$ is minimal, C is called a **regulating hull** for G. The **coregulator** of G is then the sum of all regulating hulls (inside QG). Not surprisingly, the results in [MV1] on regulating hulls and coregulators are dual to the results in [Bu1] and [La1] on regulating subgroups and regulators. Indeed the duality is just Warfield duality in the locally free setting (see [MV1]). However, the general case remains somewhat of a mystery. Other dualities for torsion–free abelian groups ([A4], [But1], [Ri2], [VW1]) are useless as they only apply in the quasi–homomorphism category. The next example shows that regulating hulls are not as plentiful as one might think. There is, of course, a dual example for regulating subgroups.

EXAMPLE 5.7: There is an ACD group G of rank 4 which is not completely decomposable and a completely decomposable group D containing G such that $|D/G| = p^2$ but D does not contain a regulating hull of G.

Let $A,B \subset S$ be subgroups of Q such that A and B have incomparable types which are strictly less than the type of S. Assume p is a prime for which $p^{-1} \notin S$. Form the completely decomposable group $D = Aa \oplus Bb \oplus Sc_1 \oplus Sc_2$, where a,b,c_1,c_2 are assumed to be elements of D of p–height 0. Let G and C be the subgroups of D given by
$$G = A(pa+c_1) \oplus B(pb+c_2) \oplus pSc_1 \oplus S(c_1+c_2) + Z(a+b),$$
$$C = A(a-c_2/p) \oplus B(b+c_2/p) \oplus pSc_1 \oplus S(c_1+c_2) \subset QD.$$
The following facts can be shown (see [MV1]). Their confluence establishes the example.
  (1) G is a subgroup of C and D.
  (2) C/G is cyclic of order p, generated by $a-c_2/p + G$.
  (3) D/G is a cyclic group of order p^2 generated by $a + G$.
  (4) G is not completely decomposable.

(5) C is a regulating hull of G.

(6) E = G + Zpa is not a regulating hull of G.

Note that G projects onto each homogeneous component of D. Thus, this condition is not sufficient to imply that D is a regulating hull for G, although it is clearly necessary.

Penultimately, we mention some recent work by J.D. Reid [R2] which unifies and expands the body of examples of almost completely decomposable groups with pathological (non Krull–Schmidt) decomposition properties.   Most of the early examples of bizarre decomposition of torsion–free groups were almost completely decomposable. Reid's insightful work reveals the basic theme underlying such examples. This theme will (we hope) be expounded by its author elsewhere in these proceedings.

Finally, we discuss briefly some work of H.P. Goeters [G] indicating that results on almost completely decomposable groups can be fruitfully generalized. Goeters has found that, in many ways, the **finitely faithful S–groups** behave like subgroups of Q. These are the torsion–free abelian groups A of finite rank such that Ext(A,A) is torsion–free. In this setting, a **type** is a quasi–isomorphism class of strongly indecomposable finitely faithful S–groups, and the set of types can be partially ordered. Goeters then defines the analogs of completely decomposable groups and Butler groups, and proves that a completely decomposable group A is quasi–isomorphic to a Butler group B if and only if for every finite set of types T, rank $\sum_{t\in T} A(\tau)$ = rank $\sum_{t\in T} B(\tau)$. Here A(τ) = $\sum\{f(X) \mid f \in Hom(X,A)\}$, where X is a finitely faithful S–group of type τ. As an example of the generality of Goeters' results, we mention that they apply to the class consisting of all proper subgroups of Q and all rank two groups which are homogeneous of type τ < type(Q) but have each rank one factor isomorphic to Q. The groups in this class are all finitely faithful S–groups, and pure subgroups of finite direct sums form the analog of Butler groups.

OPEN QUESTIONS AND PROBLEMS (conclusion).

XII. For an almost completely decomposable group G, determine the torsion quotients G/C that can occur when C is a regulating subgroup of G.

XIII. Find classes of almost completely decomposable groups which can be classified up to near isomorphism (see [DO]).

XIV. Extend the results in [AD1] to cases where the critical types are not necessarily incomparable.

XV. Investigate regulating subgroups and regulators for Butler groups in general.

If G is a Butler group, for each type τ, there is a decomposition $G(\tau) = G_\tau \oplus G^{\#}(\tau)$, with G_τ homogeneous completely decomposable of type τ. The group $\sum_\tau G_\tau$ is called a **regulating subgroup** of G and the intersection of all regulating subgroups is the **regulator** of

G. Mutzbauer has conjectured that the regulator of G is its own regulator.

XVI. Examine the cross–fertilization of ideas between almost completely decomposable groups, representations of finite posets and finitely generated modules over multiple pullback rings (see [A3] for details and a host of open questions).

LIST OF REFERENCES

[A1]   D. Arnold, Finite Rank Torsion–Free Abelian Groups and Rings, Lecture Notes in Math. 931, Springer, New York, 1982.

[A2]   ————, A class of pure subgroups of completely decomposable groups, Proc. A.M.S. 41 (1973), 37–44.

[A3]   ————, Butler groups, representations of finite posets, and modules over multiple pullback rings, Proc. of 1991 Colorado Springs conference on Methods in Rings and Modules, to appear.

[A4]   ————, A duality for quotient divisible abelian groups of finite rank, Pacific J. Math. 42(1972), 11–15.

[AD1]   D. Arnold and M. Dugas, Representations of finite posets and near–isomorphism of finite rank Butler groups, preprint.

[AD2]   ————————————, Butler groups with finite typesets and integer representations of finite posets.

[AHR]   D. Arnold, R. Hunter and F. Richman, Global Azumaya theorems in additive categories, J. Pure Appl. Algebra 16(1980), 223–242.

[ARV]   D. Arnold, F. Richman, and C. Vinsonhaler, Representations of posets and finite valuated groups, J. Alg., to appear.

[AV1]   D. Arnold and C. Vinsonhaler, Invariants for a class of torsion–free abelian groups, Proc. A.M.S. 105 (1989), 293–300.

[AV2]   ——————, Invariants for classes of indecomposable representations of finite posets, J. Alg., to appear.

[AV3]   ——————, Isomorphism invariants for abelian groups, Trans. A.M.S., to appear.

[AV4]   ——————, Duality and invariants for Butler groups, Pacific J. Math 148(1991), 1–10.

[AV5]   ——————, Quasi–isomorphism invariants for a class of torsion–free abelian groups, Houston J. Math., 15(1989), 327–340.

[AV6]   ——————, Pure subgroups of finite rank completely decomposable groups II, Lecture Notes in Mathematics 1006, Spring–Verlag, New York.

[AV7]   ——————, Representations of lattices in additive categories, preprint.

[AV8]   ——————, Representing graphs for a class of torsion–free abelian groups, Abelian Group Theory, Gordon and Breach, London, 1987, 309–322.

[AV9]   ——————————, Almost split sequences, representations of finite posets, and finite
             rank Butler groups, preprint.

[AV10]   ——————————, Quasi–isomorphism invariants for two classes of finite rank Butler
             groups, preprint.

[Ba]   R. Baer, Abelian groups without elements of finite order, Duke Math. J. 3(1937), 68–122.

[Bu1]   R. Burkhardt, on a special class of almost completely decomposable groups, Abelian
             Groups and Modules, CISM #287, Springer, New York, 1984, 141–150.

[Bu2]   ——————————, Elementary abelian extensions of finite rigid systems, Comm. Alg.
             11(13) (1983), 29–36.

[But1]   M.C.R. Butler, Torsion–free modules and diagrams of vector spaces, Proc. London
             Math. Soc. (3) 18(1968), 635–652.

[But2]   ——————————, Some almost split sequences in torsion–free abelian group theory,
             Abelian Group Theory, Gordon and Breach, New York, 1987, 91–302.

[But3]   ——————————, A class of torsion–free abelian groups of finite rank, Proc. London
             Math. Soc., 15(1965), 183–187.

[DO]   M. Dugas and E. Oxford, Invariants for a class of almost completely decomposable
             groups, preprint.

[FM]   L. Fuchs and C. Metelli, On a class of Butler groups, preprint.

[G]   H.P. Goeters, Quasi–invariants for torsion–free groups, preprint.

[GM]   P. Goeters and C. Megibben, Quasi–isomorphism and Z_2–representations for a class
             of Butler groups, preprint.

[GU1]   P. Goeters and W. Ullery, Butler groups and lattices of types, Comm. Math. Univ.
             Carolinae 31,4 (1990), 613–619.

[GU2]   ——————————————, Quasi–summands of a certain class of Butler groups, preprint

[HM]   P. Hill and C. Megibben, The classification of certain Butler groups, preprint.

[K]   S.F. Kozhukhov, On a class of almost completely decomposable torsion–free abelian
             groups, lz. Vuz. Matematika **27(10)** (1983), 29–36

[KM]   K.J. Krapf and O. Mutzbauer, Classification of almost completely decomposable
             groups, Abelian Groups and Modules, CISM #287, Springer, New York, 1984, 151–162.

[La1]   E.L. Lady, Almost completely decomposable torsion–free groups, Proc. A.M.S.
             45(1974), 41–47

[La2]   ————, Nearly isomorphic torsion free abelian groups, J. Alg. 35(1974), 235–238.

[La3]   ————, A seminar on splitting rings for torsion–free modules over Dedekind
             domains, Abelian Group Theory, Lecture Notes in Mathematics 1006, Springer, New
             York, 1983, 1–48.

[La4]   ————, Warfield duality and rank–one quasi–summands of tensor products of
             finite rank locally free modules over Dedekind domains, J. Alg. 121(1989), 129–138.

[Le]   W.Y. Lee, Co–diagonal Butler groups, Chinese J. Math. 17(1989), 259–271.

[Lw]   W. Lewis, Almost completely decomposable groups with two critical types, preprint.

[MM]   A. Mader and O. Mutzbauer, Almost completely decomposable groups with cyclic regulator quotient, preprint.

[MV1]   A. Mader and C. Vinsonhaler, Regulating hulls of almost completely decomposable groups, to appear in J. Australian Math. Soc.

[MV2]   ————————————, Torsion quotients of almost completely decomposable groups, preprint.

[MV3]   A Mader, and C. Vinsonhaler, Classifying almost completely decomposable groups, preprint.

[Re1]   J.D. Reid, Warfield duality for modules over Dedekind domains, preprint.

[Re2]   ————, Decompositions of almost completely decomposable groups, preprint.

[Ri1]   F. Richman, An extension of the theory of completely decomposable torsion–free abelian groups, Trans. A.M.S. 279(1983), 175–185.

[Ri2]   ——————, Butler groups, valuated vector spaces, and duality, Rend. Sem, Mat. Univ. Padova 72(1984), 13–19.

[S]   P. Schultz, Finite extensions of torsion–free groups, Abelian Group Theory, Proceedings, Oberwolfach Conference 1985, 333–350, Gordon and Breach, Montreux, 1987.

[VW1]   C. Vinsonhaler and W. Wickless, Dualities for torsion–free abelian groups of finite rank, Pac. J. Math. 128(1990), 474–487.

[W]   R.B. Warfield, Jr., Homomorphisms and duality for torsion–free groups, Math Z. 107(1968), 189–200.

[Y]   P. Yom, a characterization of a class of torsion-free abelian groups, preprint.

Set-Theoretic Methods:
The Uses of Gamma Invariants

PAUL C. EKLOF

Department of Mathematics
University of California, Irvine

1. Whitehead's problem generalized.

Whitehead's original problem asked:

For all $\mathbf{Z}$-modules M, does $\mathrm{Ext}(M, \mathbf{Z}) = 0$ imply M is free?

This is a problem about torsion-free abelian groups since it is possible to prove (in ZFC) that $\mathrm{Ext}(M, \mathbf{Z}) = 0$ implies M is $\aleph_1$-free, that is, every countable subgroup of M is free. As is well known, the answer to this question was proved by Shelah to be independent of ZFC. We will look at why that is the case in the context of more general questions. Instead of considering $\mathbf{Z}$-modules we can deal with arbitrary R-modules, and we can also replace $\mathbf{Z}$ by any R-module K. Thus we ask:

For all R-modules M, does $\mathrm{Ext}(M, K) = 0$ imply M is projective?

(We shall always use Ext to mean Ext_R^1.) Even for $\mathbf{Z}$-modules, we thus have a more general question (posed by Adamek and Rosický): is there *any* abelian group K such that $\mathrm{Ext}(M, K) = 0$ implies M is free (for all groups M)? We can generalize even further and ask, for any collection, $\mathcal{K}$, of modules

For all R-modules M, if $\mathrm{Ext}_R(M, K) = 0$ for all $K \in \mathcal{K}$, is M projective?

For example, if $\mathcal{K}$ is the class of torsion modules (over a domain), then a module which satisfies the hypothesis of the last question is called a *Baer module*. My aim here is not to systematically study the answers to all such questions (as R, K and $\mathcal{K}$ vary), but rather to describe some methods that can be used to investigate such questions (sometimes in ZFC, sometimes with the help of additional set-theoretic hypotheses).

43

2. κ-filtrations of modules.

Central to our method of investigation is the technique of writing a module M as the union of a continuous chain of smaller submodules. If M is a module of cardinality κ, we define

$\langle M_\alpha : \alpha \in \kappa \rangle$ is a κ-*filtration* of M if it is a chain of submodules of M, each of cardinality $< \kappa$, such that $M = \cup_{\alpha \in \kappa} M_\alpha$ and for all limit ordinals $\sigma < \kappa$, $M_\sigma = \cup_{\alpha < \sigma} M_\alpha$.

From now on, M will always denote an R-module of cardinality κ, where κ is a regular uncountable cardinal. (It will do no harm here to think of κ as being a successor cardinal, that is, one of the form $\aleph_{\nu+1}$.) Whenever I write $M = \cup_{\alpha \in \kappa} M_\alpha$, it will be understood that $\langle M_\alpha : \alpha \in \kappa \rangle$ is a κ-filtration. It will not matter *which* κ-filtration of M is used, because of the following result:

LEMMA. *If* $\langle M_\alpha : \alpha \in \kappa \rangle$ *and* $\langle M'_\alpha : \alpha \in \kappa \rangle$ *are two* κ-*filtrations of* M, *then* $\{\alpha \in \kappa : M_\alpha = M'_\alpha\}$ *is a cub in* κ.

A subset C of κ is a *cub* (closed unbounded subset) in κ if for all $X \subseteq C$, sup $X \in C \cup \{\kappa\}$ and sup $C = \kappa$. (Here and elsewhere where no other reference is given, I refer the reader to [EM] for attributions and more details.) From now on, C will always denote a cub in κ.

A module M (of cardinality κ) will be called *almost free* (resp. *almost projective*) if it has a κ-filtration consisting of free (resp. projective) modules.

3. Stationary sets: the values of Γ-invariants.

A cub is a "large" subset of κ: an appropriate analog is a set of measure 1 in a probability space. In particular, the intersection of two – or even fewer than κ – cubs is again a cub. The analogy can be made precise as follows. Let $\mathcal{F}$ be the filter on $\mathcal{P}(\kappa)$ consisting of subsets of κ which contain a cub; then $D(\kappa) = \mathcal{P}(\kappa)/\mathcal{F}$ is a Boolean algebra, whose largest element, 1, is represented by the elements of $\mathcal{F}$. The smallest element, 0, is represented by the subsets X of κ such that $X \cap C = \emptyset$ for some cub. If $S \in \mathcal{P}(\kappa)$, denote by $\tilde{S}$ its image in $D(\kappa)$. A subset S of κ is called *stationary* if $\tilde{S} \neq 0$, i.e., for every cub C, $S \cap C \neq \emptyset$; the analog of a stationary set is a set of non-zero measure. Some important properties of stationary sets and cubs are summarized in the following.

LEMMA. (*i*) *Every stationary subset of* κ *is the disjoint union of* κ *stationary subsets;* (*ii*) *the union of* $< \kappa$ *non-stationary subsets of* κ *is non-stationary;* (*iii*) *the diagonal intersection,* $\Delta\{C_\alpha : \alpha \in \kappa\}$, *of* κ *cubs* (*defined to be* $\{\beta \in \kappa : \forall \alpha < \beta(\beta \in C_\alpha)\}$) *is again a cub.*

A consequence of part (iii) is called Fodor's Lemma, or the Pressing-Down Lemma, and explains the name stationary; it says that if S is stationary in κ and $f : S \to \kappa$ such that for all $\alpha \in \kappa$, $f(\alpha) < \alpha$, then there is a stationary subset S' of S such that $f \upharpoonright S'$ is constant.

4. A Gamma invariant for projectivity.

The Gamma invariants to be defined here will act on modules of cardinality κ and take values in $D(\kappa)$. (If the cardinality of R is $\geq \kappa$, we need to consider modules generated by κ

elements, instead of modules of cardinality κ; but I will ignore that technical consideration here and always assume that $|R| < \kappa$. Later we will also look at a couple of Gamma invariants of valuation domains – see sections 14 and 15.) The Gamma invariants measure an obstruction to the module having a certain property; in this section, it is the property of being projective.

Given a module M of cardinality κ, fix a κ-filtration $\langle M_\alpha : \alpha < \kappa \rangle$ of M and let $\Gamma(M) = \tilde{S}$, where

$$S = \{\alpha \in \kappa \colon \{\beta > \alpha \colon M_\beta/M_\alpha \text{ is not projective}\} \text{ is stationary in } \kappa\}$$

(cf. [EM; IV.1.6, p. 85]). By Lemma 2 (i.e., the lemma in §2), $\Gamma(M)$ is well defined, that is, it does not depend on the choice of the κ-filtration.

THEOREM. $\Gamma(M) = 0 \Leftrightarrow M = N \oplus P$ where $|N| < \kappa$ and P is projective.

Sketch of proof: ($\Leftarrow$) By Kaplansky's result, P is the direct sum, $\bigoplus_{i \in I} P_i$, of countably-generated (projective) submodules, so there is a κ-filtration of M where each M_α has the form $N \oplus \bigoplus_{i \in I_\alpha} P_i$; then it is clear that, for this κ-filtration, $S = \emptyset$. ($\Rightarrow$) If $S \cap C = \emptyset$, then for all $\alpha \in C$ there is a cub C_α such that for all $\beta \in C_\alpha$, M_β/M_α is projective. Let C_1 be the intersection of C with the diagonal intersection of the C_α. Then C_1 is a cub, $M = \bigcup_{\alpha \in C_1} M_\alpha$, and for all $\alpha \in C_1$, if α^+ denotes the next element of C_1, M_{α^+}/M_α is projective, and if τ is the least element of C_1, $M = M_\tau \oplus \bigoplus_{\alpha \in C_1} M_{\alpha^+}/M_\alpha$. $\square$

COROLLARY. *If M is almost projective, then $\Gamma(M) = 0$ if and only if $M is projective.*

I will not get into the subject of the construction of almost projective (or almost free) modules which are not projective, except to mention that such constructions can be carried out over non-left perfect rings: in ZFC for $\kappa = \aleph_1$ and other cardinals which aren't too large; and for more cardinals assuming $V = L$. A precursor of such results was Griffith's paper [G]. For a lot more on this subject see [EM; Chaps IV and VII] and [MaS].

5. A Gamma invariant for the vanishing of Ext.

Given a collection, $\mathcal{K}$, of modules (of arbitrary size) and a module M of cardinality κ, fix a κ-filtration $\langle M_\alpha : \alpha < \kappa \rangle$ of M and define $\Gamma_\mathcal{K}(M) = \tilde{S}$ where

$$S = \{\alpha \in \kappa : \{\beta > \alpha \colon \exists K \in \mathcal{K} \text{ s.t. } \mathrm{Ext}(M_\beta/M_\alpha, K) \neq 0\} \text{ is stationary in } \kappa\}.$$

Again, this is well-defined because of Lemma 2. Note that $\Gamma_\mathcal{K}(M) \leq \Gamma(M)$. If M has a κ-filtration, $\bigcup_{\alpha \in \kappa} M_\alpha$, such that $\mathrm{Ext}(M_\alpha, K) = 0$ for all $K \in \mathcal{K}$, then it is equivalent to define S to be $\{\alpha \in \kappa \colon \exists \beta > \alpha \exists K \in \mathcal{K} \text{ s.t. } \mathrm{Ext}(M_\beta/M_\alpha, K) \neq 0\}$.

THEOREM. *If $\Gamma_\mathcal{K}(M) = 0$, then there is a submodule N of M such that $|N| < \kappa$ and for all $K \in \mathcal{K}$, $\mathrm{Ext}(M/N, K) = 0$.*

Sketch of proof (cf. [E; Lemma 2.2]): As in the proof of Theorem 4, there is a cub C_1 such that for all $\alpha < \beta$ in C_1, $\text{Ext}(M_\beta/M_\alpha, K) = 0$ for all $K \in \mathcal{K}$. Without loss of generality $C_1 = \kappa$. Let $N = M_0$ and let $L = M/N$. Then $L = \cup_{\alpha \in \kappa}(M_\alpha/N)$ is a κ-filtration of L. We will use *factor systems* from L to K, that is pairs (f, g) such that $f: L \times L \to K$ and $g: R \times L \to K$ such that for all u, v, $w \in L$ and r, $s \in R$: $f(u, v) = f(v, u)$; $f(u,v) + f(u + v, w) = f(u, v + w) + f(v, w)$; $f(u, 0) = f(0, u) = 0$; $sg(r, u) + g(s, ru) = g(sr, u)$; $g(s, u) + g(r, u) + f(su, ru) = g(s + r, u)$; and $g(r, u) + g(r, v) + f(ru, rv) = g(r, u + v) + rf(u, v)$. Then $\text{Ext}(L, K) = 0$ if and only if for every factor system (f, g) from L to K, there is a function $h: L \to K$ such that $h(0) = 0$ and for all u, $v \in L$ and $r \in R$, $f(u, v) = \delta h(u, v) \stackrel{\text{def}}{=} h(u) + h(v) - h(u + v)$ and $g(r, u) = \delta_R h(u, v) \stackrel{\text{def}}{=} rh(u) - h(ru)$. (Compare [F; p. 210] and [Mac; p.69].)

Now for any $K \in \mathcal{K}$, given a factor system (f, g) from L to K, define inductively a continuous chain of functions $h_\alpha: M_\alpha/N \to K$ such that for all $\alpha \in \kappa$, $(\delta h_\alpha, \delta_R h_\alpha) = (f \restriction (M_\alpha/N \times M_\alpha/N), g \restriction (R \times M_\alpha/N))$. If h_α has been defined, then since $\text{Ext}(M_{\alpha+1}/N, K) = 0$ (by induction), there exists $h': M_{\alpha+1} \to K$ with $(\delta h', \delta_R h') = (f \restriction (M_{\alpha+1}/N \times M_{\alpha+1}/N), g \restriction (R \times M_{\alpha+1}/N))$. Thus $h_\alpha - (h' \restriction M_\alpha/N)$ is a homomorphism: $M_\alpha/N \to K$, which extends to a homomorphism $\theta: M_{\alpha+1}/N \to K$ since $\text{Ext}(M_{\alpha+1}/M_\alpha, K) = 0$; then let $h_{\alpha+1} = h' + \theta$. □

The astute reader will have noted the difference between Theorem 4 and Theorem 5; the question of a converse to Theorem 5 will be the principal subject of the next few sections. Rather than dealing with the most general situation, we will assume the appropriate analog of "almost projective", that is, we will assume, from now on, that M has a κ-filtration, $\cup_{\alpha \in \kappa} M_\alpha$, such that for all α, $\text{Ext}(M_\alpha, K) = 0$ for all $K \in \mathcal{K}$.

6. A converse to Theorem 5, for certain $\mathcal{K}$.

Fuchs noted that under certain conditions, there is a converse to Theorem 5 provable in ZFC (cf. [EF; Lemma 9], [EFS; Lemma 9] and [FV; Lemma 3.1]).

> **THEOREM.** *Suppose $\mathcal{K}$ is closed under direct sums, and M has a κ-filtration, $\cup_{\alpha \in \kappa} M_\alpha$, such that for all α, $\text{Ext}(M_\alpha, K) = 0$ for all $K \in \mathcal{K}$. If $\Gamma_\mathcal{K}(M) \neq 0$, then $\text{Ext}(M, K') \neq 0$ for some $K' \in \mathcal{K}$.*

Sketch of proof: Without loss of generality we can suppose that there is a stationary set S such that for all $\alpha \in S$, there is $K_\alpha \in \mathcal{K}$ with $\text{Ext}(M_{\alpha+1}/M_\alpha, K_\alpha) \neq 0$. Let $K' = \oplus_{\alpha \in S} K_\alpha$ and for all α, let $K'_\alpha = \oplus_{\beta \in S \cap \alpha} K_\beta$. Define inductively a continuous chain of factor systems (f_α, g_α) from M_α to K'_α. The crucial case is when (f_α, g_α) has been defined and $\alpha \in S$. Then since $\text{Ext}(M_{\alpha+1}/M_\alpha, K_\alpha) \neq 0$ and $\text{Ext}(M_{\alpha+1}, K_\alpha) = 0$, there exists $\theta_\alpha \in \text{Hom}(M_\alpha, K_\alpha)$ which does not extend to a homomorphism on $M_{\alpha+1}$. Also, since $\text{Ext}(M_\alpha, K_\alpha) = 0$, there exists $h_\alpha: M_\alpha \to K'_\alpha$ such that $(f_\alpha, g_\alpha) = (\delta h_\alpha, \delta_R h_\alpha)$. Choose $h_{\alpha+1}: M_{\alpha+1} \to K'_{\alpha+1} = K'_\alpha \oplus K_\alpha$ such that for all $u \in M_\alpha$, $h_{\alpha+1}(u) = (h_\alpha(u), \theta_\alpha(u))$ and define $(f_{\alpha+1}, g_{\alpha+1}) = (\delta h_{\alpha+1}, \delta_R h_{\alpha+1})$. Let $(f, g) = \cup_{\alpha \in \kappa}(f_\alpha, g_\alpha)$. It is then not hard to show that there is no $h: M \to K'$ such that $(\delta h, \delta_R h) = (f, g)$. (Use the fact that for any such h there is a cub C such that for all $\alpha \in C$, $h[M_\alpha] \subseteq K'_\alpha$.) □

When $\mathcal{K}$ is the class of all torsion modules (over a domain), this result is used to characterize Baer modules (see [EF] and [EFS]); more precisely, it is used – in an inductive argument on the cardinality of the module – to deal with the case of modules of regular cardinality; Shelah's Singular Compactness Theorem (cf. [EM; §IV.3]) is used for the singular case. (See also §12 below.) Fuchs and Viljoen [FV] have studied a weaker form of Baer's problem for modules over valuation domains in which one seeks to characterize those M such that $\mathrm{Ext}(M, T) = 0$ for all bounded and divisible torsion modules T. In this case the Theorem is used with $\mathcal{K}$ equal to the class of all direct sums of a fixed bounded torsion T.

7. On solving Whitehead's Problem.

From now on, we will consider the case when $\mathcal{K}$ is a singleton $\{K\}$, and we will write Γ_K instead of $\Gamma_{\{K\}}$. Let us for a moment consider the classical Whitehead Problem, i.e., $R = \mathbf{Z} = K$. Let M be an almost free $\mathbf{Z}$-module of cardinality κ. We want to know whether $\mathrm{Ext}(M, \mathbf{Z}) = 0 \Rightarrow \Gamma(M) = 0$. By Theorem 5 we know that $\mathrm{Ext}(M, \mathbf{Z}) = 0 \Leftarrow \Gamma_{\mathbf{Z}}(M) = 0$ and by definition, $\Gamma_{\mathbf{Z}}(M) \leq \Gamma(M)$. Thus for a negative answer to the problem, it suffices to show, in this context, that the converse to Theorem 5 is false.

In order to obtain a positive solution to the problem, we need not only a converse to Theorem 5 but also a reversal of the inequality, that is we need $\Gamma_{\mathbf{Z}}(M) = \Gamma(M)$. Assuming that we have a converse to Theorem 5 (for all cardinalities) — which will be the case if $V = L$ — we proceed by induction on cardinality. The inductive hypothesis is that for all $\mathbf{Z}$-modules N of cardinality $< \kappa$, $\mathrm{Ext}(N, \mathbf{Z}) = 0 \Rightarrow \Gamma(N) = 0$. This implies that if M has (regular) cardinality κ and $\mathrm{Ext}(M, \mathbf{Z}) = 0$, then M is almost free and $\Gamma(M) = \Gamma_{\mathbf{Z}}(M)$. Hence, by the converse to Theorem 5, $\Gamma(M) = 0$, i.e., M is free. (The Singular Compactness Theorem is needed at singular cardinalities to continue the induction.)

The base case of the inductive hypothesis, that is $\mathrm{Ext}(N, \mathbf{Z}) = 0 \Rightarrow \Gamma(N) = 0$ for N of cardinality $< \aleph_1$, is provable in ZFC, so $\Gamma(M) = \Gamma_{\mathbf{Z}}(M)$ for M of cardinality $\aleph_1$. However it is consistent with ZFC that there is an M of cardinality $\aleph_2$ with $\Gamma_{\mathbf{Z}}(M) = 0 < \Gamma(M)$ (cf. [EM; XII.1.16]).

8. The uses of prediction principles: a converse to Theorem 5.

If S is a stationary subset of κ, by $\Diamond_\kappa(S)$ is meant the following principle:

> there is a family $\{W_\alpha : \alpha \in S\}$ of sets such that for each $\alpha \in S$, $W_\alpha \subseteq \alpha$, and for all $X \subseteq \kappa$, $\{\alpha \in S : W_\alpha = X \cap \alpha\}$ is stationary in κ.

$\Diamond_\kappa(S)$ is consistent with but not provable from ZFC. Gödel's Axiom of Constructibility, $V = L$, implies $\Diamond_\kappa(S)$ for all regular uncountable κ and all stationary subsets S of κ. However, it is consistent with ZFC + GCH that $\Diamond_\kappa(S)$ holds for some stationary subsets S of κ and fails for others.

> **THEOREM.** *Assume $\Diamond_\kappa(S)$ and suppose $|K| \leq \kappa$ and M has a κ-filtration, $\cup_{\alpha \in \kappa} M_\alpha$, such that for all α, $\mathrm{Ext}(M_\alpha, K) = 0$. If $\mathrm{Ext}(M, K) = 0$, then $\Gamma_K(M) \not\supseteq \tilde{S}$.*

> **COROLLARY.** *Assume $V = L$ and suppose K and M are as in the Theorem. If $\mathrm{Ext}(M, K) = 0$, then $\Gamma_K(M) = 0$.*

Thus, assuming V = L, it follows — as outlined in §7 — that every W-group (i.e., $\mathbf{Z}$-module M with $\text{Ext}(M, \mathbf{Z}) = 0$) is free. Shelah has also proved that it is consistent with ZFC that there is a non-zero ideal W of $D(\omega_1)$ such that an $\aleph_1$-free group M of cardinality $\aleph_1$ is a W-group if and only if $\Gamma(M) \in W$ (cf. [S1]). On the other hand, he has also shown that it is consistent with ZFC that there are $\aleph_1$-free groups M_1 and M_2 of cardinality $\aleph_1$ such that $\Gamma(M_1) = \Gamma(M_2)$ and M_1 is a W-group but M_2 is not (cf. [S2]). The proof uses uniformization principles like those discussed below (see also [EM; XII.3]).

The method of this section has been used to characterize Whitehead modules over other domains, assuming V = L: see [BFS].

9. Whitehead's Problem and uniformization principles.

Suppose that S is a stationary subset of κ consisting of limit ordinals of cofinality ω. If $\delta \in S$, a *ladder on* δ is a strictly increasing function $\eta_\delta: \omega \to \delta$ such that $\sup\{\eta_\delta(n): n \in \omega\} = \delta$. A *ladder system on* S is a family $\eta = \{\eta_\delta: \delta \in S\}$ of ladders on the members of S. A λ-*coloring* of a ladder system η is a family $c = \{c_\delta: \delta \in S\}$ of functions $c_\delta: \omega \to \lambda$. We say that $f: \kappa \to \lambda$ *uniformizes* the coloring c of η if for each $\delta \in S$, there exists $f^*(\delta) \in \omega$ such that $f(\eta_\delta(n)) = c_\delta(n)$ for $n \geq f^*(\delta)$.

Martin's Axiom (MA) plus ¬CH implies that for every stationary subset S of ω_1, every $\aleph_0$-coloring of a ladder system on S can be uniformized. Then the method of proof of the Theorem of the next section shows that MA + ¬CH implies there is a non-free W-group.

In fact, it can be proved (cf. [EM; §XII.3]) that there is a non-free W-group of cardinality $\aleph_1$ if and only if there is a stationary subset S of $\aleph_1$ and a ladder system η on S such that every 2-coloring of η can be uniformized. For more on uniformization and its relation to the Whitehead problem, see [EMS].

10. The failure of a converse to Theorem 5.

Let κ be the successor of a singular cardinal, ρ, of cofinality ω. By $UP_\kappa(S)$ is meant the following principle:

> S is a stationary subset of κ consisting of limit ordinals of cofinality ω and for every ladder system η on S, every cardinal $\lambda < \rho$, and every λ-coloring c of η, there is a function $f: \kappa \to \lambda$ which uniformizes c.

Shelah has shown that such principles are consistent with ZFC + GCH (cf. [S3]). On the other hand, $UP_\kappa(S)$ is not consistent with $\Diamond_\kappa(S)$, since we can use the diamond principle to predict, and then foil, any possible uniformizing function f.

From now on, whenever we assume $UP_\kappa(S)$ we will also be assuming that κ is the successor of a singular cardinal, ρ, of cofinality ω and S is a stationary subset of κ consisting of limit ordinals of cofinality ω.

> **THEOREM.** *Assume $UP_\kappa(S)$ and suppose R is not left perfect. Then there exists a module M such that $\Gamma_K(M) \supseteq \tilde{S}$, but $\text{Ext}(M, K) = 0$ for all K of cardinality $< \rho$.*

Sketch of proof (cf. [EM; XII.3.4 and XII.3.6]and [T4]): Since R is not left perfect, there is an infinite descending chain $a_0 R \supset a_0 a_1 R \supset ... \supset a_0 a_1 \cdots a_n R \supset ...$ of principal right ideals. Let $\eta = \{\eta_\delta : \delta \in S\}$ be a ladder system on S such that the range of each ladder consists of successor ordinals. Let F be freely generated by $\{x_{\nu+1} : \nu \in \kappa\} \cup \{z_{\delta n} : \delta \in S, n \in \omega\}$. Let F' be the submodule of F (freely) generated by $\{w_{\delta n} : \delta \in S, n \in \omega\}$ where $w_{\delta n} = a_n z_{\delta, n+1} - z_{\delta n} - x_{\eta_\delta(n)}$. Let $M = F/F'$. To see that $\Gamma(M) \supseteq \tilde{S}$, consider the κ-filtration of M where M_α is generated by $\{x_{\nu+1} : \nu + 1 < \alpha\} \cup \{z_{\delta n} : \delta \in S \cap \alpha, n \in \omega\}$. Then for all $\delta \in S$, $M_{\delta+1}/M_\delta$ is not projective (cf. [EM; VII.1.1]).

Now suppose that $|K| = \lambda < \rho$ (where, recall, $\kappa = \rho^+$). To show that $\mathrm{Ext}(M, K) = 0$ it suffices to show that every $\psi \in \mathrm{Hom}(F', K)$ extends to a member, θ, of $\mathrm{Hom}(F, K)$. Define a coloring $c = \{c_\delta : \delta \in S\}$ of η by: $c_\delta(n) = \psi(w_{\delta n})$. By $UP_\kappa(S)$, there is a uniformizing function $f : \kappa \to K$. Define a homomorphism θ on F by defining θ on the given basis: let $\theta(x_{\nu+1}) = -f(\nu+1)$; for $n \geq f^*(\delta)$, let $\theta(z_{\delta n}) = 0$; and for $n < f^*(\delta)$ define $\theta(z_{\delta n})$ by downward induction so that for all $n \in \omega$ we have $\theta(w_{\delta n}) = \psi(w_{\delta n})$ for all $\delta \in S$. $\square$

Now the consistency proof for $UP_\kappa(S)$ shows that S can be taken to be *non-reflecting*, i.e., for all limit ordinals $\sigma < \kappa$ (of uncountable cofinality), $S \cap \sigma$ is not stationary in σ. This easily implies that M is almost-free. But, in fact, the following argument, due to Alan Mekler, derives the result that M is almost-free from the uniformization principle as stated. To show that, for any α, M_α is free, it suffices to show there is a function f^* on limit ordinals $\delta \in S \cap \alpha$ such that for all $\delta_1 \neq \delta_2 \in S \cap \alpha$, $\{\eta_{\delta_1}(n) : n \geq f^*(\delta_1)\}$ and $\{\eta_{\delta_2}(n) : n \geq f^*(\delta_2)\}$ are disjoint. This will be the case if every monochromatic ρ-coloring of η can be uniformized. (By a *monochromatic* coloring, c, we mean each c_δ is constant.) Indeed, we just uniformize a monochromatic coloring such that c_δ and c_γ assign different colors whenever $\delta \neq \gamma$ are both $< \alpha$. Thus it remains only to prove the following.

> **LEMMA.** *Assume $UP_\kappa(S)$. Then then every monochromatic ρ-coloring of a ladder system η on S can be uniformized.*

Sketch of proof: Given a monochromatic coloring c, one first partitions the colors into $\mathrm{cf}(\rho)$ sets X_i of size $< \rho$. In the first use of uniformization, color η_δ with color i if the color of c_δ is a member of X_i. Apply $UP_\kappa(S)$ — actually we need it only for monochromatic colorings — and let (g, g^*) be the uniformizing pair. Next, for each i choose a coloring d where $d_\delta = c_\delta$ if the color of c_δ belongs to X_i and otherwise d_δ is some fixed element not in X_i. Now uniformize to get (h_i, h_i^*). Then we can uniformize the original coloring c if we let $f^*(\delta) = \max\{g^*(\delta), h_i^*(\delta)\}$ when the color of c_δ is in X_i. $\square$

11. Whitehead test modules and Ext rings.

Say that an R-module K is a *Whitehead test module* (for being projective) if for all R-modules M (of arbitrary cardinality), $\mathrm{Ext}(M, K) = 0$ implies M is projective. So Whitehead's original problem asks if $\mathbf{Z}$ is a Whitehead test module; the question of Adamek and Rosický mentioned in §1 is whether (for $R = \mathbf{Z}$) there is *any* Whitehead test module. We know that it is consistent that there *is* one (namely $\mathbf{Z}$ when $V = L$). The following is proved in [ES1]:

> **THEOREM.** *It is consistent with ZFC + GCH that for every singular cardinal ρ of cofinality ω, $UP_\kappa(S)$ is true for some (non-reflecting) stationary*

subset of $\kappa = \rho^+$ consisting of ordinals of cofinality ω.

Hence, as a consequence of Theorem 10, it is consistent with ZFC +GCH that for every non-left perfect R, there is no Whitehead test module.

Trlifaj [T5] has observed that this result is best possible in the sense that (it is provable in ZFC that) for any left perfect R there *is* a Whitehead test R-module. Trlifaj [T1, 2, 3, 4, 5] has made a study of Whitehead properties of modules over (von Neumann) regular rings. A ring R is called an *Ext ring* if every R-module is either injective or is a Whitehead test module. Trlifaj [T4] independently gave a proof of the second part of the Theorem for regular rings, from which it follows that it is consistent with ZFC that every regular Ext ring is semisimple Artinian.

12. Baer test modules.

Say that a torsion R-module T (where R is a domain) is a *Baer test module* if for all modules M of projective dimension ≤ 1, $\mathrm{Ext}(M, T) = 0$ implies M is a Baer module, i.e., $\mathrm{Ext}(B, T') = 0$ for all torsion modules T'. Assuming V = L, there is a Baer test module for any domain R (cf. [EFS; Theorem C] and [ES1; §4]). On the other hand, by the methods of §11 it can be shown that it is consistent with ZFC that for every domain R which is not a field, there is no Baer test module (cf. [ES1; Thm. 4.6]).

13. Non-standard uniserial modules.

We now change the subject, but still continue with the general theme of Gamma invariants. From now on, R will denote a valuation domain and Q its quotient field. A *uniserial R-module* is one whose submodules are totally ordered by inclusion. For simplicity of exposition we confine ourselves here to uniserial modules *of type Q/R*, that is ones which are divisible and torsion and such that the annihilator of every element is a principal ideal. Obviously Q/R is an example of such a module. The question is whether there is a *non-standard* uniserial module of type Q/R, i.e., one which is not a quotient of Q. More precisely, the question is whether there are valuation domains R for which non-standard uniserial modules of type Q/R exist. From now on, when we say "non-standard uniserial (module)" we will mean "non-standard uniserial module of type Q/R". Fuchs proved that the existence of a non-standard uniserial implies the existence of a valuation ring which is not a quotient of a valuation domain, answering a question of Kaplansky (see [FS]).

The question of the existence of non-standard uniserial modules was answered in the affirmative by Shelah (cf. [S4] and [FS]), but the model-theoretic transfer arguments he used gave no clue as to *which* R have non-standard uniserials. Osofsky [O1, 2] gave a concrete construction of a valuation domain which has a non-standard uniserial module, and made a conjecture as to when R has a non-standard uniserial. It turns out that the conjecture is independent of ZFC, as I will explain below.

14. A Gamma invariant for completeness.

From now on we will assume that R is a valuation domain of cardinality $\aleph_1$. If Q is countably generated, then every uniserial (of type Q/R) is standard, so we will assume that Q is not countably generated.

If R is "complete" in some sense (e.g. almost maximal), then there are no non-standard uniserial R-modules. We are going to introduce a couple of Gamma invariants of R which can be viewed as measuring (the lack of) "completeness". Write

$$Q = \cup_{\nu < \omega_1} r_\nu^{-1} R$$

for some sequence of elements $\langle r_\nu : \nu < \omega_1 \rangle$ such that for all $\mu < \nu$, $r_\mu | r_\nu$ and r_ν does not divide r_μ if $\mu < \nu$, i.e. $r_\mu^{-1} r_\nu$ is not a member of R. Define $\Gamma(Q/R)$ to be $\tilde{S}$, where

$$S = \{\delta \in \lim(\omega_1) : R / \cap_{\nu < \delta} r_\nu R \text{ is not complete}\}$$

where the topology on $R / \cap_{\nu < \delta} r_\nu R$ is the metrizable linear topology with a basis of neighborhoods of 0 given by the submodules $r_\nu R$ ($\nu < \delta$). (The notation used is essentially from [ES2].) It is not hard to show that $\Gamma(Q/R)$ is well-defined, that is, it depends only on R and not on the choice of the $\langle r_\nu : \nu < \omega_1 \rangle$.

As an ad hoc definition, say that R is *standard* if every uniserial R-module of type Q/R is standard. In this setting, Osofsky's conjecture says that R is not standard if and only if $\Gamma(Q/R) \neq 0$. The conjecture is true assuming V = L; the following result is proved in [ES2] using techniques from [BSa].

> **THEOREM.** (i) If $\Gamma(Q/R) = 0$, then R is standard.
> (ii) Assuming $V = L$, if $\Gamma(Q/R) \neq 0$, then R is not standard.

Any uniserial module of type Q/R is a direct limit of modules isomorphic to $r_\nu^{-1} R/R$ ($\nu \in \omega_1$), but the embeddings may not be the canonical ones. In the proof of part (ii), the embeddings are defined inductively and the predictions given by the diamond principles are used to choose the embeddings so as to foil any isomorphism of the end result with Q/R.

On the other hand, it is consistent with ZFC + GCH that there is a valuation domain R (of cardinality $\aleph_1$) such that $\Gamma(Q/R) \neq 0, 1$ and R is standard. (See [ES2]; this proof uses Axiom(S), a "proper-forcing axiom" which is consistent with GCH and implies that colorings of ladder systems on $S \subseteq \omega_1$ can be uniformized.)

In order to obtain a valuation domain R with $\Gamma(Q/R) \neq 0, 1$, one must start with a stationary and co-stationary set S of ω_1 — which can only be obtained non-constructively by an application of the Axiom of Choice — and then construct the valuation domain. If one confines oneself to "natural" valuation domains (of cardinality $\aleph_1$), which in particular have $\Gamma(Q/R) \in \{0, 1\}$, then Osofsky's Conjecture is correct (in ZFC): R is standard if and only if $\Gamma(Q/R) = 1$. (See [ES3]; the proof divides into several cases, one of which uses weak diamond, and another of which uses a construction like that of an Aronszajn tree.)

15. Another Gamma invariant for completeness.

Given a valuation domain R of cardinality $\aleph_1$, choose an ω_1-filtration of R, $R = \cup_{\alpha < \omega_1} N_\alpha$, such that each N_α is a countable subring of R. Define $\Gamma'(Q/R) = \tilde{S}'$ where

$$S' = \{\delta \in \lim(\omega_1) : \exists \langle u_\sigma : \sigma < \delta \rangle \text{ representing a Cauchy sequence in}$$
$$R / \cap_{\nu < \delta} r_\nu R \text{ and s.t. for all units } f \in R \; \exists \sigma < \delta \; \text{ s.t. } u_\sigma f \notin N_\delta \bmod r_\sigma R\}.$$

Note that $\Gamma'(Q/R) \leq \Gamma(Q/R)$ since if $R/\cap_{\nu<\delta} r_\nu R$ is complete, for any Cauchy sequence $\langle u_\sigma : \sigma < \delta \rangle$, if f is the limit of $\langle u_\sigma^{-1} : \sigma < \delta \rangle$, then f shows that $\langle u_\sigma : \sigma < \delta \rangle$ does not witness to $\delta \in S'$. The following is a theorem of ZFC (see [ES2]):

THEOREM. *If* $\Gamma'(Q/R) \neq 0$, *then* R *is not standard.*

The proof employs an argument like that in proofs using diamond principles, but here carried out in ZFC, using the N_δ to give the predictions. The converse is false: it is possible to construct R such that $\Gamma'(Q/R) = 0$ but R is not standard (see [ES3]).

16. Paradise.

This paper was delivered at a conference held in "paradise" — the island of Curaçao. I was reminded of Hilbert's assertion that mathematicians would not be driven from the paradise created for them by Cantor. In a similar fashion, I would like to claim that the results discussed above demonstrate that set-theorists — or those who employ set-theoretic methods — will not be driven from the paradise created by module theory for them to play their games in.

REFERENCES

[BSa] S. Bazzoni and L. Salce, *On non-standard uniserial modules over valuation domains and their quotients*, J. Algebra **128** (1990), 292-305.

[BFS] T. Becker, L. Fuchs and S. Shelah, *Whitehead modules over domains*, Forum Math. **1** (1989), 53-68.

[E] P. C. Eklof, **Set Theoretic Methods in Homological Algebra and Abelian Groups**, Les Presses de L'Université de Montréal (1980).

[EF] P.C. Eklof and L. Fuchs, *Baer modules over valuation domains*, Ann. Mat. Pura Appl.(IV) **150** (1988), 363-374.

[EFS] P.C. Eklof, L. Fuchs, and S. Shelah, *Baer modules over domains*, Trans. Amer. Math. Soc. **322** (1990), 547-560.

[EM] P. Eklof and A. Mekler, **Almost Free Modules**, North-Holland (1990).

[EMS] P. Eklof, A. Mekler and S. Shelah *Uniformization and the diversity of Whitehead groups*, to appear.

[ES1] P. Eklof and S. Shelah, *On Whitehead modules*, J. Algebra **142** (1991), 492-510.

[ES2] P. Eklof and S. Shelah, *On a conjecture regarding non-standard uniserial modules*, to appear in Trans. Amer. Math. Soc.

[ES3] P. Eklof and S. Shelah, *Explicitly non-standard uniserial modules*, to appear.

[F] L. Fuchs, **Infinite Abelian Groups**, vol.I, Academic Press (1970) .

[FSa] L. Fuchs and L. Salce, **Modules Over Valuation Domains**, Marcel Dekker (1985).

[FS] L. Fuchs and S. Shelah, *Kaplansky's problem on valuation rings*, Proc. Amer. Math. Soc. **105** (1989), 25-30.

[FV] L. Fuchs and G. Viljoen, *A weaker form of Baer's splitting problem over valuation domains*, Quaestiones Math. **14** (1991), 227-236.

[G] P.A. Griffith, *A note on a theorem of Hill*, Pacific J. Math. **29** (1969), 279-284.

[Mac] S. MacLane, **Homology**, Academic Press (1963).

[MaS] M. Magidor and S. Shelah, *When does almost free imply free?* (*For groups, transversal etc.*), to appear in Jour. Amer. Math. Soc.

[O1] B. L. Osofsky, *Constructing Nonstandard Uniserial Modules over Valuation Domains*, **Azu-maya Algebras, Actions, and Modules:** proceedings of a conference in honor of Goro Azu-maya's seventieth birthday, Contemporary Mathematics **124** (1992), 151–164.

[O2] B. Osofsky, *A construction of nonstandard uniserial modules over valuation domains*, Bull. Amer. Math. Society, **25** (1991), 89–97.

[S1] S. Shelah, *Whitehead groups may not be free even assuming CH, I*, Israel J. Math. **28** (1977), 193–203.

[S2] S. Shelah, *Whitehead groups may not be free even assuming CH, II*, Israel J. Math. **35** (1980), 257–285.

[S3] S. Shelah, *Diamonds, uniformization*, J. Symbolic Logic **49** (1984), 1022–1033.

[S4] S. Shelah, *Nonstandard uniserial module over a uniserial domain exists*, Lecture Notes in Math vol. 1182, Springer-Verlag (1986), pp. 135–150.

[T1] J. Trlifaj, *Whitehead property of modules*, Czech. Math. J. **36** (1986), 467–475.

[T2] J. Trlifaj, *Lattices of orthogonal theories*, Czech. Math. J. **39** (1989), 595–603.

[T3] J. Trlifaj, *Associative rings and the Whitehead property of modules*, Algebra Bericht **63**, R. Fischer (1990).

[T4] J. Trlifaj, *Von Neumann regular rings and the Whitehead property of modules*, Comment. Math. Univ. Carolinae **31** (1990), 621–625.

[T5] J. Trlifaj, *Non-perfect rings and a theorem of Eklof and Shelah*, Comment. Math. Univ. Carolinae **32** (1991), 27–32.

Modules with Distinguished Submodules
and Their Endomorphism Algebras

RÜDIGER GÖBEL

Fachbereich 6 — Mathematik und Informatik
Universität Essen, 4300 Essen 1, Germany

Introduction

It is my intention to give a survey of recent results on endomorphism algebras in the category of modules with distinguished submodules. We will give several applications to module theory and will discuss some open questions. Moreover I would like to draw attention to an interesting crossing of three main streams in mathematics, which are representation theory (and module theory in general), model theory and infinite combinatorics. The paper grew out of a survey talk I presented at the Curacao meeting on Abelian Groups.

Let R be an arbitrary commutative ring with $1 \neq 0$ and let $\mathbf{I} = (\mathbf{I}, <)$ always be a finite partially ordered set. An R_I-**module** will be an $|\mathbf{I}| + 1$ — tuple $\mathbf{F} = (F, F^i : i \in \mathbf{I})$ of a right R-module F with distinguished submodules F^i $(i \in \mathbf{I})$ such that $F^i \subset F^j$ for all $i \leq j \in \mathbf{I}$. A **morphism** $\varphi : \mathbf{F} \to \mathbf{F}'$ between two R_I-modules $\mathbf{F}, \mathbf{F}'$ is an R-homomorphism $\varphi : F \to F'$ with $F^i \varphi \subset F'^i$ for all $i \in \mathbf{I}$. This makes R_I-modules with their morphisms into an abelian category $Mod - R_I$ with sums, products, kernels, cokernels and extensions. The subcategory of all finitely generated R_I-modules is denoted by $mod - R_I$, and if R is a field then we will write K for R.

The category $mod - K_I$ has been introduced by Nazarova and Roiter [37] in terms of matrices in solving the Brauer-Thrall problem. In [22] Gabriel introduced its setting in terms of representations of quivers and proved Theorem 1 by using more algebraic arguments. Since then $mod - K_I$ has been studied intensively in the last two decades. The numerous results on $mod - K_I$ will be discussed in a nice forthcoming book by D. Simson [47], see also [6] for further references. We will not discuss the connection of $Mod - K_I$ with questions concerning its decidability and undecidability. These results related to algorithms of Turing machines answering sentences in the language of K_I-modules lead to similar patterns as Theorem 1. The results, mainly due to W. Baur, are discussed in [40,

This work is supported by the "Deutsche Forschungsgemeinschaft" under the project "Starre Systeme in der Algebra", contract no. Go 268/6-1.

55

41]. The recent influence of representation theory of p. o. sets on abelian groups, mainly on finite rank Butler groups, can be seen from the survey articles by D. Arnold [1, 2, 3], D. Arnold and C. Vinsonhaler [4].

I will concentrate on $Mod - R_I$ with a special emphasis that the modules will mainly be of infinite rank. The category $Mod - K_I$ has been studied less intensively in representation theory, nevertheless important results have been established in the last 120 years. This will be incorporated later on.

The endomorphism algebra $\mathrm{End}_R F$ of an R-module F becomes a complete Hausdorff topological R-algebra with respect to the **finite topology fin**. Making addition continuous, it is sufficient to define a basis of neighbourboods of $0 \in \mathrm{End}_R F$, which is

$$\{\mathrm{Ann}\ E : E \leq F, E\ \text{finite}\}\ , \text{where}\ \mathrm{Ann}\ E = \{\sigma \in \mathrm{End}\ F, E\ \sigma = 0\}$$

is the **annihilator** ideal of E. If $\mathbf{F}$ is an R_I-module, then

$$\mathrm{End}\mathbf{F} \subset (\mathrm{End}\ F, \mathbf{fin})\ \text{is a closed subalgebra.}$$

Is any closed subalgebra of $(\mathrm{End}\ F, \mathbf{fin})$ of the form $\mathrm{End}\ \mathbf{F}$ for a suitable $\mathbf{F}$? We will characterize all p. o. sets $\mathbf{I}$ when this will always be the case. In fact this will be the case if and only if $\mathbf{I}$ is of "infinite representation type"; see the definition below. If we choose $\mathbf{I}$ to be an antichain of length 5, this answers a question of Corner's [11]. It will be convenient to follow tradition and to denote R_I-modules by R_n-modules if $\mathbf{I}$ is an antichain of length n, cf. Ringel [43].

We will begin with a few examples from abelian groups which motivate an investigation of R_I-modules. Let R = $\mathbf{Q}$,

$$F =< e_1, \ldots., e_d, f_1, \ldots, f_d > \text{a vector space with basis}\ e_1, \ldots, f_d.$$

If $F^0 =< e_1, \ldots, e_d >_{\mathbf{Q}}$, $F^1 =< f_1, \ldots, f_d >_{\mathbf{Q}}$, $F^2 =< e_1 + f_1, \ldots, e_d + f_d >_{\mathbf{Q}}$ and $F^3 =< e_1 + f_2, \ldots, e_{d-1} + f_d >_{\mathbf{Q}}$ are the distinguished $\mathbf{Q}$-submodules of F, then it is easy to see that the $\mathbf{Q}_4$-module $\mathbf{F} = (F, F^i : i < 4)$ has endomorphism algebra $\mathbf{Q}$ and $\mathbf{F}$ must be indecomposable because $\mathbf{Q}$ has no idempotents different from 0 and 1.

The above examples for all $d \geq 1$ also show that $mod - \mathbf{Q}_4$ has infinitely many indecomposables and hence $mod - \mathbf{Q}_4$ is of infinite representation type. If $n < 4$, then $mod - K_n$ is of finite representation type and all indecomposable K_n-modules (of finite dimension) are listed in Brenner [8]. It follows from representation theory and is a result of Kleiner [31] that finite representation is independent of the field K. A new direct proof has been given by Arnold, Richman [5]. We also learn from the example above some strategy for constructing complicated R_I-modules. The first distinguished submodules F^0, F^1, F^2 are constructed "canonically" in a trivial way and only the fourth module F^3 is more complicated and kills all possible decompositions of $\mathbf{F}$. This will always be the case later on. Our interest on R_I-modules comes from abelian groups. Indecomposable $\mathbf{Q}_4$-modules can be used to construct indecomposable abelian groups. The $\mathbf{Q}_4$-structure has to be coded into the desired abelian group G. This is done by arithmetic, using four distinct primes p_0, p_1, p_2, p_3. Working over more general R-modules, primes can be replaced by comaximal multiplicatively closed subsets of non-zerodivisors.

Using our example above, we consider the free abelian subgroups

$$U^0 =< e_1, \ldots, e_d >, U^1 =< f_1, \ldots, f_d >, U^3 =< e_1 + f_1, \ldots, e_d + f_d >$$

and

$$U^4 = < e_1 + f_2, ..., e_{d-1} + f_d >$$

of $U = < e_1, ..., f_d >$. Clearly $U^i \otimes \mathbf{Q} = F^i$ and if $p_i^{-\infty} U^i$ denotes the subgroup of F^i generated by all elements x in F^i with some multiple $p_i^k x$ in U^i, then

$$G = < p_i^{-\infty} U_i : i < 4 >$$

is a torsion-free group of rank $2d$. Any $\sigma \in \text{End } G$ extends uniquely to $\bar{\sigma} \in \text{End } \mathbf{F}$ and an easy arthmetic argument, using the district primes, shows that $F^i \bar{\sigma} \subset F^i$ hence $\bar{\sigma} \in End \mathbf{F} = \mathbf{Q}$ and G must be indecomposable.

This is essentially the method by which indecomposable torsion-free abelian groups of rank 2 have been constructed by Levi [35] and Pontryagin [39]. The prime number argument could be replaced by an argument using algebraically independent elements over $\mathbf{Z}$. The existence of indecomposable abelian groups of any rank was only shown in 1974 in a celebrated paper by Shelah [46]. Hence it is very tempting to prove a theorem on R_4-modules which gives the existence of indecomposable abelian groups immediately, compare Theorem 5.

If λ is an infinite cardinal, then Theorem 5 provides a $\mathbf{Z}_4$-module $\mathbf{U} = (U, U^i : i < 4)$ with End $\mathbf{U} = \mathbf{Z}$. Moreover U is free abelian of rank λ and the subgroups U^i are summands of U. In particular End $\mathbf{F} = \mathbf{Q}$ if $\mathbf{U} \otimes \mathbf{Q} = \mathbf{F}$, which is $F = U \otimes \mathbf{Q}$ and $F^i = U^i \otimes \mathbf{Q}$. The argument above shows that $G = < p_i^{-\infty} U^i : i < 4 >$ is indecomposable of rank λ.

The following examples are more sophisticated and use the full strength of Theorem 6. The finite topology becomes crucial if we want to consider infinite decompositions of modules. The first collection of examples uses a topological ring due to Corner [12]. The second set of examples needs another topological ring from Göbel, Ziegler [29].

Ring 1 [12]: If κ is an infinite cardinal, then there exists a Hausdorff complete topological ring R with a basis of neighbourhoods of 0 consisting of κ two-sided ideals $J_i (i < \kappa)$ such that R/J_i is free abelian of rank κ and with the further property:

(*) For each $0 \neq x \in R$ there exists a summable family of κ orthogonal idempotents e_i $(i < \kappa)$ such that $x e_i \neq 0$ for all $i < \kappa$.

Ring 2 [29]: There exists a ring R satisfying the same hypothesis as **ring 1** but with the properties (**) in place of (*).

(**) (i) For each $\rho < \lambda$ there is a summable family $e_{\rho\alpha}(\alpha < \rho)$ of orthogonal idempotents $\neq 0$.

(ii) There is no orthogonal family of λ idempotents $\neq 0$.

Using ring 1 and Theorem 6, then for any infinite cardinal $\lambda \geq \kappa$ we find a κ-superdecomposable group G of cardinality λ. The group G is κ-superdecomposable if any a summand $\neq 0$ of G decomposes into a direct sum of κ non-trivial summands.

A modification of ring 1 [12] provides an abelian group G of any cardinal $\geq \kappa$ with $G = H_1 \oplus H_2 = G_1 \oplus G_2$ each G_i κ-superdecomposable and each H_i indecomposable.

Using ring 2 and Theorem 6, we see that for every infinite cardinal λ there exists an abelian group G of cardinality λ such that the following holds

(i) We can find decompositions $G = \oplus_{i \in \kappa} G_i$ $(G_i \neq 0)$ for any $\kappa < \lambda$

(ii) G does not decompose into λ non-trivial summands.

The case of cf $\lambda = \omega$ is particular interesting. Observe that realization theorems as above hold for all cardinals $\geq |R|$. Realization theorems which drive from what we [14] called **"Shelah's Black Box"** have the defect that some cardinals of the constructed modules $\geq |R|$ are missing: if $cf|R| = \omega$, then the constructed abelian groups are of size $\geq |R|^{\aleph_0} > |R|$, for instance. We also want to mention that a fine tuning for coding the distinguished submodules (as above) into a completely decomposable group of rank $\aleph_0$ leads to Butler groups of countable rank with prescribed (topological) endomorphism ring R whenever R^+ is a countable Butler group which is p-reduced for four primes p, cf. [17]. Similar to [42], the "p. o. set-theorems" 5 and 6 can be applied to establish realization theorems of endomorphism rings of Butler groups. In order to describe our final results we will need Kleiner's [31] list of p. o. sets. This list $\mathbf{K}$ consists of the following five p. o. sets.

$$(1,1,1,1)\ (2,2,2)\ (1,3,3)\ (N,4)\ (1,2,5).$$

The ordering (n,k) is the disjoint union of two chains n and k of length n and k respectively. The ordering N is a set of four elements $\{a,b,c,d\}$ where $a < b, c < d$ and $c < b$.

In 1975 Kleiner [31] characterized the orderings of finite representation type. These orderings admit only finitely many nonisomorphic indecomposable K_I-modules for some field K. The following result on finite dimensional K_I-modules is due to Kleiner [31, p.607, Theorem 1].

Theorem 1. *Let $I = (I, \leq)$ be a partially ordered set. The following are equivalent.*
(1) $\mathbf{I}$ *is of infinite representation type.*
(2) There is an ordering from $\mathbf{K}$ that can be embedded into $\mathbf{I}$.

Further proofs of this result can be found in Simson [47] and Kerner [30]. In a subsequent paper [32] Kleiner determined all indecomposable representations of p. o. sets of finite type. This list of the indecomposable representations was corrected by Arnold and Richman [5] and Simson [47].
If $mod - K_I$ is of finite representation type we also say that $\mathbf{I}$ is of finite representation type. A modernized proof of Kleiner's result, which is also based on Nazarova and Roiter's "derivative" of p. o. sets [37], may be found in Simson [47].
Passing to $Mod - K_I$ (of infinite dimension), the following theorem, proved independently by Ringel, Tachikawa [45] and Simson [48] illustrates our starting point, cf. [6, Theorem 4.1].

Theorem 2: *If $\mathbf{F}$ is an indecomposable K_I-module and $\mathbf{I}$ is of finite representation type, then $\dim_K F$ must be finite.*

We see that over p. o. sets $\mathbf{I}$ of finite representation type and $\dim_K F = \infty$ it will never be possible to realize End $\mathbf{F} = K$ because End $\mathbf{F}$ will have non-trivial orthogonal projections. It is our aim to realize **any** R-algebra A as End $\mathbf{F} = A$ for a suitable R_I-module $\mathbf{F}$. Then the p. o. set $\mathbf{I}$ **must** be of infinite representation type, as follows from Theorem 2.
First we restrict ourselves to antichains $\mathbf{I}$ and recall a major step, due to Corner [11] (1969), towards our anticipated result. If A is an R-algebra, then $\|A\|$ will denote the minimal cardinal of generators for A (as an R-algebra) and A_R denotes the R-module A.

Proposition 3: *If A is an R-algebra and $\lambda \geq \|A\|$ is an infinite cardinal, then we can find an R_5-module*

$$\mathbf{F} = (\oplus_\lambda A_R, F^0, F^1, F^2, F^3, F^4)\ with\ \text{End}\ \mathbf{F} = A.$$

The cardinal restrictions on $\|A\|$, λ in Corner's original result [11] have been eliminated in Franzen, Göbel [19] using techniques from Shelah [46]. Theorem 2 shows that the category of R_5-modules in Proposition 3 cannot be replaced by R_n-modules for $n < 4$. This leads to the question whether $Mod - R_4$ can be substituted. Already Corner [11] in 1966 remarks *"it is not known whether the trivial algebra can be realized by 4 subspaces of an infinite-dimensional vector space, even in the countable case: indeed, I know of no algebra that can be shown not to be realizable by 4 subspaces of some infinite-dimensional vector space."* The answer to his second remark will be given for any commutative base ring R : There isn't any!

Moreover, the answer to his first remark is hidden in an over 100 year old paper on algebraic reduction of bilinear forms by Kronecker [34] published in the "Sitzungsberichte der königlich preussischen Akademie der Wissenschaften zu Berlin" on November 27th 1890. As Kronecker mentions on *p.*1225, he had finished his paper already in 1874 but he did not publish it at that time because he felt that the applied methods were not pure algebraic and getting too much into analysis. He changed his mind 16 years later! In this paper, known in representation theory in connection with the Kronecker modules (see Ringel [44]), Kronecker realizes $\mathbf{Q}$ as endomorphism algebra of a $\mathbf{Q}_4$-module $\mathbf{F}$ of countable dimension. In fact, Kronecker provides a countable, orthogonal family of such modules $\mathbf{F}_i$ with

$$\mathrm{Hom}(\mathbf{F}_i, \mathbf{F}_j) = \delta_{ij}\mathbf{Q} \ (i, j \in \omega)$$

where the Kronecker symbol δ_{ij} represents the obvious maps. This result is a crucial starting point for the more general Theorem 5. It is also used in Ringel's interesting paper [43, *p.*407] which provides a

Proposition 4: *There is a full and exact embedding*
$Mod - K_5 \to Mod - K_4$.

Using Proposition 3, this result ensures that in case of vector spaces, all K-algebras A can be realized over $Mod - K_4$. In particular we can find arbitrarily large indecomposable K_4-modules. It remains to consider the following generalizations:
(i) replace the field K by a ring R
(ii) replace 4 by a p. o. set I of infinite representation type
(iii) determine the algebraic structure of the R_I-modules realizing an R-algebra A
(iv) derive a topological isomorphism (equality) $\mathrm{End}_R\mathbf{F} = A$ in order to get the examples at the beginning.

Main results

The first step of the constructions in $Mod - R_4$ imitates Kronecker's idea. However, we must overcome difficulties arising from zero-divisors in a general ring R. This is ensured by passing from R to the polynomial ring $R[x]$. The role of primes is taken by a countable sequence $p_i \in R[x]$ of monic comaximal polynomials p_i (p_i, p_j are comaximal if we can find q_i, $q_j \in R[x]$ with $p_iq_i + p_jq_j = 1$). Then let S be the multiplicatively closed subset of $R[x]$ generated by the $p_i's$. Localizing at S we derive $R[x] \subseteq S^{-1}R[x]$ and we have unique partial fraction representation in $S^{-1}R[x]$ in terms of $\frac{1}{p_i}'s$. If $\alpha \subseteq \omega = \{0, 1, ..., \}$, then let

$$L_\alpha = \{f \cdot p_{i_o}^{-1} \cdot \ ... \ \cdot p_{i_n}^{-1} : f \in R[x], i_o, \ ... \ , i_n \in \alpha \ \text{distinct}\}.$$

It is easy to see that $L_\alpha \subseteq S^{-1}R[x]$ is an S-torsion-free $R[x]$-submodule and Baer's type arguments as in Fuchs [20, pp. 107-111], show the following.

If $\varphi : L_\alpha \to L_\beta$ is an $R[x]$-homomorphism, then φ is multiplication with (1φ) and $\varphi = 0$ if $\alpha \backslash \beta$ is infinite, cf. Göbel, May [28]. If d_i denotes the degree of p_i and

$$U_i = \bigoplus_{0 \le j < d_i} \frac{x^j}{p_i} R,$$

then $U_\alpha = \oplus_{i \in \alpha} U_i \subset L_\alpha$ is the building block for the generalized Kronecker module. If $L_\alpha = U_\alpha \oplus R[x] \supseteq W_\alpha = U_\alpha \oplus 1R$, then the R_4-modules $\mathbf{F}_\alpha (\alpha \subset \omega)$ are as follows. $F_\alpha = U_\alpha \oplus W_\alpha$, $F_\alpha^0 = U_\alpha \oplus 0$, $F_\alpha^1 = 0 \oplus W_\alpha F_\alpha^2 = <(u,u) : u \in U_\alpha>$ and the "complicated" submodule

$$F_\alpha^3 = <(u, ux) : u \in U_\alpha>.$$

Changing the indexing set, we obtain an orthogonal family $\mathbf{F}_\alpha (\alpha \in \omega)$ of R_4-modules with

$$\mathrm{Hom}(\mathbf{F}_\alpha, \mathbf{F}_\beta) = \delta_{\alpha\beta} R.$$

The next step is the construction of a countable "fully" rigid system. The notion of a fully rigid system is due to Corner [13] (published recently but known for several years).

Definition: *Let λ be a cardinal and A be an R-algebra over a commutative ring R. An A-fully rigid system on a set λ is a faithful family $M_\alpha (\alpha \subseteq \lambda)$ of A-modules with $M_\alpha \subseteq M_\beta$ for all $\alpha \subseteq \beta \subseteq \lambda$ and*

$$\mathrm{Hom}_R(M_\alpha, M_\beta) = A \ \text{if} \ \alpha \subseteq \beta \ \text{and} \ \mathrm{Hom}_R(M_\alpha, M_\beta) = 0 \ \text{if} \ \alpha \not\subseteq \beta.$$

Similarly fully rigid systems can be defined in any category, in particular for the category $Mod - R_I$, which can we used in Theorem 5.

Choosing an antichain (with respect to inclusion) of the power set of λ, every fully rigid system on λ leads to a rigid system on 2^λ with

$$\mathrm{Hom}_R(M_i, M_j) = \delta_{ij} A \ \text{for all} \ i, j \in 2^\lambda.$$

The converse is not true.
The following three observations are crucial hints for the construction of an R-fully rigid system on ω. It is known that such a $\mathbf{Q}$-system over ω exists for torsion-free abelian groups of rank 2. In order to derive Proposition 4, Ringel [43, p.405 − 406] considers elements in $Ext(\mathbf{F}_\alpha, \mathbf{F}_\beta)$ with $\mathbf{F}_\alpha, \mathbf{F}_\beta$ Kronecker modules in $Mod - K_4$. Moreover, Fuchs [21, p.107, condition (iii)] uses the hypothesis $Ext(..., ...) \neq 0$ to derive the existence of arbitrarily large indecomposable modules over certain commutative rings [21, p.110 − 111]. By the same token, building extensions of our original Kronecker modules, we are able to convert our rigid family $\mathbf{F}_\alpha$ of R_4-modules into a fully rigid system over ω. The final arguments in $Mod - R_4$ are Shelah's combinatorial methods to derive the general realization Theorem 5 below for an antichain $\mathbf{I}$ of length 4, see Göbel, May [28]. It remains to consider the other four p. o. sets in Kleiner's list $\mathbf{K}$, which was investigated in Böttinger, Göbel [6]. The analogue of Kronecker's module in case $\mathbf{I} = (N, 4)$ is the following R_I-module.
Let $U = U_\omega$ be from above and $V = U \oplus 1R \subseteq S^{-1} R[x]$.
Let $V \to V^i (v \to v^i)$ denote copies (of elements) of V for $i < 5$. Similarly let $U^i \subseteq V^i$. Then we define

$$F = V^0 \oplus U^1 \oplus U^2 \oplus U^3 \oplus U^4$$

$$F^0 = U^3 \oplus U^4 \subseteq F^1 = U^1 \oplus U^2 \oplus U^3 \oplus U^4$$

$$F^1 \supset F^2 = < u^1 + u^2 + u^3 + u^4 : u \in U > \subseteq F^4$$

$$F^4 = < xu^0 + u^2, \, u^0 + u^4, \, u^1 + u^2 + u^3 + u^4 : u \in U >$$

which represents "N" and

$$F^5 = V^0 \subseteq F^6 = V^0 \oplus U^1 \subseteq F^7 = V^0 \oplus U^1 \oplus U^2 \subseteq F^8 = V^0 \oplus \bigoplus_{i=1}^{3} U^i$$

which represents "4".

Similarly we get a rigid family of orthogonal $R_{(N,4)}$-modules which then is treated as the case $Mod - R_4$, cf. Böttinger, Göbel [6]. The structure of the other generalized Kronecker modules over a p. o. set $\mathbf{I}$ from Kleiner's list are inspired by the construction in the appendix by Dlab and Ringel [15]. Finally we derive the following result.

Theorem 5. *Let $\mathbf{I} = (\mathbf{I}, \leq)$ be a partially ordered set and λ an infinite cardinal. The following are equivalent*

(1) $\mathbf{I}$ is of infinite representation type

(2) There is an ordering from $\mathbf{K}$ that can be embedded into $\mathbf{I}$.

(3) For all commutative rings R, for all R-algebras A with $\lambda \geq \|A\|$ and for $F = \oplus_\lambda A$ there is a family of R_I-modules $\mathbf{F}_\alpha = (F, F_\alpha^i : i \in \mathbf{I})(\alpha \subseteq \lambda)$ with $Hom(\mathbf{F}_\alpha, \mathbf{F}_\beta) = A$ and $\mathbf{F}_\alpha \subseteq \mathbf{F}_\beta$ if $\alpha \subseteq \beta$ and $Hom(\mathbf{F}_\alpha, \mathbf{F}_\beta) = 0$ if $\alpha \not\subseteq \beta$.

Moreover, the submodules F_α^i in (3) are R-summands of F_α.

Remarks: The case $(1) \to (2)$ is of course Kleiner's theorem, and $(3) \to (1)$ (take $R = K$) follows from Ringel-Tachikawa and Simson (Theorem 2). If $\mathbf{I}$ is an antichain of length 4, then Theorem 5 is from Göbel, May [28] and if $\mathbf{I}$ is an antichain of length 5 the result is (up to some cardinal restrictions lifted in Franzen, Göbel [19]) due to Corner [11].

In order to derive examples mentioned above we must strengthen Theorem 5 and derive a new result which is based on the finite topology of endomorphism rings. The result follows from the construction of $\mathbf{F}$ in Theorem 5 (Böttinger, Göbel [6]) and Göbel, Ziegler [29, pp. 494 – 495]. We are only interested in a topological generalization of "$(1) \to (3)$" in Theorem 5, which is stated as our next

Theorem 6: *Let $\mathbf{I}$ be a p. o. set of infinite representation type and A be a complete Hausdorff R-algebra over a commutative ring R whose topology has a basis $N_j (j \in \mathbf{J})$ of right ideals. If λ is an infinite cardinal with $|\mathbf{J}|, |A/N_j| \leq \lambda$ for all $j \in \mathbf{J}$, then there exists an R_I-module $\mathbf{F} = (\oplus_\lambda(\oplus_{j \in J} A/N_j), F^i : i \in \mathbf{I})$ such that $A = End_R\mathbf{F}$ is a topological equality.*

Epilogue or Spätlese

Theorem 5 can be used to derive new results in module theory as seen by our previous examples. Such a treatment can be axiomatized as observed recently by Corner [13]. His result can now be sharpened replacing 5 by 4. A proof of the following corollary follows by inspection of the proof in Corner [13] substituting our Theorem 5 at all relevant places.

Corollary 7: *Let A be an R-algebra with an A-fully rigid system N_α ($\alpha \subseteq 4 = \{0,1,2,3\}$) on 4 and let λ be any infinite cardinal $\geq |N_4|$. Then there exists an A-fully rigid system of R-modules of cardinality λ on λ.*

Remark: Observe that the hypothesis concerning an A-fully rigid system on 5 in Corner's [13] hypothesis needs the knowledge of 32 particular modules, while 16 modules suffice by Corollary 7. Using Corollary 7, a recent result of Fuchs [21, $p.110$, Theorem 6] can be sharpened replacing 6 by 5. Fuchs [21] is based on Corner [13], and his Lemma 4 must be replaced by Corollary 7 to obtain the following interesting application.

Corollary 8: *Let $\mathcal{X}$ be a class of R-modules which is closed under extensions, direct sums and images under homomorphisms between modules in $\mathcal{X}$. Suppose we can find a rigid system of five modules X_0, X_1, X_2, X_3, X_4 in $\mathcal{X}$*

$$(i.e.\ \operatorname{Hom}_R(X_i, X_j) = \delta_{ij} R \text{ for all } i, j < 5)$$

with $R \subseteq \operatorname{Ext}_R^1(X_i, X_0)$ for $1 \leq i < 5$. For every infinite cardinal $\lambda \geq max_{i<5} |X_i|$ we can find an R-fully rigid system of R-modules in $\mathcal{X}$ of cardinality λ on λ.

If R has only the trivial idempotents 0, 1, then all members of the R-fully rigid system are obviously indecomposable. Examples for particular rings which lead to arbitrarily large indecomposable divisible torsion modules are discussed in Fuchs [21].

Question A: *Find an example showing that 5 in Corollary 8 can not be replaced by 4.*

Finally we want to discuss the minimality of 4 in Theorem 5 in case of antichains. The minimality follows from Theorem 1 which holds for fields K. The following condition on an R-algebra A is quite general but excludes the possibility for R to be a field. Let $S \leq R$ be a multiplicatively closed subset of non-zero divisors such that $\bigcap_{s \in S} As = 0$ (*A is S-reduced*) and $sa = 0$ implies $a = 0$ (*A is S-torsion-free*). Then we can show that 2 is minimal.

Theorem 9: (Böttinger, Göbel [7]) *There exists an R_2-module $\mathbf{F} = (F, F^0, F^1)$ where F is an S-pure submodule of $\prod_\lambda$ containing $\oplus_\lambda A$ and*
$$\operatorname{End} \mathbf{F} = A.$$

Project B: *Replace the commutative ring R by a ring and prove similar results as above.*

Problem C: *Find a proof of Theorem 9 which is based on the arguments giving Theorem 5. This would cover a larger cardinal spectrum because the present proof uses Shelah's Black Box.*

Question D: *Can F be free in Theorem 9?*

References

1. D. Arnold, Notes on Butler groups and balanced extensions, Boll. Un. Mat. Ital. A (6) **5** (1986) 175 – 184.

2. D. Arnold, Representations of partially ordered sets and abelian groups, in: *Proceedings of the Abelian Group Conference at Perth*, Contemp. Math., Vol.**87** (Amer. Math. Soc., R. I. 1988) 91 – 109 .

3. D. Arnold, C. Vinsonhaler, Invariants of indecomposable representations of finite posets, Journ. Alg. **147** (1992) 245 – 264.

4. D. Arnold, C. Vinsonhaler, Finite rank Butler groups, a survey on recent results, this volume

5. D. Arnold, F. Richman, Field-independent representations of partially ordered sets, to appear in Forum Math.

6. C. Böttinger, R. Göbel, Endomorphism algebras of modules with distinguished partially ordered submodules over commutative rings, Journ. Pure and Appl. Algebra **76** (1991) 121 – 141.

7. C. Böttinger, R. Göbel, Modules with two distinguished submodules, this volume

8. S. Brenner, Endomorphism algebras of vector spaces with distinguished sets of subspaces, J. Algebra, **6** (1967) 100 – 114.

9. S. Brenner, On four subspaces of a vector space, J. Algebra, **29** (1974) 587 – 599.

10. S. Brenner, M. C. R. Butler, Endomorphism rings of vector spaces and torsion free abelian groups, J. London Math. Soc., **40** (1965) 183 – 187.

11. A. L. S. Corner, Endomorphism algebras of large modules with distinguished submodules, J. Algebra, **11** (1969) 155 – 185.

12. A. L. S. Corner, On the existence of very decomposable abelian groups, pp. 354 – 357 in *"Abelian Group theory"*, ed. R. Göbel, L. Lady, A. Mader Springer LNM Vol. 1006 (1983).

13. A. L. S. Corner, Fully rigid systems of modules, Rend. Sem. Mat. Univ. Padova,**82** (1989) 55 – 66.

14. A.L.S. Corner, R. Göbel, Prescribing endomorphism algebras, a unified treatment, Proc. London Math. Soc.(3), **50** (1985) 447 – 479.

15. V. Dlab, C. M. Ringel, Indecomposable representations of graphs and algebras, Mem. Amer. Math. Soc. **173**, AMS Providence, 1976.

16. Y. A. Drozd, Coxeter transformations and representations of partially ordered sets, Funct. Anal. and Appl., **8** (1974) 219 – 225.

17. M. Dugas, B. Thomé, Countable Butler groups and vector spaces with four distinguished subspaces, Journ. Alg. **138** (1991) 249 – 272.

18. P. Eklof, E. Mekler, *Almost free modules, set-theoretic methods*, North-Holland, Amsterdam – New York 1990.

19. B. Franzen, R. Göbel, The Brenner-Butler-Corner-Theorem and its applications to modules, 209 – 227 in: *"Abelian Group Theory"*, ed. R. Göbel and E. A. Walker, Gordon and Breach (1986) 209 – 227.

20. L. Fuchs, *Infinite abelian groups,* Vol. 2, Academic Press, New York 1973.

21. L. Fuchs, Large indecomposable modules in torsion theories, Aequationes Math. **34** (1987) 106 – 111.

22. P. Gabriel, Représentations indécomposables des ensembles ordonnés, Séminaire P. Dubreil, Paris (1972 – 1973) 1301 – 1304.

23. P. Gabriel, Unzerlegbare Darstellungen I, Manuscripta Math. **6** (1972) 71 – 103.

24. P. Gabriel, Indecomposable representations II, Symp. Math. **11** (1973) 81 – 104.

25. P. Gabriel, L. A. Nazarova, A. V. Roiter, Représentations indécomposables, un algorithme, C. R. Acad. Sci. Paris Ser. I **307** (1988) 701 – 706.

26. I. M. Gelfand, V. A. Ponomarev, Problems of linear algebra and classification of quadruples of subspaces in a finite-dimensional vector space, in: *Hilbert Space operators*, Tihany 1970, Colloq. Math. János Bolyai, Vol. 5 (North-Holland, Amsterdam – New York) 163 – 237.

27. R. Göbel, Vector spaces with five distinguished subspaces, Resultate Math., **11** (1987) 211 – 228.

28. R. Göbel, W. May, Four submodules suffice for realizing algebras over commutative rings, J. Pure and Appl. Algebra **65** (1990) 29 – 43.

29. R. Göbel, M. Ziegler, Very decomposable abelian groups, Math. Z. **200** (1989) 485 – 496.

30. O. Kerner, Partially ordered sets of finite representation type, Comm. Algebra **9** (8) (1981) 783 – 809.

31. M. M. Kleiner, Partially ordered sets of finite type, J. Soviet Math. **3** (1975) 607 – 615.

32. M. M. Kleiner, On the exact representations of partially ordered sets of finite type, J. Soviet Math. **3** (1975) 616 – 628.

33. M. M. Kleiner, Pairs of partially ordered sets of tame representation type , Linear Algebra Appl. **104** (1988) 103 – 115.

34. L. Kronecker, Algebraische Reduction der Schaaren bilinearer Formen, Sitzungsberichte der königl. preuss. Akademie der Wiss. Berlin (1890) 1225 – 1237.

35. F. Levi, Arithmetische Gesetze im Gebiet diskreter Gruppen, Rend. Circ. Mat. Palermo **35** (1912) 225 – 236.

36. L. A. Nazarova, Partially ordered sets of infinite type, Izv. Akad. Nauk SSR, Ser. Mat., **39** (1975) 963 - 991 = Math. USSR izv., **9** (1975) 911 – 938.

37. L. A. Nazarova, A. V. Roiter, Representations of partially ordered sets, J. Soviet. Math. **3** (1975) 585 – 607.

38. L. A. Nazarova, A. V. Roiter, Representations of bipartite completed posets, Comment. Math. Helvetici **63** (1988) 498 – 526.

39. L. Pontryagin, The theory of topological commutative groups, Ann. Math. **35** (1934) 535 — 569.

40. M. Prest, *Model theory and modules*, London Math. Soc. **130** (1988), Cambridge Univ. Press

41. M. Prest, Model theory and representation type of algebras, pp. 219 – 260 in "*Logic Colloquium '86*", ed. F. R. Drake, J. K. Truss, North-Holland 1988.

42. K. M. Rangaswamy, C. Vinsonhaler, Butler groups and representations of infinite rank, in preparation.

43. C. M. Ringel, Infinite-dimensional hereditary algebras, Symp. Math., **23** (1979) 321 – 412.

44. C. M. Ringel, *Tame Algebras and Integral Quadratic Forms*, Lecture Notes in Mathematics Vol. **1099** (Springer, 1984).

45. C. M. Ringel, H. Tachikawa, QF-3rings, J. reine angew. Math.,**272** (1975) 49 – 72.

46. S. Shelah, Infinite abelian groups, Whitehead problem and some constructions, Israel J. Math., **18** (1974) 243 – 256.

47. D. Simson, *On representations of partially ordered sets*, monograph, to appear Gordon and Breach London 1992.

48. D. Simson, Functor categories in which every flat object is projective, Bull. Acad. Pol. Ser. Math. **22** (1974) 375 – 380.

On the Structure of Torsion-free Groups
of Infinite Rank

PAUL HILL *

Department of Mathematics
Auburn University, Alabama, U.S.A.

1. INTRODUCTION.

In response to the kind invitation of the organizers of the Curacao Conference, this is a survey paper on the structure of torsion-free groups of infinite rank. In attempting to survey a subject of such broad scope, one encounters at once two basic questions: where should the survey begin, and what should be its boundaries?

In regard to the first question, I am not sure that the *Alice in Wonderland* advice is feasible here, but if we were to begin at the beginning then certainly we would start with Reinhold Baer's famous 1937 paper entitled "Abelian groups without elements of finite order". In view of the fact that this paper by Baer [Ba] and other important contributions to the subject prior to 1973 are treated in excellent expository fashion in $[F_1]$ $[F_2]$, $[F_3]$ and $[G_1]$, we shall make reference here to such works almost exclusively for background and historical purposes only. In fact, we shall concentrate on results established in the last decade, which has been a very active period for torsion-free groups. (Needless to say, "group" here means "abelian group".) For a moment, however, let us return to Baer's paper.

A group is simply presented if it can be presented by generators and relations where each relation involves at most two generators. It is not a difficult exercise to show that a torsion-free group is simply presented if and only if it is completely decomposable, that is, if and only if it is isomorphic to a co-product of subgroups of $\mathbb{Q}$. Therefore, for torsion-free groups, the terms "simply presented" and "completely decomposable" are interchangeable. Actually, Baer [Ba] called these groups "completely reducible". Whatever they are called, Baer completely determined the structure of these groups, but his goal was more ambitious. He wanted to be able, likewise, to describe separable groups. A torsion-free group G is separable if each finite set is contained in a completely decomposable direct summand. Baer's investigation of separable groups presented something of a paradox: success would imply failure. There was little realistic hope, particularly at that time, of classifying separable

*Supported by NSF grant DMS-90-96243.

groups without showing that they are completely decomposable, but in this case separable groups would lose their identity. Baer, of course, knew that although countable separable groups are completely decomposable, in general they are not. Thus, he perhaps never had much expectation of classifying separable groups in the way he had done for completely decomposable groups using numerical invariants like rank $(G(\sigma)/G(\sigma^*))$. (Notation not explained herein agrees more or less with conventional usage, but we prefer $G(\sigma^*)$ to the older usage of $G^*(\sigma)$ because our notation serves as a constant reminder that the fully invariant subgroups $G(\sigma)$ and $G(\sigma^*)$ are of the same ilk.) As for as I know, Baer never seriously pursued the topic again. Surprisingly enough, except for the summand problem, not much more is known today about separable groups than in 1937. This fact is illustrated by the following problem.

Problem 1. *Can separable groups that appear as pure subgroups of completely decomposable groups be classified using the Baer invariants, rank $(G(\sigma)/G(\sigma^*))$, and other numerical invariants?*

We have mentioned the summand problem for separable groups. Baer proved, in his initial paper, that summands of separable groups are again separable for the homogeneous case, but the general case was not settled until 1970 by Fuchs [F_4]. Since 1970, however, there have been many generalizations of this result including the same result for a class of groups called k-groups, which are more general than separable groups. We will discuss this and other generalizations in section 5.

We turn now to address the second of our fundamental questions about the nature of this survey. What should the survey encompass? As we have already indicated, the survey will concentrate on results of the last decade especially those belonging to the second half, but as our initial discussion would indicate we will attempt to properly connect these investigations to more classical results and thereby hopefully provide some degree of continuity between classical and more recent results. There is another restriction related to the subject matter of this survey. Since there are several other survey papers being presented to this conference on different areas of torsion-free groups, a result that (in the judgement of the author) falls more into one of these other categories will generally be omitted here although it may contribute also to the structure of infinite rank torsion-free groups. For example, one could claim that the extensive work by U. Albrecht ([A_1], [A_2]) dealing with A-projective groups has something to do with structure, but it probably has more to do with endomorphism rings and/or homological methods if all of these topics are to be distinguished. Be mindful that this is only one example among many that we do not include in our survey. In fact, it is with perhaps more guilt that we omit a discussion of realization theorems such as those found in [C-G], [D-G_1] and [D-G_2] that extend and generalize the classical results of Corner [C] because these results have a more direct relation with what we think of as structure.

Having given some indications of what we are excluding, it is time to say what we have included in this selective survey. I have broken down recent advances concerning the structure of infinite rank torsion-free groups into five broad categories:

(1) Free (Extremely Simply Presented) Groups and Relative Structure.
(2) Butler Groups of Infinite Rank.
(3) Refined Concepts of Heights and Purity with Applications.
(4) k-Groups: A Generalization of Separable Groups.
(5) Simply Presented Torsion-free Groups and Relative Structure.

We shall accentuate the positive, but at the same time by giving a number of counterexamples and other negative results we have tried to define the limits for which certain results are true. Finally, there are ten open problems that are stated for the purpose of stimulating further research.

2. Freeness and Relative Structure.

In an attempt to tie the past to the present, let us recall that for many years (roughly from 1940 unitl about 1980) the main attack on the structure of torsion-free groups of infinite rank was directed toward establishing various criteria for freeness. Probably the three most famous problems in the history of abelian groups concerned precisely this question. The three problems referred to are the following.

(1) The Bounded Sequence Problem. Is the group consisting of the bounded sequences of integers a free group?
(2) Baer's Problem. Does $Ext(G,T) = 0$ for all torsion T imply that G is free?
(3) Whitehead's Problem. Does $Ext(G,\mathbb{Z}) = 0$ imply that G is free?

Solutions to the first two of these problems can be found in $[F_3]$, and P. Eklof [E] has given an excellent expository account of the third problem.

Early on, in the 1930's, Pontryagin had given a rather definitive criterion for freeness of countable groups, and I found a generalization in $[H_1]$ that applied to arbitrary groups. P. Griffith $[G_2]$ proved that if the torsion-free group G is an extension of a free group by a simply presented p-primary group, then G must be free. Other criteria for freeness include having a discrete norm [L], [S], [Z]. For a detailed treatment of the freeness or almost freeness of a group, especially results that involve special set-theory hypotheses, the reader should consult [E-M]. We also point to the paper presented to this conference by M. Magidor.

Without reservation, we can say that the structure theory of free groups is at least as well advanced as any other area of torsion-free groups. As we have mentioned, many criteria for freeness have been established. Moreover, it is the case that significant progress has been made in recent years concerning relative structure of free groups. In particular, we know according to the following theorem when two subgroups of a free group are equivalent. Note in passing that there is a sharp contrast here between free groups and completely decomposable groups. Recall that two subgroups H and K of G are equivalent if there is an automorphism of G that maps H onto K. The definitive result concerning the equivalence of subgroups of free groups is the following.

Theorem 2.1. $[H\text{-}M_3]$. *Two subgroups H and K of a free group are equivalent if and only if (1) $G/H \cong G/K$, and (2) $\dim(H + pG/pG) = \dim(K + pG/pG)$ for each prime p.*

In simplistic terms, two subgroups of a free group are equivalent unless there is a fairly obvious reason why they are not. We mention that J. Erdös [Er] proved the above theorem for *pure* subgroups H and K. This was, it seems, the first example of an equivlance theroem for torsion-free groups unless one counts the invariant factor theorem for finitely generated groups as an equivalence theorem. Since a free resolution of a torsion-free group is automatically pure, Erdös presented a case for the classification of all torsion-free groups, but his result now is considered to be a translation of the structure of torsion-free groups to another (but equally complicated) problem. While the above theorem, by no means, determines the structure of all abelian groups, it does have significant consequences. For example the theorem provides a new proof of the stacked bases theorem $[H\text{-}M_3]$. Moreover, it provides a solution to the long-standing Problem 36 in $[F_1]$.

A more dramatic consequence of the above theorem and a companion existence theorem is a complete classification, using numerical invariants, of the free resolutions of any abelian group. The existence theorem (Theorem 4.2 in $[H\text{-}M_3]$) is a little too technical to state here, but we can say that it gives the exact conditions under which there is a free resolution

$$(R) \qquad\qquad 0 \to H \to G \to A \to 0$$

of an abelian group A that has prescribed and predetermined $\dim(H + pG/pG)$ for the various primes p.

If $A = \oplus A_i$ in the free resolution (R), we say that the free resolution (R) *splits* over the decomposition $A = \oplus A_i$ if we can write $G = \oplus G_i$ with $\eta(G_i) = A_i$ where η is the map $G \to A$ in (R). Note that this implies that $H = \oplus(H \cap G_i)$. Hence, for example,

$$0 \to pq\mathbb{Z} \to \mathbb{Z} \to \mathbb{Z}/pq\mathbb{Z} \to 0$$

cannot split over $\mathbb{Z}/pq\mathbb{Z} = \mathbb{Z}/p\mathbb{Z} \oplus \mathbb{Z}/q\mathbb{Z}$ for distinct primes p and q.

Recall that the stacked bases problem (solved by Cohen and Gluck) due to Kaplansky inquires whether or not free groups H and G must have stacked bases whenever G/H is a direct sum of cyclic groups. An affirmative answer is provided by a corollary of the classification of free resolutions mentioned above. A generalizaton of the stacked bases theorem is the following.

Theorem [H-M_3]. *Let m be an infinite cardinal and suppose that $A = \oplus A_i$ where $|A_i| < m$ for each i (belonging to an arbitrary index set). If (R) is a free resolution of A, then there is a direct decomposition $A = \oplus B_j$ such that $|B_j| < m$ for each j and such that (R) splits over the decomposition $A = \oplus B_j$. Moreover, if m is uncountable, then we can choose the B_j's so that every free resolution of A splits over the B_j's.*

We should remark that there are limitations on how far one can go in the direction of the stacked bases theorem. For example, there is a free resolution of a simply presented mixed group A that does not split over any decomposition $A = \oplus A_i$ of A where A_i has torsion-free rank 1 for each i.

Although we know a lot about the relative structure of free groups, the following problem remains open.

Problem 2. *If H and K are (not necessarily pure) subgroups of the free group G and if $\phi : G/H \rightarrowtail G/K$ is an isomorphism between the respective quotients, find necessary and sufficient conditions on ϕ in order that it can be lifted to an automorphism of G.*

We introduced this section by discussing the many tests for freeness that are now available. Lest we leave the impression that one can always tell when a torsion-free group is free, we conclude this section with an open question concerning freeness. The following is called the simple (but not conclusive) test for equivalence, which is abbreviated $S.T.E.$ (Simple Test for Equivalence), for subgroups H and K of G.

   $(S.T.E.)$ $H \cong K$ and $G/H \cong G/K$.

Problem 3. *Suppose that G is a homogeneous torsion-free group of type zero. If $S.T.E.$ suffices for the equivalence of pure subgroups, must G be free?*

3. BUTLER GROUPS OF INFINITE RANK.

There are several equivalent characterizations of Butler groups of finite rank that diverge in the infinite rank case. In particular, the following are equivalent for torsion-free groups G of finite rank (but not for infinite rank); see, for example, [B], [A] and [A-V].

   (i) G is a pure subgroup of a completely decomposable group of finite rank.
   (ii) G is a homomorphic image of a completely decomposable group of finite rank.
   (iii) Bext $(G, T) = 0$ for all torsion groups T.

The several different but equivalent descriptions of Butler groups of finite rank lead to some ambiguity as to what a Butler group of infinite rank ought to be. We might say that infinite rank torsion-free groups satisfying any one of these conditions lose much of the flavor of Butler groups of finite rank because the condition of finite rank yields gratuitous side conditions. Condition (iii) has been used primarily as the definition of a Butler group of infinite rank (see, for example, [D], and [D-R_2]), but an alternate concept was introduced in [B-S] and [A-H]. In order to respect precedent and, at the same time, to avoid confusion we introduce the following terminology; actually, this is only a slight variance from [F-M].

Definition 3.1. *An infinite rank torsion-free group G is said to be a Butler group of type 1 if $Bext\,(G,T) = 0$ for all torsion T; whereas, G is said to be a Butler group of type 2 if G is the union of a smooth chain $0 = G_0 \equiv G_1 \subseteq \cdots \subseteq G_\alpha \subseteq \cdots$ of pure subgroups G_α such that $G_{\alpha+1} = G_\alpha + B_\alpha$ where B_α is a Butler group of finite rank. (Butler groups of type 2 are also known as B_2-groups.)*

It was shown in [B-S] that type 1 and type 2 Butler groups coincide for a countable G, and this result was extended in [D-H-R] to groups of cardinality $\aleph_1$. At the present time (August, 1991), it is not known whether the two types of infinite rank Butler groups are actually different or not. It was shown, however, in [B-S] (see also [F-M]) that every Butler group of type 2 is of type 1. As we have indicated, the converse remains open. This problem, whether or not explicitly formulated, has been around for awhile; it is described in [F-M] as being one of the major open problems (in infinite rank Butler groups).

Problem 4. *Must a type 1 Butler groups also be of type 2?*

Under the continum hypothesis (CH), it was shown in [D-H-R] that Problem 4 has an affirmative answer whenever $|G| \leq \aleph_\omega$. A pleasant structural feature of Butler groups of type 2 is that they possess an Axiom 3 characterization. Recall that a group G is said to satisfy Axiom 3 with respect to a subgroup property $\mathcal{P}$ if there exists a collection $\mathcal{C}$ of $\mathcal{P}$-subgroups of G that satisfies the following conditions.

(0) $0 \in \mathcal{C}$.
(1) If $A_i \in \mathcal{C}$, then $\sum A_i \in \mathcal{C}$.
(2) If B is a countable subgroup of G, there exists a countable A in $\mathcal{C}$ with $A \supseteq B$.

Albrecht and Hill [A-H] call a subgroup H of G *decent* if for each finite subset S of G there exists a finite number of rank 1 subgroups A_i of G such that $H + \sum A_i$ is pure in G and contains S. (Clearly, $\sum A_i$ can be replaced by a Butler group B of finite rank.)

Theorem 3.2. [A-H]. *A torsion-free group G is a Butler group of type 2 if and only if G satisfies Axiom 3 with respect to decent subgroups.*

The following concept may make the notions of "nice" subgroup and Axiom 3 more appealing to those interested in the homological aspects of the subject.

Definition 3.3. *A subgroup property $\mathcal{P}$ belongs to the genus **nice** if property $\mathcal{P}$ can be described by stipulating that certain maps extend.*

We remark that in the original definition and application of Axiom 3, which was applied to primary groups, the only subgroup property $\mathcal{P}$ considered was that of being a nice subgroup. Incidentally, it is not a completely trivial result that nice subgroups of primary groups belong to the genus **nice**.

In reference to Theorem 3.3, we remark that there is no loss of generality in assuming that the decent subgroups involved are pure since any torsion-free group satisfies Axiom 3 with respect to pure subgroups [H_2]. Moreover, it was demonstrated in [D-R_2] that a pure subgroup of a countable Butler group is decent if and only if it satisfies the Torsion Extension Property (T.E.P.). Since decent subgroups of uncountable Butler groups are also closely linked to T.E.P. subgroups in [D-H-R] and [F-M] and since T.E.P. subgroups manifestly belong to the genus **nice**, we are led naturally to the following problem.

Problem 5. *Do infinite rank Butler groups (or either type) satisfy Axiom 3 with respect to a class of subgroups that belong to the genus* **nice**?

The next question is essentially due to Dugas and Thomé [D-T]. Recall that a subgroup H of G is separable if for each coset $g + H$ there are countable number of elements $h_1, h_2, \ldots$ in H such that given $h \in H$, then $|g + h| \leq |g + h_n|$ for some n.

Problem 6. *Do Butler groups satsify Axiom 3 with respect to subgroups that are separable subgroups (in the sense of Hill)?*

Other results on infinite rank Buter groups are contained in section 6.

4. Refined Concepts of Heights and Purity with Applications.

One is of course familiar with the concept of the height-matrix of an element in a mixed group $[F_3]$. But heretofore it has been thought that a height-sequence suffices for a torsion-free group since the height of nx is uniquely determined by that of x. I submit, however, that a height-matrix for an element in a torsion-free group can be useful. This height-matrix is of course different from the familiar one in the mixed case. This matrix is a $\mathbb{Q}_0 \times \mathbb{P}$ matrix where $\mathbb{Q}_0$ is the set of nonnegative rational numbers and $\mathbb{P}$ is the set of positive rational primes. For definiteness, let $\mathbb{Q}_0$ be well ordered with zero as the initial element.

Definition 4.1. *Let x be an element of the torsion-free group G. The height-matrix, M_x^G, of x in G is a $\mathbb{Q}_0$ by $\mathbb{P}$ matrix. More precisely, if $(r, p) \in \mathbb{Q}_0 \times \mathbb{P}$, the (r, p)-entry of M_x^G is denoted by $M_x^G(r, p)$ and is defined as follows:*

$$M_x^G(r, p) = \left\{ \begin{array}{l} - \text{ (read ``blank'') if } rx \text{ does not exist.} \\ \sup\{|y|_p^G : x = y + z \text{ with } z \in G(|rx|^*) \text{ if } rx \text{ does exist.}\} \end{array} \right\}$$

It is convenient here to adopt a special convention for the height of zero; namely, $|0|_p^G = \infty^+$ for any prime p. Moreover, if $s = (\infty, \infty, \ldots, \infty, \ldots)$ or $(\infty^+, \infty^+, \ldots, \infty^+, \cdots)$, then $G(s^*)$ is understood to be zero. Observe that the entries of the matrix M_x^G must belong to the set $[0, \omega] \bigcup \{\infty, \infty^+, -\}$. As usual, if the group G is clear from the context, we will write M_x in place of M_x^G and of course $|x|_p$ in place of $|x|_p^G$. Notice that the first row of M_x is simply the height-sequence of x.

Since this is a survey paper, it is perhaps best left to forthcoming research papers to explore the full ramifications of the refined concept of height defined by the above matrix. It seems to me as if the preceding or some similar manifestation of the idea of refined height is likely an essential requirement for certain new developments concerning the structure of torsion-free groups. These words, incidentally, are not entirely prophetic but are in part already historical; indeed, many of the recent advances depend on derivatives of this kind of refined concept of height such as "*-purity", "weak *-purity", "*-valued coproduct", and "primitive element".

Any refined or modified notion of height carries with it a corresponding notion of purity. For a long time there have been various concepts of purity. However, those that seem most relevant to the structure of torsion-free groups are relatively new and include the following: strong purity, Σ-purity, *-purity, and weak *-purity. Strong purity is aptly named since it implies all the others listed; in fact, accordingly to [H-M_2], we have the implications:

$$\text{strong purity} \Rightarrow \Sigma\text{-purity} \Rightarrow *\text{-purity} \Rightarrow \text{weak } *\text{-purity} \Rightarrow \text{purity.}$$

None of the preceding implications is reversible. Recall that the subgroup H of G is strongly pure if, for each $h \in H$, there is a homomorphism $\phi_h : G \to H$ that maps h onto itself. Clearly, strong purity is of the genus **nice**. The role of strong purity (see [J-R] and [N]) has largely been subsumed by Σ-purity [H-M_2].

A subgroup H of G is Σ-pure if whenever $h = g_1 + g_2 + \cdots + g_n$ with $g_i \in G(s_i)$, then $h = h_1 + h_2 + \cdots + h_n$ where $h_i \in H(s_i)$. We mention here that a Σ-pure subgroup of a completely decomposable group or even of a separable group is actually strongly pure, but this is not true in general or even for k-groups. Our main application of Σ-purity applies to k-groups and will appear in section 5. We note that Σ-purity is simple and natural, and it is probably the most significant of the four refinements of purity listed here.

A pure subgroup H of G is said to be $*$-pure if $H \bigcap G(s^*) = G(s^*)$ and $H \bigcap G(s^*, p) = H(s^*, p)$, where $G(s^*, p)$ is the abbreviated notation for $G(s^*) + G(ps)$. If we only have that the pure subgroup H of G satisfies $H \bigcap G(s^*, p) = H \bigcap G(s^*) + H \bigcap G(ps)$, then H is called weakly $*$-pure. Here is a criterion that uses $*$-purity for a group to be completely decomposable.

Theorem 4.2. [H-M_2]. *If G is the union of a countable number of $*$-pure subgroups G_n each of which is completely decomposable, then G itself is completely decomposable.*

It is perhaps worth mentioning that weak $*$-purity seems more applicable to Butler groups - at least as the theory has now been developed. The next result provides some sufficient conditions for weak $*$-purity.

Theorem 4.3. [H-M_4]. *Let H be a pure subgroup of the torsion-free group G. If $G(\sigma)/H(\sigma) + G(\sigma^*)$ is torsion free for each type σ, then H is weakly $*$-pure. In particular, if G is completely decomposable then a pure subgroup H is weakly $*$-pure if $H \bigcap G(\sigma) \subseteq G(\sigma^*)$ for all σ. Moreover, a balanced subgroup H of a completely decomposable group G is weakly $*$-pure if and only if $(G/H)(\sigma^*)$ is pure in $(G/H)(\sigma)$ for each type σ.*

In §6, we shall present an equivalence theorem for weakly $*$-pure subgroups of completely decomposable groups, but first we look at k-groups.

5. k-GROUPS.

In the introduction, we mentioned Baer's notion of separable groups. A natural and important generalization of these groups are k-groups.

Definition 5.1. *A torsion-free group G is called a k-group if each finite subset is contained in a completely decomposable $*$-pure subgroup.*

Earlier we promised an application of Σ-purity in the section on k-groups. Indeed, we have the following.

Theorem 5.2. [H-M_2]. *A Σ-pure subgroup of a k-group is again a k-group.*

The theorem above includes most, if not all, of the summand theorems of this type including those listed below.

Corollary 5.3. [F_4] *A summand of a separable group is itself separable.*

Corollary 5.4. [H-M_1]. *A summand of a k-group is a k-group.*

Corollary 5.5. [N]. *A strongly pure subgroup of a separable group is again separable.*

In his initial study of torsion-free groups, Baer [Ba] introduced the notion of a primitive element. It turns out, however, that Baer's definition of primitive element is too stringent. For example, an element that purely generates a direct summand in a completely decomposable group may not be a primitive element according to Baer's definition. But this is precisely the property that motivates the Hill-Megibben concept of a primitive element, which has proved to be not only relevant but fundamental [H-M_1] and [H-M_2].

Definition 5.6. *An element x in a torsion-free group G is said to be non-primitive (or to have non-primitive height) if $M_x(r,p) > |x|_p$ for some prime p not relevant to the rational number r, where $|x|_p$ is finite. Otherwise, x is primitive. Equivalently, x is primitive if and only if each nonblank row of the matrix M_x is simply the height sequence of x.*

We remark that a nonzero rational multiple of x is primitive if and only if x itself is primitive; see [H-M_1]. Moreover, if s is the height sequence of x, then each element of the coset $x + G(s^*)$ is also primitive. If G is separable (in particular, if G is completely decomposable) then x is primitive if and only if $\langle x \rangle_*$ is a direct summand.

Primitive elements are often used in conjunction with $*$-valued coproducts. A coproduct $\oplus A_i$ of subgroups A_i of G is called a $*$-valued coproduct if it is true that whenever $x = \sum a_i$ with $a_i \in A_i$ then $x \in F$ if and only if $a_i \in F$ whenever F is any one of the fully invariant subgroups $G(s), G(s^*)$, or $G(s^*, p)$ where s is a height sequence and p is a prime.

An alternate definition of a torsion-free k-group is one that has the property that each finite subset is contained in a $*$-valued coproduct of the groups generated by (a finite number of) primitive elements. Indeed, this is the original definition [H-M_1].

We can use k-groups to define an important class of subgroups for torsion-free groups called *knice* subgroups. Actually, it is better here to consider pure knice subgroups.

Definition 5.7. *A pure knice subgroup of a torsion-free group G is a balanced subgroup H for which G/H is a k-group.*

The following result demonstrates an even closer relationship between pure knice subgroups and k-groups.

Theorem 5.8. [H-M_1]. *A pure knice subgroup H of a torsion-free group G is a k-group if and only if G is a k-group.*

Separable groups and k-groups are linked by our next result.

Theorem 5.9. [H-M_1]. *The torsion-free group is separable if and only if (1) G is a k-group, and (2) each pure knice subgroup of finite rank is a summand.*

Recall, for a torsion-free group, that G is completely decomposable if and only if it is simply presented. We can use pure knice subgroups to complete the trinity which for a long time has been established in the torsion case.

Theorem 5.10. [H-M_1]. *The following are equivalent for a torsion-free group G.*

(a) *G is completely decomposable.*
(b) *G is simply presented*
(c) *G satisfies Axiom 3 with respect to pure knice subgroups.*

Remark. *The purity in the above theorem is essentially free becasue every torsion free group satisfies Axiom 3 with respect to pure subgroups [H_2].*

The preceding result naturally gives rise to the following problem.

Problem 7. *Do pure knice subgroups of torsion-free groups belong to the genus* **nice**?

6. Simply Presented Torsion-free Groups and Relative Structure.

As was mentioned in the previous section, completely decomposable, simply presented, and Axiom 3 groups (with respect to pure knice subgroups) are one and the same class of torsion-free groups. Even though the structure of these groups was completely determined by Baer [Ba] (in terms of the numerical invariants rank $G(\sigma)/G(\sigma^*)$), there are many recent advances and remaining open questions about their subgroups and relative structure.

First, we will consider the question concerning when a subgroup H of a torsion-free simply presented group is again simply presented. By way of introduction, we recall the famous Baer-Kulikov-Kaplansky theorem that states that a direct summand of a simply presented torsion-free group is again simply presented. In view of the result that a Σ-pure subgroup of a k-group is again a k-group, it is natural to ask whether a Σ-pure subgroup of a simply presented torsion-free group is again simply presented. The answer is "no".

Theorem 6.1. [H-M_2]. *A strongly pure subgroup of simply presented torsion-free group need not be simply presented.*

An example demonstrating this is

$$G(A_1, A_2, \ldots, A_n, \ldots) \rightarrowtail \oplus A_i \twoheadrightarrow \mathbb{Q},$$

where the A_i's run over those subgroups of $\mathbb{Q}$ having no ∞ in their type.

In this connection, the following result establishes a necessary condition for a pure subgroup of any torsion-free group to be simply presented.

Theorem 6.2. [A-H]. *Let H be a pure subgroup of a torsion-free group G. In order for H to be completely decomposable it must be a separable subgroup (in the sense of Hill).*

M. Dugas and K. Rangaswamy have just announced an almost definitive solution to the problem concerning when a pure subgroup of a simply presented torsion-free group is again simply presented.

Theorem 6.3. [D-R_3]. *Let H be a pure subgroup of the torsion-free simply presented group G. Then H is simply presented if G satisfies Axiom 3 over H with respect to subgroups that are pure, separable (in the sense of Hill), and have the property that each countable subgroup is contained in a countable and pure simply presented subgroup.*

The above result presents us with the following problem.

Problem 8. *Do the subgroups satisfying the conditions of the preceding theorem belong to the genus **nice**? Alternately, is there a better Axiom 3 system than the one given?*

If we consider balanced subgroups of a torsion-free simply presented group G, we know from [A-H] or [D-H-R] that they are Butler groups of type 2 if $|G| \leq \aleph_1$. However, for larger groups the situation is somewhat ambiguous.

Theorem 6.4. *(i)* [D-H-R]. *Under CH (continuum hypothesis), a balanced subgroup of a torsion-free group G of cardinality not exceeding $\aleph_\omega$ is a Butler group (of type 2).*

(ii) [D-T]. *Assume $\neg CH$. There is a balanced subgroup of a torsion-free simply presented group of cardinality $\aleph_2$ that is not a Butler group (of type 1 or 2).*

One of the most fundamental problems for torsion-free groups is the following.

Problem 9. *Find necessary and sufficient conditions in order for two pure subgroups H and K of a torsion-free simply presented group G to be equivalent.*

Recall again that we say that two subgroups are equivalent if there is an automorphism of the group that maps one subgroup onto the other. This means that the subgroups are indeed equivalent as subgroups. Equivalence theorems form part of the foundation of relative structure, an aspect of abelian group theory which has grown in significant proportions in recent years. The following is a partial answer to Problem 9.

Theorem 6.5. [H-M_4]. *Let H and K be weakly $*$-pure subgroups of the torsion-free simply presented group G. Then H and K are equivalent if and only if*

(1) *there is an isomorphism $\phi : G/H \rightarrowtail G/K$ that respects heights (computed in G).*

and

(2) *$rank\,(H \cap G(\sigma)/H \cap G(\sigma^*)) =\ rank\,(K \cap G(\sigma)/K \cap G(\sigma^*))$ for all types σ.*

I think that the key to Problem 9 is the next problem, which is a restatement of Problem 2 in the more general context of homogeneous simply presented torsion-free groups as opposed to free groups.

Problem 10. *If H and K are subgroups (not necessarily pure) of a homogeneous, simply presented, torsion-free group G such that $G/H \cong G/K$, find necessary and sufficient conditions on an isomorphism $\phi : G/H \rightarrowtail G/K$ in order that it can be lifted to an automorphism π of G.*

Although the above problem is basically an infinite rank problem, it has been reduced to the finite rank case. This, among other things, suggests that there may be more in common between infinite rank and finite rank torsion-free groups than we sometimes think. Perhaps the distinction is often unnecessary and the traditional barrier between the two could be at least partially removed.

REFERENCES

[A] D. Arnold, Pure subgroups of finite rank completely decomposable groups, in Abelian Group Theory, Lecture Notes in Math., Vol. 874, Springer-Verlag, New York, 1981.

[A_1] U. Albrecht, Endomorphism rings and A-projective torsion-free abelian groups, in Abelian Group Theory (Proceedings of the Honolulu Conference), Lecture Notes in Math., Vol. 1006, Springer-Verlag, New York, 1983.

[A_2] U. Albrecht, A note on locally A-projective Abelian groups, Pacific Jour. Math., 120 (1985), 1-17.

[A-H] U. Albrecht and P. Hill, Butler groups of inifinite rank and Axiom 3, Czech. Math. J. 37 (1987), 293-309.

[A-V] D. Arnold and C. Vinsonhaler, Pure subgroups of finite rank completely decomposable groups II, in Abelian Group Theory, Lecture Notes in Math., Vol. 1006 Springer-Verlag, New York, 1983.

[B] L. Bican, Splitting in abelian groups, Czech. Math. J. 28 (1978), 356-364.

[B-S] L. Bican and L. Salce, Butler groups of infinite rank, Lecture Notes in Math. 1006 Springer (1983), 171-189.

[Ba] R. Baer, Abelian groups without elements of finite order. Duke Math. J., 3 (1937), 68-122.

[C] A. L. S. Corner, Every countable reduced torsion-free ring is an endomorphism ring, Proc. London Math. Soc. 13 (1963), 687-710.

[C-G] A. L. S. Corner and R. Göbel, Prescribing endomorphism Algebras - a unifed treatment, Proc. London Math. Soc. 50, 447-479.

[D] M. Dugas, On some subgroups of infinite rank Butler groups, Rend. Sem. Math. Univ. Padova 79 (1988), 153-161.

[D-G_1] M. Dugas and R. Göbel, Every cotorsion-free algebra in an endomorphism algebra, Math. Z. 181, 451-470.

[D-G_2] M. Dugas and R. Göbel, Torsion-free abelian groups with prescribed finitely topologized endomorphism rings, Proc. Amer. Math. Soc. 90 (1984), 519-527.

[D-H-R] M. Dugas, P. Hill, and K. Rangaswamy, Butler groups of infinite rank II, Transactions Amer. Math. Soc. 320 (1990), 643-664.

[D-R_1] M. Dugas and K. Rangaswamy, On torsion-free abelian k-groups, Proc. Amer. Math. Soc. 99 (1987), 403-408.

[D-R_2] M. Dugas and K. Rangaswamy, Infinite rank Butler groups, Trans. Amer. Math. Soc. 305 (1988) 129-142.

[D-R_3] M. Dugas and K. Rangaswamy, Separable pure subgroups of completely decomposable torsion-free abelian groups, to appear.

[D-T] M. Dugas and B. Thomé, The functor Bext under the negation of CH, to appear.

[E] P. Eklof, Whitehead's problem is undecideable, Amer. Math. Monthly, 83 (1976), 775-788.

[E-M] P. Eklof and A. Mekler, Almost Free Modules, North-Holland, Amsterdam, 1990.

[Er] J. Erdös, Torsion-free factor groups of free abelian groups and a classification of torsion-free abelian groups, Publ. Math. Debrecen, 5 (1957), 172-184.

[F_1] L. Fuchs, Abelian Groups, Publ. House Hungarian Acad. of Sci., Budapest, 1958.

[F_2] L. Fuchs, Infinite Abelian Groups, Vol. 1, Academic Press, New York, 1970.

[F_3] L. Fuchs, Infinite Abelian Groups, Vol. 2, Academic Press, New York, 1973.

[F_4] L. Fuchs, Summands of separable abelian groups, Bull. London Math. Soc. 2 (1970), 205-208.

[F-M] L. Fuchs and C. Metelli, Countable Butler groups, to appear.

[G_1] P. Griffith, Infinite Abelian Groups, Univ. of Chicago Press, Chicago, 1970.

[G_2] P. Griffith, Extensions of free groups by torsion groups, Proc. Amer. Math. Soc. 24 (1970), 677-679.

[H_1] P. Hill, On the freeness of abelian groups: A generalization of Pontryagin's theorem, Bull. Amer. Math. Soc. 76 (1970), 1118-1120.

[H_2] P. Hill, The third axiom of countability for abelian groups, Proc. Amer. Math. Soc. 82 (1981), 347-350.

[H_3] P. Hill, Isotype subgroups of totally projective groups, in Lecture Notes in Math., Vol. 874, Springer-Verlag, New York, 1981.

[H-M_1] P. Hill and C. Megibben, Torsion-free groups, Trans. Amer. Math. Soc. 295 (1986), 735-751.

[H-M_2] P. Hill and C. Megibben, Pure subgroups of torsion-free groups, Trans. Amer. Math. Soc. 303 (1987), 765-778.

[H-M_3] P. Hill and C. Megibben, Generalizations of the stacked bases theorem, Trans. Amer. Math. Soc. 312 (1989), 377-402.

[H-M_4] P. Hill and C. Megibben, Equivalence theorems for torsion-free groups, to appear.

[J-R] S. Janakiraman and R. Rangaswamy, Strongly pure subgroups of abelian groups, Lecture Notes in Math., Vol. 574, Springer-Verlag, New York, 1977, pp. 57-64.

[L] J. Lawrence, Countable abelian groups with a discrete norm are free, Proc. Amer. Math. Soc. 90, 352-354.

[N] L. Nongxa, Strongly pure subgroups of separable torsion-free abelian groups, Trans. Amer. Math. Soc. 290 (1985), 363-373.

[S] J. Steprāns, A characterization of free abelian groups, Proc. Amer. Math. Soc. 93, 347-349.

[Z] F. Zorzitto, Discretely normed abelian groups, Aequationes Math. 29, 172-174.

RESEARCH ARTICLES

Almost Split Sequences and Representations of Finite Posets

D. ARNOLD[1]
Department of Mathematics
Baylor University

C. VINSONHALER[2]
Department of Mathematics
University of Connecticut

INTRODUCTION

The theory of **Rep(k,S)**, the category of finite-dimensional representations of a finite poset S over a field k, is applicable to both finite rank Butler groups [Ar] and [AD] and finite valuated groups [ARiV]. Elements of Rep(k,S) are denoted by $U = (U,U(s)|s \in S)$, where U is a finite dimensional k-vector space, each U(s) is a subspace of U, and if $s \leq t$ in S, then U(s) is contained in U(t).

This paper is devoted to an examination of almost split sequences and related topics in Rep(k,S) as a potential tool for abelian group theorists. Almost split sequences are known to exist in Rep(k,S), [BM] as summarized in [Bu1]. These sequences arise from an embedding of Rep(k,S) into a category of finitely generated modules over a finite dimensional 1-Gorenstein k-algebra kS, the incidence algebra of S, and the existence of almost split sequences in this category of modules.

---

[1] Research supported, in part, by NSF grant # DMS-9101000
[2] Research supported, in part, by NSF grant # DMS-9022730

Much of this paper is expository, intended for readers unfamiliar with the theory of almost split sequences of finitely generated modules over artin algebras. Some fundamental properties of almost split sequences in Rep(k,S) are given in Section 1. These results are standard fare for finitely generated modules over artin algebras [AuR] and Krull-Schmidt categories with exact sequences (of which Rep(k,S) is an example) from the point of view of irreducible morphisms [R]. Self-contained proofs, being relatively easy, are included.

Constructions, entirely within Rep(k,S), of Coxeter correspondences C^+, C^-:Rep(k,S) → Rep(k,S), as given in [ArV1], are summarized at the beginning of Section 2. These correspondences coincide with modifications of the traditional correspondences DTr, respectively TrD, (as defined in [AuR]) in the category of finitely generated kS-modules given in [BM]. The fact that the correspondences coincide is not obvious, see [ArV1]. It now follows from [BM], that if U in Rep(k,S) is indecomposable, then there are almost split sequences in Rep(k,S) of the form

$$0 \to C^+(U) \to B \to U \to 0$$

if U is not projective and

$$0 \to U \to D \to C^-(U) \to 0$$

if U is not injective.

Known existence proofs do not immediately give rise to explicit constructions. Briefly, an almost split sequence of kS-modules with M as a right-hand term arises as a socle element of $Ext^1(M,DTr(M))$ regarded as a module over the opposite of the endomorphism ring of M modulo the ideal of endomorphisms that factor through projectives [AuR], also see [Bu3]. This construction must then be translated to the image of Rep(k,S), as in [BM], and pulled back to Rep(k,S).

We give a construction of almost split sequences in Rep(k,S) constructed from the right-hand term within Rep(k,S) (Theorem 2.2). This construction is predicated on the fact that almost split sequences exist, we do not have an independent proof of existence. Specifically, if U is indecomposable non-projective, there is an almost split sequence

$$0 \to C^+(U) \to B \overset{g}{\to} U \to 0 \qquad\qquad (*)$$

the pullback of an injective envelope

$$0 \to C^+(U) \to W \overset{g}{\to} L \to 0 \qquad\qquad (**)$$

of $C^+(U)$ in Rep(k,S) and a natural embedding U → L (Theorem 2.2).

An injective envelope (**) of $C^+(U)$ arises from a projective cover $0 \to K \to \text{DIM } U \xrightarrow{g} U \to 0$ of U in Rep(k,S), which is easily constructed from U. Moreover, W, respectively L, is defined on the underlying vector space DIM U, respectively U; and $g:W \to L$ coincides with $g:\text{DIM } U \to U$ as a vector space morphism. The point is that almost split sequences can now be constructed entirely within Rep(k,S) from the right-hand term without recourse to a complicated chain of translations from the theory of finitely generated modules over artin algebras.

The almost split sequences $0 \to U \to B \to V \to 0$ in Rep(k,S) with V, respectively U, 1-dimensional can be completely described (Example 2.3). If, in addition, V and U are not projective or injective, then U, respectively V, and B are included in the classes of representations $G(\chi)$, respectively $G[\chi]$, classified by invariants in [ArV1].

Results on almost split sequences in Rep(Q,S) can be immediately translated to a quasi-homomorphism category of finite rank Butler groups by a well known category equivalence. Let **T** be a finite distributive lattice of types and **B**$_T$ the quasi-homomorphism category of finite rank Butler groups with typeset in T. There is a category equivalence $F:B_T \to \text{Rep}(Q,S)$ for $S = JI(T)^{op}$, the opposite of the poset of join-irreducible elements of T, given by $F(G) = (QG, QG(\tau):\tau \in S)$ [Bu1,Bu2]. The properties of almost split sequences given in Theorem 2.4, when translated to B$_T$, demonstrate the special nature of these extraordinary sequences of groups. The utility of almost split sequences to the theory of Butler groups remains unclear.

Some explicit non-trivial examples of almost split sequences in Rep(k,S) can now easily be constructed for the case $S = \Lambda_n$, a poset of n pairwise incomparable elements (Section 3). These examples demonstrate the efficacy of our constructions. Posets of this form arise from those B$_T$'s with T a finite Boolean algebra and Λ_n the set of atoms of T.

An unsolved problem for Butler groups is the classification by group-theoretic invariants of $C^{+r}G$ and $C^{-r}G$ for the special class of strongly indecomposable G's of the form G[A] or G(A) by group-theoretic invariants. Only the case $r = 0$ is resolved [ArV2], as noted above. The representation analog of this problem is the classification of (indecomposable) representations $C^{+r}U$ and $C^{-r}U$ for indecomposable U's of the form $G[\chi]$ or $G(\chi)$ up to isomorphism. One approach to a solution is to find invariants for the U's that are preserved by C^+ and C^-.

The dimension vector, dim U, plays a significant role in the representation version of this problem, where **dim U** = $(x, x_s | s \in S)$, $x = \dim_k U$, and $x_s = \dim_k(U(s)/\Sigma\{U(t) | t < s\})$. For example, if S has finite representation type, then the dimension vector completely classifies indecomposable representations in Rep(k,S) [D] or [BM]. However, dim C^+U and dim C^-U need not be directly computable from dim U for an arbitrary finite poset S.   A key observation for the case that S has finite representation type is that if U is an indecomposable representation in Rep(k,S) or in Rep(k,S^{op}) that is **exact** (each $x_s \neq 0$), then U has a **good projective cover** (there is a projective cover $0 \rightarrow K \rightarrow N \rightarrow U \rightarrow 0$ of U in Rep(k,S) with K **trivial,** i.e. each K(s) = 0) [BM]. In this case, dim C^+U and dim C^-U are computable from dim U.

Those posets S such that each indecomposable U in Rep(k,S) (exact or not) has a good projective cover are described in Theorem 4.4. T. Giovannitti in [G] gives a description of those finite posets such that each representation has projective dimension ≤ 1.   Such posets include, properly, those posets given in Theorem 4.4.

As an application of Theorem 4.4, we have Corollary 4.5: If S is the disjoint union of poles, then dim U is a complete isomorphism invariant for $U \in$ Rep(k,S) of the form C^+X or C^-X with X a 1-dimensional representation. A special case of this result is the known classification of all indecomposable **preinjectives** ($C^{+r}X$, X indecomposable injective) and indecomposable **preprojectives** ($C^{-r}X$, X indecomposable projective) in Rep(k,S) by their dimension vectors.

1. PROPERTIES OF ALMOST SPLIT SEQUENCES

We write **Rep(S)** for Rep(k,S) and **U** = $(U, U(s) | s \in S)$. A **morphism** $f:(U, U(s)) \rightarrow (V, V(s))$ in Rep(S) is a k-homomorphism $f:U \rightarrow V$ with f(U(s)) contained in V(s) for each $s \in S$. Categorical properties of the preabelian category Rep(S) are spelled out in [ArV1]. We mention only that a sequence $0 \rightarrow U \rightarrow V \rightarrow W \rightarrow 0$ is exact in Rep(S) if and only if it is an exact sequence of vector spaces with $0 \rightarrow U(s) \rightarrow V(s) \rightarrow W(s) \rightarrow 0$ exact for each s in S.

A representation U is indecomposable in Rep(S) if and only if the finite dimensional k-algebra **End(U)**, the endomorphism ring of U, is local. Consequently, Rep(S) is a Krull-Schmidt category.

A morphism $f:U \to V$ in Rep(S) is a **splitting monomorphism** if there is $g:V \to U$ with $gf = 1_U$ and a **splitting epimorphism** if there is $g:V \to U$ with $fg = 1_V$. An exact sequence $0 \to U \xrightarrow{f} V \xrightarrow{g} W \to 0$ in Rep(S) is **almost split** if:(i) it is not split; (ii) if $h:Y \to W$ is not a splitting epimorphism, then there is $\alpha:Y \to V$ with $g\alpha = h$; and (iii) if $h:U \to Y$ is not a splitting monomorphism, then there is $\alpha:V \to Y$ with $\alpha f = h$.

Call an exact sequence $0 \to U \to V \to W \to 0$ in Rep(S) **left almost split** if (i) and (iii) hold and **right almost split** if (i) and (ii) do.

PROPOSITION 1.1: Assume $0 \to U \xrightarrow{f} V \xrightarrow{g} W \to 0$ is almost split.

(a) The endomorphism rings of U and W are local rings, in particular both U and W are indecomposable representations.

(b) If $0 \to U' \xrightarrow{f'} V' \xrightarrow{g'} W \to 0$ is right almost split with U' indecomposable, then there is an isomorphism of exact sequences:

$$0 \to U' \xrightarrow{f'} V' \xrightarrow{g'} W \to 0$$
$$\downarrow \alpha \quad \downarrow \alpha \quad \downarrow 1$$
$$0 \to U \xrightarrow{f} V \xrightarrow{g} W \to 0$$

(c) If $0 \to U \xrightarrow{f'} V' \xrightarrow{g'} W' \to 0$ is left almost split with W' indecomposable, then there is an isomorphism of exact sequences:

$$0 \to U \xrightarrow{f'} V' \xrightarrow{g'} W' \to 0$$
$$\downarrow 1 \quad \downarrow \alpha \quad \downarrow \alpha$$
$$0 \to U \xrightarrow{f} V \xrightarrow{g} W \to 0$$

Proof: (a) Let h and h' be non-units in the endomorphism ring of W. Then h and h' are not splitting epimorphisms, since End(W) is artinian. Choose β, $\beta':W \to V$ with $g\beta = h$ and $g\beta' = h'$. Then $\beta + \beta':W \to V$ with $g(\beta+\beta') = h + h'$. Since g is not a splitting epimorphism, $h + h'$ is not a unit of End(W). This proves that End(W) is a local ring. The proof that End(U) is a local ring is similar.

(b) Since $g':V' \to W$ is not a splitting epimorphism, there is $\alpha:V' \to V$ with $g\alpha = g'$. Similarly, there is $\beta:V \to V'$ with $g'\beta = g$, since $0 \to U' \xrightarrow{f'} V' \xrightarrow{g'} W \to 0$ is right almost split.

Resulting is a commutative diagram

$$0 \to U' \xrightarrow{f'} V' \xrightarrow{g'} W \to 0$$
$$\downarrow \gamma \quad \downarrow \beta\alpha \quad \downarrow 1$$
$$0 \to U' \xrightarrow{f'} V' \xrightarrow{g'} W \to 0$$

If $\gamma \in \text{End}(U')$ is a unit, then $\alpha: U' \to U$ is an isomorphism, since U is indecomposable. In this case, $\alpha: V' \to V$ is also an isomorphism, as desired. Since U' is indecomposable, we may suppose that γ is a nilpotent endomorphism of U', say $\gamma^n = 0$. Then $(\beta\alpha)^n$ induces $h': W \to V'$ splitting $V' \xrightarrow{g'} W \to 0$, since $g'(\beta\alpha)^n = g'$ and g' an epimorphism implies that $g'h'$ is a unit of $\text{End}(W)$. This contradicts the hypothesis that g' is not a splitting epimorphism.

The proof of (c) is dual to that of (b).

A representation U is a **subrepresentation** of V if U is a subspace of V with $U(s)$ a subspace of $V(s)$ for each s in S and a **pure subrepresentation** if $U(s) = U \cap V(s)$ for each s; equivalently U is a kernel of a representation morphism $V \to W$ for some W. If U is a subspace of V and $V = (V,V(s)) \in \text{Rep}(S)$, then $\mathbf{U}* = (U, U \cap V(s))$ is a pure subrepresentation of V. Moreover, $V/\mathbf{U}* = (V/U,(V(s)+U)/U) \in \text{Rep}(S)$ is a cokernel of $U \to V$. If U and U' are subrepresentations of V, define $\mathbf{U} + \mathbf{U'} = (U+U',U(s)+U'(s))$ and $\mathbf{U} \cap \mathbf{U'} = (U \cap U',U(s) \cap U'(s))$.

PROPOSITION 1.2: Assume that

$$0 \to U \xrightarrow{f} V \xrightarrow{g} W \to 0 \qquad\qquad (1.2.1)$$

is an almost split sequence in $\text{Rep}(S)$ and that

$$0 \to U \xrightarrow{f'} V' \xrightarrow{g'} W \to 0 \qquad\qquad (1.2.2)$$

is a non-split exact sequence in $\text{Rep}(S)$. The following statements are equivalent:

(a)   (1.2.2) is left almost split;

(b)   If U' is a non-zero pure subrepresentation of U, then

$$0 \to U/U' \xrightarrow{f'} V'/f'(U') \xrightarrow{g'} W \to 0$$

is split exact;

(c)   (1.2.2) is almost split;

(d)   If W' is a proper (pure) subrepresentation of W, then

$$0 \to U \xrightarrow{f'} (g')^{-1}(W') \xrightarrow{g'} W' \to 0$$

is exact;

(e)   (1.2.2) is right almost split.

Proof: (a) → (b)   Since U' is non-zero, the natural morphism h:U → U/U' is not a splitting monomorphism. Thus, by (a), there is α:V' → U/U' with $\alpha f'$ = h. Now α sends f'(U') to ker h = U', and so induces a splitting of U/U' $\xrightarrow{f'}$ V'/f(U').

(b) → (c)   Since (1.2.2) is not split, g':V' → W is not a splitting epimorphism. Thus, (1.2.1), being almost split, guarantees the existence of an α:V' → V with $g\alpha$ = g'. If ker α = 0, then α' = $f^{-1}\alpha f'|_U$ is not nilpotent, hence must be a unit of End(U) by Proposition 1.1.a. In this case, α:V' → V is an isomorphism, and so induces an isomorphism of exact sequences:

$$\begin{array}{ccccccccc}
0 \to & U & \xrightarrow{f'} & V' & \xrightarrow{g'} & W & \to 0 \\
 & \downarrow \alpha' & & \downarrow \alpha & & \downarrow 1 & \\
0 \to & U & \xrightarrow{f} & V & \xrightarrow{g} & W & \to 0
\end{array}$$

Thus, (1.2) is almost split, as desired.

Otherwise, ker α' is a pure non-zero subrepresentation of U so there is, by (b), a split exact sequence

$$0 \to U/\text{ker } \alpha' \xrightarrow{f'} V'/f'(\text{ker } \alpha') \xrightarrow{g'} W \to 0.$$

As α sends V'/ker α isomorphically to a subrepresentation of V and $g\alpha$ = g' is an epimorphism, this split exact sequence induces a splitting of (1.2.1), a contradiction.

(c) → (a) is clear. The equivalence of (c), (d), and (e) is proved by a dual argument.

COROLLARY 1.3: Assume that there is an almost split sequence

$$0 \to U \xrightarrow{f} V \xrightarrow{g} W \to 0 \qquad\qquad (1.3.1)$$

and that either End(U) or End(W) is a division algebra.   If $0 \to U \xrightarrow{f'} V' \xrightarrow{g'} W \to 0$ is an exact sequence that is not split exact, then it is almost split.

Proof: We   show   that   if   End(W)   is   a   division   algebra,   then
$0 \to U \overset{f'}{\to} V' \overset{g'}{\to} W \to 0$ is almost split.   A dual argument handles the case that
End(U) is a division algebra.

Note that f': U → V' is not a splitting monomorphism.   Thus, there is
α:V → V' with αf = f', since (1.3.1) is almost split.   Then α induces
β: W → W with βg = g'α.   Moreover, β ≠ 0, otherwise α induces a splitting of f
of (1.3.1), a contradiction.   Consequently, β is a unit of the division algebra
End(W).   Therefore, α:V → V' is an isomorphism inducing an isomorphism of
exact sequences

$$0 \to U \overset{f}{\to} V \overset{g}{\to} W \to 0$$
$$\downarrow 1 \quad \downarrow \alpha \quad \downarrow \beta$$
$$0 \to U \overset{f'}{\to} V' \overset{g'}{\to} W \to 0$$

This shows that the bottom row is also almost split, as desired.

2. COXETER CORRESPONDENCES AND CONSTRUCTION
OF ALMOST SPLIT SEQUENCES

We   begin   with   a   summary,   without   details,   of   some   constructions
given in [ArV1].   Recall that U in Rep(S) is trivial if U(s) = 0 for each s in S
and define U to be **cotrivial** if U(s) = U for each s in S.   Note that U has no
trivial summands if and only if U = Σ{U(s)|s∈ S} and no cotrivial summands if
and only if ∩{U(s)|s∈ S} = 0.

Projectives   relative   to   exact   sequences   in   Rep(S)   are   finite   direct
sums of 1-dimensional representations U such that either U is trivial or
there is an s with U(t) = U for each s ≤ t and U(t) = 0 for all other t.   Dually,
injectives   relative   to   exact   sequences   in   Rep(S)   are   finite   direct   sums   of
1-dimensional representations U such that either U is cotrivial or there is
an s with U(t) = 0 for each t ≤ s and U(t) = U for all other t.   Equivalently, U
in Rep(S) is projective with no trivial summands if and only if there is a
vector space decomposition U = ⊕{U_s|s∈ S} with each U(s) = ⊕{U_t|t≤s}.

Moreover, U is injective with no cotrivial summands if and only if there is a vector space decomposition $U = \oplus\{V_s | s \in S\}$ with each $U(s) = \oplus\{V_t | s \not\leq t\}$.

Given U with no trivial summands, define **DIM U** $= \oplus\{U^+(s) | s \in S\}$, where $U^+(s) = U(s)/\Sigma\{U(t) | t < s\}$ and **(DIM U)(s)** $= \oplus\{U^+(t) | t \leq s\}$. Then DIM U is projective. There is a projective cover $0 \to K \to$ DIM U $\overset{g}{\to} U \to 0$ of U in Rep(S) given by $g = \oplus g_s$, where each g_s is a vector space splitting of $U(s) \to U^+(s)$. Define **W** $= (DIM\ U, W(s))$, with each $W(s) = \oplus\{U^+(t) | s \not\leq t\}$ and **C^+U** $= K_*$, the purification of K in W. Then $0 \to C^+U \to W \overset{g}{\to} L \to 0$ is an injective envelope of C^+U, where $L = (U, L(s))$ and $L(s) = g(W(s)) = \Sigma\{U(t) | s \not\leq t\}$ for each s.

Dually, for a U with no cotrivial summands, define **CODIM U** $= \oplus\{U^-(s) | s \in S\}$, where **$U^-(s)$** $= (\cap\{U(t) | t > s\})/U(s)$ and **(CODIM U)(s)** $= \oplus\{U^-(t) | s \not\leq t\}$. Then CODIM U is injective and $f:U \to$ CODIM U is an injective envelope of U. Define **W'** $= (CODIM\ U, W'(s))$ with $W'(s) = \oplus\{U^-(t) | t \leq s\}$, **L'** $= f(U)_* = (f(U), f(U) \cap W'(s))$, and **$C^-U$** to be the cokernel of L' in W'. In particular, $C^-U = (CODIM\ U)/U$ with $(C^-U)(s) = (W'(s) + f(U))/f(U)$ for each s in S. Furthermore, $0 \to L' \to W' \to C^-U \to 0$ is a projective cover of C^-U.

The correspondences C^+ and C^- are well defined with the following properties: C^+ and C^- preserve indecomposables and finite direct sums; $C^+U = 0$ if and only if U is projective; $C^-U = 0$ if and only if U is injective; if U has no injective summands, then $C^+C^-U \approx U$; and if U has no projective summands, then $C^-C^+U \approx U$ [ArV1].

COROLLARY 2.1: [ArV1 via BM] Suppose that U is an indecomposable representation in Rep(S).

(a) If U is not projective then there is an almost split sequence
$$0 \to C^+(U) \to B \to U \to 0$$

(b) If U is not injective then there is an almost split sequence
$$0 \to U \to D \to C^-(U) \to 0.$$

THEOREM 2.2: Let U be an indecomposable in Rep(S) that is not projective, $0 \to K \to \text{DIM } U \overset{g}{\to} U \to 0$ a projective cover of U, and $0 \to C^+(U) \to W \overset{g}{\to} L \to 0$ an injective envelope of $C^+(U)$ in Rep(S) as given above.

(a) Inclusion i: U → L is a morphism.

(b) The representation pullback

$$0 \to C^+(U) \to B \overset{g}{\to} U \to 0 \qquad\qquad (2.2.1)$$

$$\downarrow \quad\quad \downarrow \quad \downarrow\, i$$

$$0 \to C^+(U) \to W \overset{g}{\to} L \to 0$$

of g:W → L and i:U → L is an almost split sequence in Rep(S).

Proof: (a) To show that U is a subrepresentation of L, recall that $L(s) = \Sigma\{U(t)|s \not\leq t\}$. Since U is indecomposable, U(s) is contained in L(s); otherwise the vector space complement of $U(s) \cap L(s)$ in U(s) is a proper representation summand of the indecomposable representation U, a contradiction (see [ARiV] for details).

(b) Since $L = (U, \Sigma\{U(t):s \not\leq t\})$, it follows that the pullback $B = (W, B(s))$ can be defined on the vector space W by $B(s) = g^{-1}(U(s)) \cap W(s)$. First assume that (2.2.1) is split exact, say h:U → B such that $gh = 1_U$. Let $Y = (h(U), h(U) \cap W(s)) = (h(U))_*$, the purification of h(U) in W. Then g:Y → L = (U,L(s)) is a representation isomorphism, as $g(h(U)\cap W(s)) = U \cap g(W(s)) = U \cap L(s) = L(s)$ for each s in S. Since $C^+(U) \cap Y = 0$ and B = W as vector spaces, it follows that $C^+(U)$ is a summand of the injective representation W, in which case $C^+(U)$ is injective. Consequently, $0 = C^-C^+(U) \approx U$, a contradiction. Thus, (2.2.1) is not split exact.

In view of Corollary 2.1 and Proposition 1.2, it is sufficient to show that if V is a proper pure non-projective indecomposable subrepresentation of U, then $0 \to C^+(U) \to g^{-1}(V) \overset{g}{\to} V \to 0$ is split exact. By induction on rank U, the pullback sequence $0 \to C^+(V) \to B_V \overset{g_V}{\to} V \to 0$ is almost split, noting that if rank V = 1, then this sequence is almost split by Corollaries 2.1.a and 1.3. Inclusion of V in U induces inclusion $L_V \to L_U$. This gives a commutative diagram of injective envelopes;

$$0 \to C^+(V) \to W_V \overset{g_V}{\to} L_V \to 0$$
$$\downarrow h \quad\quad \downarrow h \quad\quad \downarrow$$
$$0 \to C^+(U) \to W_U \overset{g}{\to} L_U \to 0$$

since V contained in U induces DIM U $\to$ DIM V and, therefore, $h:W_V \to W_U$. Moreover, h induces a commutative diagram of pullbacks:

$$0 \to C^+(V) \to \quad B_V \quad \overset{g_V}{\to} V \to 0$$
$$\downarrow h \quad\quad \downarrow h \quad\quad \downarrow =$$
$$0 \to C^+(U) \to g^{-1}(V) \overset{g}{\to} V \to 0.$$

Now $h:C^+(V) \to C^+(U)$ is not a split monomorphism, since $C^+(U)$ is indecomposable and $V \approx C^-C^+V$ is a proper subrepresentation of $U \approx C^-C^+U$. But the top row is almost split by induction, so there is $\alpha:B_V \to C^+(U)$ such that α, restricted to $C^+(V)$, coincides with h. Then, $\beta:V \to g^{-1}(V)$, defined by $\beta(v) = h(b) - \alpha(b)$ for some $b \in B_V$ with $g_V(b) = v$ is a well defined morphism. Moreoover, $g\beta(v) = gh(b) - g\alpha(b) = g_V(b) = v$ for each $v \in V$, and the bottom row is split exact, as desired.

Let $\mathbf{P} = (k, P(s)=k)$ be a rank-1 cotrivial representation. Isomorphism classes of 1-dimensional representations are in 1-1 correspondence with subrepresentations of P [ARiV], called **types**. The set of types form a distributive lattice under sum and intersections. Given $\chi = (X_1,...,X_n)$ an n-tuple of types, define $\mathbf{G(\chi)}$ to be the kernel of the morphism $X_1 \oplus ... \oplus X_n \to P$ induced by inclusion of the X_i's in P. Basic properties of representations of the form $G(\chi)$ are given in [ArV1].

EXAMPLE 2.3: Let S be a finite poset and X a non-projective type. Given an injective envelope $0 \to C^+(X) \to W \to L \to 0$ of $C^+(X)$, let $\chi = (X_1,...,X_n)$ be an n-tuple of types with $W \approx X_1 \oplus ... \oplus X_n$, and define $\chi' = (X_1,...,X_n,X)$. The unique, up to isomorphism, almost split sequence in Rep(S) with X as a right hand term is

$$0 \to G(\chi) \to G(\chi') \to X \to 0 \qquad\qquad (2.3.1)$$

where the first map is inclusion and the second is projection onto X.

Proof: Observe that $C^+(X) \approx G(\chi)$ is indecomposable by [ArV1, Corollary 2.5]. Then (2.3.1) is the pullback sequence given in Theorem 2.2, noting that g:W → L and inclusion i:X → L induces g ⊕ i:W ⊕ X → L with kernel isomorphic to G(χ'). Uniqueness is guaranteed by Proposition 1.1.b.

For representations U and X with X a type, define **U(X)** = ($\cap$\{U(s)|X(s)≠0\})* and **U[X]** = (Σ\{U(s)|X(s)=0\})*, recalling that if V is a subspace of U, then V* = (V,V$\cap$U(s)) $\in$ Rep(S) is the purification of V in U (see [ARiV, Lemma 1.2] for an equivalent definition). Let M be a (non-empty) set of types and define **U(M)** = (Σ\{U(X)|X$\in$ M\})* and **U[M]** = ($\cap$\{U[X]|X$\in$ M\})*. These representations arise in the classification of indecomposable representations U = G(χ) and their duals U = G[χ] by invariants in [ArV1]; the invariants are dim$_k$U[M], respectively dim$_k$U(M) for non-empty subsets M of P. Moreover, they are direct translations of G(M) and G[M] for a finite rank Butler group G and a finite set M of types (isomorphism classes of subgroups of Q).

Note that if Y = Σ\{X|X$\in$ M\}, then U(Y) = $\cap$\{U(X)|X$\in$ M\}. Similarly, if Y' = $\cap$\{X|X$\in$ M\}, then U[Y'] = Σ\{U[X]|X$\in$ M\}. Furthermore, (U⊕V)(M) = U(M) ⊕ V(M) and (U⊕V)[M] = U[M] ⊕ V[M], as can be seen by considering (U⊕V)(X) and (U⊕V)[X] for each X $\in$ M.

We also have that (U/U[M])[M] = 0 and U(M)(M) = U(M). To see this, recall that (U/U[M])(s) = (U(s)+U[M])/U[M] for each s in S. Hence, for X in M, (U/U[M])[X] = Σ\{(U/U[M])(s)| X(s) = 0\} = U[X]/U[M]. It now follows that (U/U[M])[M] = $\cap$\{(U/U[M])[X]|X$\in$ M\} = 0. An analogous argument shows that U(M)(M) = U(M).

THEOREM 2.4: Let M be a non-empty set of types and 0 → V → B → U → 0 an almost split sequence in Rep(S).
(a) If U(M) ≠ U, then 0 → V(M) → B(M) → U(M) → 0 is split exact.
(b) If V[M] ≠ 0, then 0 → V/V[M] → B/B[M] → U/U[M] → 0 is split exact.

Proof: (a) Since U(M) ≠ U, inclusion i: U(M) → U is not a splitting epimorphism. Thus, there is h:U(M) → B with gh = i. Note that B(M) contains image h. This gives a split exact sequence
0 → K → B(M) → U(M) → 0 with K = V $\cap$ B(M) = V(M).

(b) follows from (a) and an application of the duality
σ: Rep(S) → Rep(S^{op}) induced by vector space duality (see [ArV1]).

3. ALMOST SPLIT SEQUENCES IN REP(Λ_n)

Write $\Lambda_n = \{1,2,...,n\}$, a poset of $n \geq 3$ pairwise incomparable elements. Then $P_0, P_1, ... , P_n$ are the indecomposable projectives in Rep(Λ_n), where $P_j = (k,P_j(i))$ with $P_j(j) = k$ and $P_j(i) = 0$ if $i \neq j$ for $1 \leq i \leq n$ and $0 \leq j \leq n$. Moreover, $Q_0 = P$, $Q_1, ... , Q_n$ are the indecomposable injectives, where $Q_j = (k,Q_j(i))$ with $Q_j(j) = 0$ and $Q_j(i) = k$ if $i \neq j$.

It is well known, dating back to [GP], that the correspondences C^+ and C^-: Rep(Λ_n) $\rightarrow$ Rep(Λ_n) are functors. Also, if $O \rightarrow U \rightarrow V \rightarrow W \rightarrow O$ is an exact sequence in Rep(Λ_n), then $0 \rightarrow C^+U \rightarrow C^+V \rightarrow C^+W \rightarrow O$ is exact if U and V have no projective summands and $0 \rightarrow C^-U \rightarrow C^-V \rightarrow C^-W \rightarrow O$ is exact provided that V and W have no injective summands.

EXAMPLE 3.1: Let X be a rank-1 representation of Λ_n. If X is not projective and $r \geq 1$, there is an almost split sequence
$O \rightarrow C^{+(r+1)}X \rightarrow C^{+r}G(\chi') \rightarrow C^{+(r-1)}G(\chi') \rightarrow 0$, where $\chi = (Q_i | X \not\leq Q_i)$ and $\chi' = (X,\chi)$.

Proof: By Example 2.3 for $S = \Lambda_n$, there is an almost split sequence
$0 \rightarrow G(\chi) = C^+X \rightarrow G(\chi') \rightarrow X \rightarrow 0$. Applying C^{+r} preserves exactness, noting that $C^{+r}X$ and $C^{+(r-1)}G(\chi')$ have no projective summands.

Example 3.1 and its dual are mild generalizations of known almost split sequences having an indecomposable preinjective, respectively preprojective, as a right-hand, respectively left-hand term.

EXAMPLE 3.2: Assume that $U = (U,U_1,...,U_n,U_{n+1}) \in$ Rep(k,Λ_{n+1}) is indecomposable with $U = U_1 \oplus ... \oplus U_n$ and $U_{n+1} \cap U_i = 0$ for $i \leq n$.
(a) $C^+U = (U_{n+1},V_1,...,V_n,0)$ with $V_i = U_{n+1} \cap (\oplus\{U_j : j \neq i\})$ for $i \leq n$.
(b) There is an almost split sequence $0 \rightarrow C^+(U) \xrightarrow{f} B \xrightarrow{g} U \rightarrow 0$
where $B = (U \oplus U_{n+1}, B_1,...,B_n,U_{n+1})$, $B_i = \{(u_1,...,u_{n+1}) : u_i=0, \Sigma u_j \in U_i\}$ for $i \leq n$, $f(x) = (x,-x)$, and g is induced by the inclusions $U \rightarrow U$ and $U_{n+1} \rightarrow U$.

Proof: (a) A projective cover of U is given by $0 \rightarrow K \rightarrow$ DIM $U \rightarrow U \rightarrow 0$, where DIM $U = (U \oplus U_{n+1},U_i)$, and g:DIM $U \rightarrow U$ is induced by inclusion of the U_i's in $U = U_1 \oplus ... \oplus U_n$. Then the exact sequence $0 \rightarrow C^+(U) \xrightarrow{f} B \xrightarrow{g} U \rightarrow 0$

defines C^+U, where $W = (DIM\ U, \oplus\{U_j | j \neq i\})$ and $L = (U, L_i = U)$. Since ker g is the purification of $\{(x, -x) \in U \oplus U_{n+1} | x \in U_{n+1}\}$ in DIM U, it follows that $C^+(U)$ may be chosen as described.

    (b)   is an application of Theorem 2.2, since this sequence can be shown directly to be a pullback of $G:W \to L$ and inclusion of U in L, recalling that the middle term B_U of this pullback is of the form $(W, g^{-1}(U_i) \cap W(i))$.

    With P_i as in the first paragraph of this section, define $P_{ij} = P_i + P_j$, a type. Then $U = G[P_{12}, P_3, P_4]$ and $V = G[P_1, P_2, P_{34}]$ are indecomposables in $Rep(\Lambda_4)$ with the same dimension vector $(2,1,1,1,1)$ but U and V are not isomorphic. As a consequence of (a) of the following example, U (and similarly V) is neither preprojective nor preinjective.

EXAMPLE 3.3: Let $U = G[P_{12}, P_3, P_4]$ and $V = G[P_1, P_2, P_{34}] \in Rep(\Lambda_4)$.
(a) $C^+U \approx V$, $C^+V \approx U$, and so $C^{+2}U \approx U$ and $C^{+2}V \approx V$.
(b) There is an almost split sequence $0 \to V \xrightarrow{f} B \xrightarrow{g} U \to 0$, where $B \approx (k^4, ke_2 \oplus k(e_3+e_4), ke_1 \oplus k(e_3+e_4), k(e_2-e_4) \oplus k(e_1-e_4), k(e_2-e_3) \oplus k(e_1-e_3))$ and $\{e_1, e_2, e_3, e_4\}$ is a standard basis for k^4. Moreover, B is indecomposable.
(c) There is an almost split sequence $0 \to U \to C^+B \to V \to 0$.

Proof: (a) Write $U = (kx \oplus ky, k(x+y), k(x+y), kx, ky)$. Then there is a projective cover $0 \to K \to DIM\ U \to U \to 0$ with $DIM\ U = (k^4, ke_1, ke_2, ke_3, ke_4)$, $g:DIM\ U \to U$ induced by $g(e_1) = g(e_2) = x + y$, $g(e_3) = x$, $g(e_4) = y$ and $K = ker\ g = (k(-e_2+e_3+e_4) \oplus k(-e_1+e_3+e_4), K(i)=0)$, a trivial representation. Then $W = (k^4, W(i) = \oplus\{ke_j | j \neq i\}) \approx Q_1 \oplus Q_2 \oplus Q_3 \oplus Q_4$. It now follows that $C^+U = (K, k(-e_2+e_3+e_4), k(-e_1+e_3+e_4), k(e_1-e_2), k(e_1-e_2)) \approx V$, recalling that C^+U is the purification of K in W. A similar argument shows that $C^+V \approx U$.
    (b) Applying Theorem 2.2 shows that $B = (W, (g^{-1}U(s)) \cap W(s))$ is as given. Finally, B is indecomposable, being in the list of indecomposables given by S. Brenner in [Br,(ii),p.58].
    (c) follows from (a), (b), and the fact that C^+ preserves exactness in this case.

4. GOOD PROJECTIVE COVERS

There are alternate definitions of C^+ and C^- given in [ArV1, Proposition 1.3]. Vector space duality induces an exact contravariant duality $\sigma : \text{Rep}(S) \to \text{Rep}(S^{op})$, where $\sigma(U) = (U^*, U(s)^\perp)$, $U^* = \text{Hom}_k(U, k)$, and $U(s)^\perp = \{f \in U^* | f(U(s)) = 0\} \approx (U/U(s))^*$. Define $\rho = \sigma C^+ : \text{Rep}(S) \to \text{Rep}(S^{op})$. Then $C^+ U \approx \sigma\rho(U)$ if U has no trivial summands and $C^- U \approx \rho\sigma(U)$ if U has no cotrivial summands. The σ and ρ given here coincide with σ and ρ given in [D].

The following lemma shows that dim $\rho(U)$ is computable from dim U for an arbitrary finite poset S.

LEMMA 4.1: Let $U \in \text{Rep}(S)$ with no projective summands and let dim $U = (x, x_s | s \in S)$. Then dim $\rho(U) = (y, x_s | s \in S)$, where $y = \Sigma\{x_s | s \in S\} - x$. In particular, if U is an exact representation (each $x_s \neq 0$), then so is $\rho(U)$.

Proof: Let $0 \to C^+(U) \to W \to L \to 0$ be an injective envelope of $C^+(U)$. Applying σ gives an exact sequence $0 \to \sigma(L) \to \sigma(W) \to \rho(U) \to 0$ with $\sigma(W)$ projective. In fact, this is a projective cover of $\rho(U)$ [ArV1, Proposition 1.3]. A computation of dimensions completes the proof, recalling that $L = (U, L(s))$, $W = (\text{DIM } U, W(s))$, $W(s) = \oplus\{U^+(t) | s \nleq t\}$, $\sigma(W)(s) \approx \oplus\{U^+(t) \ t \geq s\}$, and $x_s = \dim_k U^+(s)$.

As a result of Lemma 4.1 and the preceding remarks, the computation of dim $C^+(U)$ and dim $C^-(U)$ from dim U is reduced to computing dim $\sigma(U)$ from dim U. In view of the definition of σ, $\dim_k U = \dim_k U(s) + \dim_k \sigma(U)(s)$ for each s in S. However, computing dim $\sigma(U)$ from dim U is more delicate, recalling that the sth coordinate of dim V is $\dim_k(V(s)/\Sigma\{V(t):t<s\})$.

Projective covers (defined in [ArV1]) are unique up to isomorphism, so if there is a good projective cover of U (as defined in the introduction), then each projective cover of U is a good projective cover. Moreover, the class of representations having good projective covers is closed under finite direct sums and summands, as projective covers of finite direct sums of representations are direct sums of projective covers of the summands.

LEMMA 4.2: Suppose that $U \in$ Rep(S) has no trivial summands, dim U = (x, x_s), and $0 \rightarrow K \rightarrow$ DIM $U \rightarrow U \rightarrow 0$ is a projective cover of U. Then the following are equivalent:
(a) K is a trivial representation;
(b)   (DIM U)(s) $\rightarrow U(s)$ is an isomorphism for each s in S;
(c) $\dim_k U(s) = \Sigma\{x_t | t \leq s\}$ for each s in S;

Proof: A routine consequence of the definitions.

PROPOSITION 4.3: [BM, Proposition 5.5] Assume that U is in Rep(S), both U and $\sigma(U)$ have good projective covers, dim U = (x, x_s), and dim $\sigma(U)$ = (x, z_s). Then for each $s \in S$:
$$\Sigma\{x_t | t \leq s\} + \Sigma\{z_t | t \geq s\} = x.$$
In particular, dim $\sigma(U)$ can be computed from dim U in this case.

Proof: By Lemma 4.2, $\dim_k U(s) = \Sigma\{x_t | t \leq s\}$ and $\dim_k \sigma(U)(s) = \Sigma\{z_t | t \geq s\}$, recalling that $\sigma(U) \in$ Rep(k, S^{op}). Finally, as noted above, $x = \dim_k U = \dim_k U(s) + \dim_k \sigma(U)(s)$. We can, therefore, compute dim $\sigma(U)$ from dim U inductively, observing that if s is minimal in S, then $z_s = x - x_s$.

THEOREM 4.4: Let S be a finite poset. Then the following are equivalent:
(a)   Each representation of S has a good projective cover;
(b)   Each rank-1 representation has a good projective cover;
(c)   S is a disjoint union of rooted trees.

Proof: (c) $\rightarrow$ (a)   The finite poset S is a disjoint union of rooted trees if and only if each element of S has at most one immediate predecessor, where **t is an immediate predecessor of s in S** if $t < s$ and $t \leq t' \leq s$ in S implies that $t' = t$ or s. Let U be a representation of S with dim U = (x, x_s). It is sufficient to prove by induction on the number of predecessors of s in S, that $\dim_k U(s) = \Sigma\{x_t | t \leq s\}$ (Lemma 4.2). If s is minimal in S then $\dim_k U(s)$ = x_s, as desired. Otherwise, s has exactly one immediate predecessor t. Thus, $\dim_k U(s) = x_s + \dim_k U(t)$ and the proof is complete by induction, as t has fewer predecessors.

(a) $\rightarrow$ (b) is clear. For (b) $\rightarrow$ (c), assume that s in S has two immediate predecessors r and t. Define a rank-1 representation $X = (k, X(v))$ by $X(v) = k$ if $v \geq r$ or $v \geq t$ and $X(v) = 0$ otherwise. Then DIM $U = X(r) \oplus X(t)$ and so there is a projective cover $0 \rightarrow K \rightarrow X(r) \oplus X(t) \rightarrow X \rightarrow 0$. Since $K(s) \approx k$, K is not a trivial representation and so X does not have a good projective cover.

A **pole** is a linearly ordered finite poset.

COROLLARY 4.5: Suppose that S is a disjoint union of poles.
(a) Each representation of S and S^{op} has a good projective cover.
(b) The dimension vector is a complete isomorphism invariant for representations of S of the form $C^{+r}X$, or $C^{-r}X$ for X a rank-1 representation of S.

Proof: (a) Apply Theorem 4.4. (b) Observe that dim $C^{+}U$ and dim $C^{-}U$ can be computed from dim U by Lemma 4.1, Proposition 4.3, and the fact that $C^{+}U \approx \sigma\rho(U)$ and $C^{-}U \approx \rho\sigma(U)$. Moreover, rank-1 representations are classified by their dimension vector.

LIST OF REFERENCES

[AD]   D. Arnold and M. Dugas, *Representations of finite posets and near isomorphism of finite rank Butler groups*, preprint.
[Ar]   D. Arnold, *Representations of finite posets and abelian groups*, Abelian Group Theory, Cont. Math. 87, A.M.S., 1989, 91-110.
[ARiV] D. Arnold, F. Richman, and C. Vinsonhaler, *Representations of finite posets and valued groups*, J. Alg., to appear.
[ArV1] D. Arnold and C. Vinsonhaler, *Invariants for classes of indecomposable representations of finite posets*, J. Alg., to appear.
[ArV2] _____, *Isomorphism invariants for abelian groups*, Trans. A.M.S., to appear.
[AuR]   M. Auslander and I. Reiten, *Representation theory of artin algebras IV*, Comm. Alg. 5(1977), 443-518.
[BM]   R. Bautista and R. Martinez, *Representations of partially ordered sets and 1-Gorenstein Artin algebras*, Proc. of 1978 Antwerp Conf. on Ring Theory, Marcel Dekker, 1979, 385-433.
[Br]   S. Brenner, *On four subspaces of a vector space*, J. Alg. 29(1974),587-599.
[Bu1]   M. C. R. Butler, *Some almost split sequences in torsion-free abelian group theory*, Abelian Group Theory, Gordon and Breach, 1987, 291-302.
[Bu2]   _____, *Torsion-free modules and diagrams of vector spaces*, Proc. London Math. Soc., (3) 18(1968), 635-652.
[Bu3]   _____, *The construction of almost split sequences*, I, Proc. London Math. Soc. (3) 40(1980), 72-86.
[D]   Y. A. Drozd, *Coxeter transformations and representations of partially ordered sets*, Funct. Anal. and Appl. 8(1974), 219-225.
[G]   T. Giovannitti, *Pure subgroups of projective diagrammatic groups*, Comm. Alg., to appear.
[GP]   I. Gelfand and V. Ponomarev, *Problems of linear algebra and classification of quadruples in a finite dimensional vector space*, Coll. Math. Soc., Bolyai 5(1970), Tihany, 163-237.
[R]   C. Ringel, *Tame Algebras and Integral Quadratic Forms*, Lecture Notes in Math., Springer, New York, 1984.

Modules with Two Distinguished Submodules

Claudia Böttinger and Rüdiger Göbel

Fachbereich 6 , Mathematik und Informatik
Universität Essen, 4300 – Essen 1, Germany
e-mail: mat1ØØ @ DEØHRZ1A.BITNET

§ 1. Introduction

Let R be a commutative ring with $1 \neq 0$ and $I = (I, \leq)$ a partially ordered set. A sequence $\mathbf{G} = (G, G^i : i \in I)$ consisting of an R-module G and distinguished submodules G^i with $G^i \subseteq G^j$ for all $i \leq j$ is called an R_I-module or an R-representation of I or an RI-space. Morphisms between two of these objects $\mathbf{G}$ and $\mathbf{G}'$ are R-homomorphisms of G into G' which map G^i into G'^i for each $i \in I$. We get a natural category $R_I - Mod$ which was studied by many authors, see references in [BG].

Connections and applications of R_I-Mod to module theory can be found in [A] and [FG] respectively. This leads to the question, which partially ordered sets I allow any prescribed R-algebra A to be the endomorphism algebra of some R_I-module G, say End $G \cong A$. The answer was given in [BG], using **Kleiner's "critical" list K** which consists of five well known orderings $(1,1,1,1), (2,2,2), (1,3,3), (N,4), (1,2,5)$, where numbers are the obvious linear orderings, ", " denotes disjoint unions and N denotes an order $\{a, b, c, d : b < a, d < a, d < c\}$. If one of these orderings can be embedded into I, then the following holds:

(*) *If λ is an infinite cardinal and A an R-algebra with a generating set not exceeding λ and $M = \bigoplus_\lambda A$, there are R_I-module structures*

$$\mathbf{M}_\alpha = (M, M_\alpha^i : i \in I) \ (\alpha \subseteq \lambda) \ on \ M$$

with $\mathrm{Hom}(\mathbf{M}_\alpha, \mathbf{M}_\beta) = A$ *if* $\alpha \subseteq \beta$ *and* $\mathrm{Hom}(\mathbf{M}_\alpha, \mathbf{M}_\beta) = 0$ *if* $\alpha \not\subseteq \beta$.

Special cases of this result are due to Corner, Ringel, Brenner, Butler and Göbel, May. They can be traced back to Kronecker, see [C], [R], [GM], [BG] for details. The converse of (*) also holds for fields R by a result due to Simson [S] and independently to Ringel, Tachikawa [RT]: *If no member of Kleiner's list embeds into I, then the field $A = R$ cannot be represented as the endomorphism algebra of an R_I-vector space $\mathbf{M}$ of infinite dimension.*

Hence we wonder whether this minimality of Kleiner's list **K** is restricted to fields. What is the critical list $\mathbf{K}_A$ corresponding to fixed R and A? An old result of [DG] or [CG] applies immediately. We recall the relevant definitions. Let S be a fixed multiplicatively closed subset of R containing no zero-divisors. We also assume $1 \in S$. An R-module G is S-torsion-free if $ms = 0$ $(m \in M,\ s \in S)$ implies $m = 0$. Moreover, G is S-reduced if $\bigcap_{s \in S} sG = 0$. Then S defines the S-topology on G taking $\{sG : s \in S\}$ as a basis of neighbourhoods of $0 \in G$. We may consider the S-completion $\hat{G}$ of G. The module G is called *cotorsion-free* if $\operatorname{Hom}(\hat{R}, G) = 0$, see [G], and [EM] for a discussion of the case $R = \mathbf{Z}$.

If $|S| = \aleph_o$ and A is a cotorsion-free R-algebra, then we can find an R-module G with $\operatorname{End} G \cong A$, cf [DG], [CG].
In this case $\mathbf{K}_A = \emptyset$.

On the other hand, a result of Kaplansky [K, p.52] can be reformulated as $\mathbf{K}_{J_p} \neq \emptyset$:

Any torsion-free module over the p-adic integers J_p of rank ≥ 2 has non-trivial summands.

Moreover, it is easy to check that $\{(1)\} \neq \mathbf{K}_{J_p}$. A similar result holds, if we restrict to separable, torsion-free abelian groups. We will determine $\mathbf{K}_{J_p}$ and $\mathbf{K}$ in the latter case of separable, torsionfree groups as a very special corollary of our main theorem.

Observe that we do not require S to be countable, which brings in the special flavour of what we would like to call *"local topologies"*. The new concept forces us to reorganize substantially the application of the Black Box, cf. [CG], [EM]. Our basic notion of a "strong ordinal" will relate to particular elements $s \in S$. This dependence on S can be made uniform if $|S| = \aleph_0$. We are able to avoid the difficulties which arise in [GS] in this way.

MAIN THEOREM. *Let (R, S) be as above. If A is any R-algebra with A_R an S-reduced, S-torsion-free R-module and $|A| \leq \lambda$, then we can find an R_2-module $\mathbf{G} = (G,\ G^o,\ G^1)$ such that $|G| = \lambda^{\aleph_0}$ and $\operatorname{End} \mathbf{G} \cong A$.*

Remarks. If $I = n$ is an ordinal, we will abuse notation and equip the underlying set $\{0, ..., n-1\}$ with an order making all elements pairwise incomparable.
G can be replaced by a fully rigid system, cf. [C1], [BG] or [GM].
Let $\mathcal{C}$ be a class of R-modules closed under taking S-pure submodules and containing the R-module $\prod_{\lambda}^{\aleph_0} A = \{f : \lambda \to A,\ [f]\ \text{countable}\ \}$, where $[f] = \{i \in \lambda,\ f(i) \neq 0\}$ denotes the support of f. Then G^o, G^1, G above can be chosen in $\mathcal{C}$.
The class $\mathcal{C}$ of all separable, torsion-free abelian groups with $R = A = \mathbf{Z}$ has the required properties. We can apply our Main Theorem to derive *the existence of a separable, torsion-free abelian group G (a pure subgroup of a product $\mathbf{Z}^\kappa$) with $|G| = \lambda^{\aleph_0}(\lambda\ infinite)$ and two distinguished subgroups G^0, G^1 such that* $\operatorname{End}(G; G^0 G^1) = \mathbf{Z}$.

If $R = A = \mathbf{J_p}$, then it follows from Kaplansky's result [K, p. 52] that G/G^o and G/G^1 must be divisible and $G^o \cap G^1 = 0$ for $\operatorname{End}(G; G^0 G^1) = \mathbf{J_p}$. This of course can be seen from our construction below as well:

If G/G^o is not divisible for instance, then we can find an endomorphism $\sigma \neq 0$ of G which sends a cyclic summand $x\mathbf{J_p}$ of G into G^1 and its complement, containing G^o, onto 0. Hence σ is not scalar multiplication by a p-adic integer on a suitable "canonical submodule" P which is predicted by the Black Box at some strong ordinal α, hence $\sigma = \sigma_\alpha$ on P is an endomorphism which is killed during the construction by enlarging G^o_α. The contradiction shows that G/G^o must be divisible.
It is also interesting to mention that this is in contrast to some earlier construction. If G is a rigid $\mathbf{J_p}$-module with $\operatorname{End} G = \mathbf{J_p} \oplus \operatorname{Fin}(G)$ as in [CG] or in [DG] and $G^o = \langle g_\alpha; \alpha \in \lambda^* \rangle$

is the subgroup generated by the "skeleton" which looks very much like G^o above, then G/G^o will not be divisible in this case! Hence the Black Box must *automatically* do some more work for us in the category of R_2-modules to ensure divisibility.

Observe that $|\text{End } G| = \lambda^{\aleph_0}$ and $|\text{End } (G; G^o)| = \lambda^{\aleph_0}$ for any $G^o \subseteq G$, hence the Main Theorem does not hold for R_1-modules and 2 is minimal.

The following Corollary is an immediate consequence of our Main Theorem.

Corollary. *Let $I = (I, \leq)$ be a partially ordered set and λ an infinite cardinal. The following are equivalent.*
(1) I is not linearly ordered.
(2) For all R-algebras A over a commutative ring R with a multiplicatively closed set S such that A_R is S-reduced and S-torsion-free of cardinal $\leq \lambda$, we can find a family $\mathbf{G}_\alpha$ $(\alpha \subseteq \lambda)$ of S-reduced and S-torsion-free R_I-modules of rank $\lambda^{\aleph_0}$ with $\text{Hom}(\mathbf{G}_\alpha, \mathbf{G}_\beta) = A$ if $\alpha \subseteq \beta$ and $\text{Hom}(\mathbf{G}_\alpha, \mathbf{G}_\beta) = 0$ if $\alpha \not\subseteq \beta$.

In this case $\mathbf{K}_A = \{(1,1)\}$. It would be interesting to determine $\mathbf{K}_A$ for other algebras, in particular for A's which are not S-reduced or S-torsion-free. Is $\mathbf{K}_A = \{(1,1,1)\}$ or $\mathbf{K}_A = \{(1)\}$ possible?

§ 2. Algebraic, Topological and Combinatorial Preliminaries

Let R be a commutative ring with 1 and S a fixed multiplicatively closed subset containing no zero-divisors and satisfying $\bigcap_{s \in S} Rs = 0$. We also assume $1 \in S$; cf. [AM].

Next we consider a fixed R-algebra A which is S-reduced and S-torsion-free. Naturally we may assume that $sA \neq A$ for all $s \in S \setminus \{1\}$.

Let λ be a fixed cardinal with $|A| \leq \lambda$ and let $T = {}^{\omega >}\lambda$ be the tree on λ. Every $\sigma \in T$ is a function $\sigma : n \to \lambda$ for some n with

$$n = \{0, ..., n-1\} = dom(\sigma) = \ell(\sigma),$$

which is the length of σ. Moreover, $\sigma \subseteq \rho$ $(\sigma, \rho \in T)$ (as relations) if and only if $\rho \lceil dom(\sigma) = \sigma$, i. e. ρ extends σ. *Branches* of T are maps $v : \omega \to \lambda$ which we identify with the linearly ordered subset $\{v \lceil n : n \in \omega\}$ of T. We will write $Br(T)$ for all branches of T. If $\kappa = cf \, \lambda$ is the cofinality of λ we fix a strictly increasing, continuous map $\rho : \kappa + 1 \to \lambda + 1$ with $\rho(0) = 0$ and $\rho(\kappa) = \lambda$.

This is used to define the norm $\|X\|$ of a subset $X \subseteq T$, by setting $\|X\| = \min\{\alpha < \kappa, X \subseteq {}^{\omega >}\rho(\alpha)\}$.

Let $A^T = \prod_{\sigma \in T} \sigma A = \{f : T \to A\}$ denote the cartesian product on T, where σ represents the 1 of an isomorphic copy σA of A and $f_\sigma \in A$. The product A^T contains the direct sum $B = \bigoplus_{\sigma \in T} \sigma A = A^{(T)}$ consisting of all $f \in A^T$ of finite support $[f] = \{\sigma \in T : f_\sigma \neq 0\}$.

We will work with an intermediate A-module $\hat{B} = \{f \in A^T : \|[f]\| \leq \aleph_0\}$. Every element $f \in \hat{B}$ can be written as a countable sum $f = \sum_{\sigma \in [f]} \sigma f_\sigma$ and the norm extends naturally to $\|f\| = \|[f]\|$. Similarly $\|X\| = \|\bigcup_{x \in X} [x]\|$ is well defined for all subsets $X \subseteq \hat{B}$ with $|X| < \kappa$. If $\kappa = \aleph_0$, then $\| \, . \, \|$ needs a few changes which can be seen in [GS] or [FG]. They are of minor interest for this paper, so we leave them to the reader.

If M is a R-module and $N \subseteq S$ is multiplicatively closed, we have a natural subfunctor $\nu = \nu_N$ of the identity taking M to $\nu M = \bigcap_{s \in N} sM$.
Its cokernel is $\bar{M}^N = M/\nu M$ with elements $\bar{x}^N = x + \nu M$ for $x \in M$ and $\bar{X}^N = \{\bar{x}^N : x \in X\}$ for any subset X of M. Moreover, endomorphisms φ of M induce endomorphisms $\bar{\varphi}^N$ of $\bar{M}^N$. Normally we will omit the exponent N as soon as N is fixed.
The following lemma is easy but crucial.

Lemma 2.1. *If M is an S-reduced, S-torsion-free R-module and if $N \subseteq S$ is multiplicatively closed, then $\bar{M}$ becomes an $\bar{N}$-reduced, $\bar{N}$-torsion-free $\bar{R}$-module.*

Proof: Changing the ring, the R-module M becomes an $\bar{R}$-module $\bar{M}$. If $\bar{x} \in \nu \bar{M}$, then $x \in sM + \nu M = sM$ for all $s \in N$, hence $x \in \nu M$ and $\bar{x} = \bar{0}$. The module $\bar{M}$ is $\bar{N}$-reduced. Suppose $\bar{t}\bar{x} = \bar{0}$ for some $t \in N$. Hence $tx \in \nu M = \bigcap_{s \in N} sM$. We may replace s by st because $t \in N$ and N is multiplicatively closed. From $tx \in \bigcap_{s \in N} tsM$ and S-torsion-freeness we conclude $x \in \nu M$ and $\bar{x} = \bar{0}$. The module $\bar{M}$ is $\bar{N}$-torsion-free. ∎

Next we refine the support $[x]$ of elements in $\hat{B}$. If $x = \sum_{\sigma \in [x]} \sigma \alpha_\sigma \in \hat{B}$, and $N \subseteq S$ is multiplicatively closed, then $\bar{x}\lceil \sigma = \bar{a}_\alpha \in \bar{A}$ and $[\bar{x}] = [x]_N = \{\sigma \in T : \bar{a}_\sigma \neq \bar{0}\}$. Clearly $[\bar{x}] \subseteq [x]$ and $\|\bar{x}\| := \|[\bar{x}]\| \leq \|x\|$. We call $\|\bar{x}\|$ the N-*norm* of x , which can obviously be extended to subsets of $\hat{B}$ with cofinality less than $cf \lambda$. We are only interested in cyclically generated subsets N of S, which are all subsets $N = \{s^n : n \in \omega\}$ for $s \in S$.
We write $\bar{x}^N = \bar{x}^s = (x)^{-s}$, $[x]_N = [x]_s, \bar{\varphi}^N = \bar{\varphi}^s$ and $\nu_N M = s^\omega M$ in such cases.
If $v \in Br(T)$ and $k \in \omega$, we define an s-*branch element* as $v^{s,k} = \sum_{\sigma \in v, \ell(\sigma) \geq k} s^{\ell(\sigma)k} \sigma \in \hat{B}$ and let $v^s = v^{s,0}$.
The set $\{v^{s,k} : k \in \omega\}$ constitutes a divisor chain modulo B in some obvious way. The following observation follows from the definitions and is crucial as well.

Observation 2.2. *Using the notation above, we have*

$$[v^{s,k}] = [v^{s,k}]_s = \{\sigma \in v : \ell(\sigma) \geq k\}.$$

Proof: If $(v^{s,k})\lceil \sigma = s^{\ell(\sigma)-k} + s^\omega \hat{A}$ for $\sigma \in v$, $\ell(\sigma) \geq k$ and $\sigma \notin [v^{s,k}]_s$, then $s^{\ell(\sigma)-k} \in s^\omega A \subseteq sA$ and $s^{\ell(\sigma)-k} = s^{\ell(\sigma)-k+1} \cdot a$ for some $a \in A$. We derive $1 = sa$ from S-torsion-freeness of A, and s is a unit of A, which was excluded. We have $[v^{s,k}] \subseteq [v^{s,k}]_s$ and the rest follows at once. ∎

An easy application of (2.1) implies the important

Corollary 2.3. *For $a \in A$, $s \in S$ and $v \in Br(T)$ the following conditions are equivalent*
(1) $\bar{a}^s = \bar{0}^s$
(2) $(v^{s,k})^{-s}\bar{a} = \bar{0}^s$
(3) $[(v^{s,k})^{-s}\bar{a}]$ is finite.

Proof: We only have to show that (3) implies $\bar{a}^s = 0$. If $[(v^{s,k})^{-s}\bar{a}]$ is finite, then $\sigma s^{\ell(\sigma)-k}\bar{a} = \bar{0}$ for any $\sigma \in v$ and $\ell(\sigma)$ large enough. However $s^{\ell(\sigma)-k} \in N = \{s^n : n \in \omega\}$ and $\bar{A}^s$ is $\bar{N}$-torsion-free by Lemma 2.1, so $\bar{a} = \bar{0}$.

Lemma 2.4. *Let D be an A-submodule of $\hat{B}$ containing B and that $[d] \neq T$ for all $d \in D$. Moreover, let every $d \in D$ be contained in an S-pure cyclic A-module cA of D with Ann $c = 0$. If $\varphi \in End\, D \setminus A$, then there exist $x \in D$, $s \in S$ with $\bar{x}^s\bar{\varphi}^s \notin \bar{x}^s\bar{A}^s$.*

Proof: First we show that (*) holds.

(*)If $d\varphi \in dA$ for all elements $d \in D$, then $\varphi \in A$.

We can find $a_d \in A$ such that $d\varphi = da_d$. If d, $e \in D$ are pure with $Ann\ d = Ann\ e = 0$, then it is easy to find $f \in D$ such that the pairs (d, f) and (e, f) are A-independent. We use $da_d + fa_f = (d + f)\varphi = (d + f)a_{d+f} = da_{d+f} + fa_{d+f}$ to derive $a_d = a_{d+f} = a_f$. A similar argument shows $a_e = a_{e+f} = a_f$ and $a_d = a_e(= a)$. The endomorphism φ is scalar multiplication by a on pure elements d of D with $Ann\ d = 0$.

If e is any element in D we can find $\sigma \in T \setminus [e]$ and $e + \sigma$, σ are pure with trivial annihilator. We have $e\varphi + \sigma a = e\varphi + \sigma\varphi = (e + \sigma)\varphi = (e + \sigma)a = ea + \sigma a$ and $e\varphi = ea$ follows. The endomorphism φ is scalar multiplication on D.

Next we will show:

(**) If $\varphi \in End\ D \setminus A$ and $[d\varphi] \subseteq [d]$ for all $d \in D$, then (2.4) holds.

There exists $g \in D$ with $g\varphi \notin gA$ by assumption on φ and (*) and we also find $\sigma \in T \setminus [g]$ by assumption on D. Clearly $\sigma\varphi = \sigma a_\sigma$ for some $\alpha_\sigma \in A$ follows from $[\sigma\varphi] \subseteq [\sigma] = \{\sigma\}$. Let $x = g + \sigma$ and choose $s \in S$ such that s does not divide $g(\varphi - a_\sigma)$. Such an $s \in S$ must exist, because $\hat{B}$ is S-reduced. Observe that $g\varphi \notin gA$, hence
$$0 \neq g(\varphi - a_\sigma) \in D \subseteq \hat{B}.$$
We claim that (x, s) satisfies (2.4). Otherwise
$$x\bar{\varphi}^s \in \bar{x}^s\bar{A}^s \text{ and } \bar{x}\bar{\varphi}^s = \bar{x}^s\bar{a}^s,$$
substituting $g + \sigma$ for x gives
$$\bar{0} = (\bar{g} + \bar{\sigma})\bar{\varphi} - (\bar{g} + \bar{\sigma})\bar{a} = \bar{g}(\bar{\varphi} - \bar{a}) + \bar{\sigma}(\bar{a}_\sigma - \bar{a}).$$
Clearly $[g(\varphi - a)] \subseteq [g\ \varphi] \cup [g\ a] \subseteq [g]$ and $\bar{\sigma}(\bar{a}_\sigma - \bar{a}) = 0$ follows. We have $\bar{a}_\sigma = \bar{a}$ from Lemma 2.1. Moreover $\bar{0} = \bar{g}(\bar{\varphi} - \bar{a}) = \bar{g}(\varphi - \bar{a}_\sigma) \neq \bar{0}$ as s does not divide $g(\varphi - a_\sigma)$, a contradiction. We derive $\bar{x}\bar{\varphi}^s \notin \bar{x}^s\bar{A}^s$.

Finally let $x \in D$ with $[x\ \varphi] \nsubseteq [x]$. There exists $\sigma \in [x\varphi] \setminus [x]$ and clearly $x\varphi\lceil\sigma = \sigma y_\sigma \neq 0$. We have $s \in S$ such that $\bar{y}_\sigma^s \neq \bar{0}^s$, as A is S-reduced. On the other hand $x\lceil\sigma = 0$ and obviously $\bar{x}\bar{\varphi}^s \notin \bar{x}\bar{A}$. ∎

The construction of the desired R_2-module G is based on Shelah's Black Box applied to $\hat{B}$. We repeat the slightly modified and adjusted definitions and the necessary combinatorial results, cf [CG], [GS].

A **canonical** submodule of $\hat{B}$ is a countably generated A-submodule P of $\hat{B}$ satisfying the following conditions:

(a) $[P] \subseteq P \subseteq \hat{P}$, where $\hat{P} = \prod_{\sigma\in[P]} \sigma A \subseteq \hat{B}$, $\|[P]\| \leq \aleph_o$.

(b) P is S-pure in $\hat{P}$, i.e. $sP = s\hat{P} \cap P$ for all $s \in S$.

A **trap** is a sequence (f, P, φ, s) with $f : {}^{\omega>}\omega \to T$ a tree embedding, $\varphi \in Hom_R(P, \hat{P})$ and $s \in S$ such that the following holds

(i) $Im f \subseteq [P] \subseteq T$ are subtrees of T.

(ii) $cf\ (\|P\|) = \omega$

(iii) $\|v\| = \|P\|$ for all $v \in Br(Im f)$.

A trivial modification of the proof of the Black Box given in [CG, pp. 475 – 478] leads to the following.

Black Box 2.5. *There exists an ordinal λ^* of cardinality $\lambda^{\aleph_0}$ and a sequence of traps $(f_\alpha, P_\alpha, \varphi_\alpha, s_\alpha)(\alpha < \lambda^*)$ such that the following holds for α, $\beta < \lambda^*$:*

(i) $\beta \leq \alpha \Rightarrow \|P_\beta\| \leq \|P_\alpha\|$

(ii) $\beta \neq \alpha \Rightarrow Br(Im\ f_\alpha) \cap Br(Im\ f_\beta) = \emptyset$

(iii) $\beta + 2^{\aleph_o} \leq \alpha \Rightarrow Br(Im\ f_\alpha) \cap Br([P_\beta]) = \emptyset$

(iv) If $s \in S$, $X \subseteq \hat{B}$ a countably generated A-submodule and $\varphi \in \text{Hom}(D, \hat{B})$ for some $D \subseteq \hat{B}$ with $X + B \subseteq D$, then there exists $\alpha < \lambda^*$ such that $X \subseteq P_\alpha$, $\varphi\lceil P_\alpha = \varphi_\alpha\lceil D$ and $s = s_\alpha$.

§ 3. Proof of the MAIN THEOREM

(3.1) The construction of $\mathbf{G} = (G, G^o, G^1)$ with $G^o, G^1 \subseteq G \subseteq \hat{B}$.

The R_2-module $\mathbf{G}$ will be constructed as the union $\mathbf{G} = \bigcup_{\mu < \lambda^*} \mathbf{G}_\mu$ of an ascending sequence of R_2-modules $\mathbf{G}_\mu = (G_\mu, G^o_\mu, G^1_\mu)(\mu < \lambda^*)$ where λ^* is taken from the Black Box. At the same time we define a set \$ of strong ordinals. Let $\mathbf{G_o} = (B, 0, 0)$, $\$ \cap 0 = \emptyset, 0^\bullet = 0$ and suppose $\mathbf{G}_\alpha$, $\$ \cap \alpha$, $\alpha^\bullet \in \{0, 1\}$ are defined for all $\alpha < \mu$. If μ is a limit, then $\mathbf{G}_\mu = \bigcup_{\alpha < \mu} \mathbf{G}_\alpha$ and $\$ \cap \mu$ are defined by continuity. If $\mu = \alpha + 1$ is a successor we will distinguish cases depending on the predictions of the Black Box at α. Suppose there is an s-branch element $v_\alpha = v^s$ for $s = s_\alpha$, for some $v \in Br(Im\ f_\alpha)$, $\alpha^\bullet \in \{0, 1\}$ and there is some $x \in G_\alpha$ with $\|x\| < \|v_\alpha\|$ such that the following holds

(i) If $g_\alpha = x + v_\alpha$, $G^{\alpha^\bullet}_{\alpha+1} = < G^{\alpha^\bullet}_\alpha, g_\alpha A >$, $G^j_{\alpha+1} = G^j_\alpha$ for $j \in \{0, 1\} \setminus \{\alpha^\bullet\}$ and $G_{\alpha+1} = < G_\alpha, g_\alpha A >_* \subseteq \hat{B}$ is the S-purification of $< G_\alpha, g_\alpha A >$, then $(g_\alpha \varphi_\alpha)^{-s} \notin (G^{\alpha^\bullet}_\alpha)^{-s}$.

(ii) If $\beta < \alpha$ was a strong ordinal, then there exists $t = t(\beta, \alpha) \in S$ with $(g_\beta \varphi_\beta)^{-t} \notin (G^{\beta^\bullet}_{\alpha+1})^{-t}$.

If (i) and (ii) are possible, we let $\mathbf{G}_{\alpha+1} = (G_{\alpha+1}, G^o_{\alpha+1}, G^1_{\alpha+1})$ and $\$ \cap \alpha + 1 = \$ \cap \alpha \cup \{\alpha\}$ as above, and call α a *strong ordinal*.

Otherwise, we choose g_α and $\mathbf{G}_{\alpha+1}$ as in (i), (ii) without the requirement $\bar{g}_\alpha \bar{\varphi}^s_\alpha \notin (G^{\alpha^\bullet}_\alpha)^{-s}$, and call α *weak*. If α is neither weak nor strong, we call α *useless* and choose $\mathbf{G}_{\alpha+1} = \mathbf{G}_\alpha$, $\$ \cap \alpha + 1 = \$ \cap \alpha$. However, we will show in the next Proposition that this case does not occur.

Remark 3.2. *If $x \in \hat{B}, X \subseteq \hat{B}$ and $s, t \in S$, then $\bar{x}^{st} \in \bar{X}^{st}$ implies $\bar{x}^s \in \bar{X}^s$.*

Proof: $\bar{x}^{st} \in \bar{X}^{st} \iff x \in X + (st)^\omega \hat{B}$, and $st\hat{B} \subseteq s\hat{B}$ implies $X + (st)^\omega \hat{B} \subseteq X + s^\omega \hat{B}$, hence $\bar{x}^s \in \bar{X}^s$.

Proposition 3.3. *There are no useless ordinals $\alpha < \lambda^*$.*

This Proposition follows immediately from a more general Lemma which we use for the proof of the Main Theorem as well.

Lemma 3.4. *Suppose $\mathbf{G}_\alpha$ is defined as in (3.1) and $s \in S$. For each $v \in Br(Im f_\alpha)$ let $x_v \in G_\alpha$, $\|x_v\| < \|v\|$ and define $g_v = x_v + v^s$. Then there exists $v \in Br(Im f_\alpha)$ such that $(g_\beta \varphi_\beta)^{-t} \notin (G^{\beta^\bullet}(v))^{-t}$ for some $t = t(\beta, s) \in S$ for all strong $\beta < \alpha$, where $\mathbf{G}(\mathbf{v}) = (G(v), G^o(v), G^1_\alpha)$, $G^1(v) = G^1_\alpha$ and $G(v) = < G_\alpha, g_v A >_* \subseteq \hat{B}$, $G^o(v) = < G^o_\alpha, g_v A >$.*

Proof: Suppose that (3.4) does not hold. Let $v \in Br(Im f_\alpha)$. Then we find

some strong $\beta < \alpha$ with $(g_\beta \varphi_\beta)^{-u} \in (G^o(v))^{-u}$ for all $u \in S$. (*)

We have $(g_\beta \varphi_\beta)^{-t} \notin (G^{\beta^\bullet})^{-t}$ for some $t \in S$ by induction. Consequently $\beta^\bullet = 0$. In particular, since S is multiplicatively closed, we have $(g_\beta \varphi_\beta)^{-st} \in (G^o(v))^{-st}$ from (*) for $u = st \in S$. We may write

$(g_\beta \varphi_\beta)^{-st} = \bar{h}^{st} + (\bar{x}^{st}_v + \bar{v}^s)\bar{a}^{st}$ for some $h \in G^o_\alpha$ (**)

To ease notation, let $\bar{y} = \bar{y}^{st}$. From the construction and $h, x \in G_\alpha$ we find $n \in \omega$ such that $\sigma \notin [\bar{h}] \cup [\bar{x}]$ for all $\sigma \in v$ with $\ell(\sigma) \geq n$. Let $\sigma \in v$ be of length $\geq n$. Using $v^s = \sum_{\sigma \in v} \sigma s^{\ell(\sigma)}$ we derive

$$(g_\beta \varphi_\beta)^- \lceil \sigma = \bar{h} \lceil \sigma + \bar{v}^s \bar{a} \lceil \sigma = (\bar{v}a) \lceil \sigma = (s^{\ell(\sigma) \cdot a})^-$$

If $\sigma \notin [(g_\beta \varphi_\beta)^-]$, then $(s^{\ell(\sigma)})^- = \bar{0}$ which is equivalent to saying $s^{\ell(\sigma)} a \in (st)^\omega A$ and $s^{\ell(\sigma)} a \in s^i t^i A$ for all $i \in \omega$. Hence $a \in s^{i-\ell(\sigma)} t^i A \subseteq s^{i-\ell(\sigma)} t^{i-\ell(\sigma)} A = (st)^{i-\ell(\sigma)} A$ for all $i \geq \ell(\sigma)$, and $a \in (st)^\omega A$ or equivalently $\bar{a} = \bar{0}$.

Equation (**) reduces to $(g_\beta \varphi_\beta)^{-st} = \bar{h}^{st} \in (G_\alpha^o)^{-st}$. Remark 3.2 gives $(g_\beta \varphi_\beta)^{-t} \in (G_\alpha)^{-t}$, contradicting our choice of t. We conclude that $\{\sigma \in v, \ell(\sigma) \geq n\} \subseteq [(g_\beta \varphi_\beta)^-] \subseteq [P_\beta]$, which is a subtree of T, and all of v must be in $[P_\beta]$. Hence $v \in Br([P_\beta])$.

The Black Box implies $\beta < \alpha < \beta + 2^{\aleph_o}$, where $\beta = \beta(v)$. We obtain less than $2^{\aleph_o}$ choices for β. Since $|Br(Im f_\alpha)| = 2^{\aleph_o}$, there are two distinct branches $v, w \in Br(Im f_\alpha)$ with $\beta = \beta(v) = \beta(w)$. We have $(g_\beta \varphi_\beta)^- = \bar{h}_v + (\bar{x}_v + \bar{v}^s) \bar{a}_v$ and $(g_\beta \varphi_\beta)^- = \bar{h}_w + (\bar{x}_w + \bar{w}^s) \bar{a}_w$ from (**). Subtracting both equations yields $\bar{0} = (\bar{h}_v - \bar{h}_w) + (\bar{x}_v + \bar{v}) \bar{a}_v + (\bar{x}_w + \bar{w}) \bar{a}_w$. Using norm arguments we conclude $\bar{a}_v = \bar{a}_w = \bar{0}$ and $(g_\beta \varphi_\beta)^{-st} \in (G^o)_\alpha^{-st}$. Remark 3.2 gives the final contradiction $\bar{g}_\beta \bar{\varphi}_\beta^t \in (G_\alpha^o)^{-st}$. $\blacksquare$

Proof of the Main Theorem: The modules G^o, G^1, G are clearly A-modules and A acts faithfully on these modules. We may identify $A \subseteq \text{End } \mathbf{G}$ by scalar multiplication. In order to show equality we assume $\varphi \in \text{End } \mathbf{G} \setminus A$ and we will derive a contradiction.

The module G satisfies the assumptions of D in (2.4).

By Lemma 2.4 we can find an $s \in S$, $x \in G$ such that $\bar{x}\bar{\varphi} \notin \bar{x}\bar{A}$ where "$-$" is taken with respect to s. Using the Black Box we also find $\alpha < \lambda^*$ such that $x \in P_\alpha \cap G$, $\varphi_\alpha \lceil G = \varphi \lceil P_\alpha$, $s = s_\alpha$ and $\|x\|, \|x\varphi\| < \|P_\alpha\|$.

If α is strong, then $(g_\alpha \varphi_\alpha)^- \notin (G_{\alpha+1}^\bullet)^-$ while $g_\alpha \in G^{\alpha^\bullet}$. Hence $g_\alpha \varphi = g_\alpha \varphi_\alpha \notin G^{\alpha^\bullet}$ and $\varphi \notin \text{End } \mathbf{G}$. It remains to show that α is strong.

Consider any $v \in Br(Im f_\alpha)$ and let $v^s = \sum_{\sigma \in v} s^{\ell(\sigma)} \sigma$ for $s = s_\alpha$.

First we show

$$(\bar{v}^s + \epsilon \bar{x})\bar{A} \cap (\bar{G}_\alpha^0 + \bar{G}_\alpha^1) = \bar{0} \text{ for any } \epsilon \in \{0, 1\}. \tag{*}$$

If $(\bar{v}^s + \epsilon \bar{x})\bar{a} \in \bar{G}_\alpha^0 + \bar{G}_\alpha^1$ for some $a \in A$, $\epsilon \in \{0, 1\}$, then

$$(\bar{v}^s + \epsilon \bar{x})\bar{a} = \sum_{\beta < \alpha, \beta \in E} \bar{g}_\beta \bar{a}_\beta \text{ with } a_\beta \in A,$$

is a sum over a finite set E. There exists an element $\sigma \in v$, $\ell(\sigma)$ large enough, such that $\sigma \notin \bigcup_{\beta \in E} [g_\beta] \cup [x]$. We derive $(\bar{v}^s + \epsilon \bar{x})\bar{a} \lceil \sigma = \bar{0}$ and $s^{\ell(\sigma)} \bar{a} = \bar{0}$. Lemma 2.1 applies and $\bar{a} = \bar{0}$ follows. This shows (*).

We want to find $\epsilon = \epsilon_v \in \{0, 1\}$ and $v^* \in \{0, 1\}$ such that

$$(\bar{v}^s + \epsilon \bar{x})\bar{\varphi} \in (< G_\alpha^{v^*}, (v^s + \epsilon x)A >)^-. \tag{**}$$

Otherwise $\bar{v}^s \bar{\varphi} \in (< G_\alpha^o, v^s A >)^- \cap (< G_\alpha^1, v^s A >)^-$ and $(\bar{v}^s + \bar{x})\bar{\varphi} \in (< G_\alpha^o, (v^s + x)A >)^- \cap (< G_\alpha^1, (v^s + x)A >)^-$.

$\bar{G}_\alpha^i \subseteq \bar{G}_\alpha^0 + \bar{G}_\alpha^1$ and (*) imply at once $\bar{v}^s \bar{\varphi} \in \bar{v}^s \bar{A}$ and $(\bar{v}^s + \bar{x})\bar{\varphi} \in (\bar{v}^s + \bar{x})\bar{A}$, hence $\bar{v}^s \bar{\varphi} = \bar{v}^s \bar{a}$ and $(\bar{v}^s + \bar{x})\bar{\varphi} = (\bar{v}^s + \bar{x})\bar{a}$ for $a, b \in A$. Subtracting both equations yields

$\bar{x}\bar{\varphi} = \bar{v}^s(\bar{b} - \bar{a}) + \bar{x}\bar{b}$ and $\bar{a} = \bar{b}$ follows from norm arguments. We derive $\bar{x}\bar{\varphi} = \bar{x}\bar{a} \in \bar{x}\bar{A}$, a contradiction. Condition (**) holds.

Applying Lemma 3.4 shows, that all requirements (i) (ii) of the construction can be fulfilled if we choose $g_\alpha = v^s + \epsilon x$ from (**) for a suitable $v \in Br(Im\ f_\alpha)$. The ordinal α turns out to be strong and $\alpha \in \$$ as desired. ■

§ 4. References

[A]   D. Arnold, Representations of partially ordered sets and abelian groups, in *Proceedings of the Abelian Group Conference at Perth,* Contemp. Math. Vol. **87**, A.M.S., (1988) 91 – 109.

[AS]   M.F. Atiya, I.G. MacDonald, *Introduction to commutative algebra,* Addison-Wesley, London 1969.

[BG]   C. Böttinger, R. Göbel, Endomorphism algebras of modules with distinguished partially ordered submodules over commutative rings, J. Pure Appl. Algebra **76** (1991) 121 – 141.

[C]   A.L.S. Corner, Endomorphism algebras of large modules with distinguished submodules, J. Algebra **11** (1969) 155 – 185.

[C1]   A.L.S. Corner, Fully rigid systems of modules, Rend. Sem. Mat. Padova **82** (1989) 55 – 66.

[CG]   A.L.S. Corner, R. Göbel, Prescribing endomorphism algebras, a unified treatment, Proc. London Math. Soc.(3), **50** (1985) 447 – 479.

[DG]   M. Dugas, R. Göbel, Every cotorsion-free algebra is an endomorphism algebra, Math. Z. **181**(1982) 451 – 470.

[EM]   P. Eklof, E. Mekler, *Almost free modules, set-theoretic methods,* North-Holland, Amsterdam – New York 1990.

[FG]   B. Franzen, R. Göbel, The Brenner-Butler-Corner theorem and its applications to modules, in *Abelian Group Theory,* Gordon and Breach, London (1987), 209 – 227.

[FG1]   B. Franzen, R. Göbel, Prescribing endomorphism algebras, the cotorsion-free case, Rend. Sem. Mat. Padova **80** (1988) 215 – 241.

[G]   R. Göbel, On stout and slender groups, J. Algebra **35** (1975) 39 – 55.

[GM]   R. Göbel, W. May, Four submodules suffice for realizing algebras over commutative rings, J. Pure Appl. Algebra **65** (1990) 29 – 43.

[GS]   R. Göbel, S. Shelah, Modules over arbitrary domains II, Fund. Math. **126** (1986) 217 – 243.

[K]   I. Kaplansky, *Infinite abelian groups,* The University of Michigan Press, Ann Arbor, 1971.

[R]   C.M. Ringel, Infinite-dimensional representations of finite-dimensional hereditary algebras, Symp. Math. **23** (1979) 321 – 412.

[RT]   C.M. Ringel, H. Tachikawa, QF-3 rings, J. reine angew. Math. **272** (1975) 49 – 72.

[S]   D. Simson, Functor categories in which every flat object is projective, Bull. Acad. Polon. Ser. Math. **22** (1974) 375 – 380.

Groups Associated with Valuations

H.H. BRUNGS

Department of Mathematics, University of Alberta
Edmonton, Alberta, Canada   T6G 2G1

1.   A subring B of a skew field F is called a valuation ring of F if $x \in F \backslash B$ implies $x^{-1} \in B$. We say (F, B) is a valued skew field. The lattice of right ideals as well as the lattice of left ideals of a valuation ring are totally ordered, i.e. such a ring is a chain domain with F as its skew field of quotients.

Let W be a totally ordered set and ∞ an element larger than every element in W. Further assume that there exists a mapping $v : F \to W \cup \infty$ with

i) $v(x) = \infty$ iff $x = 0$;
ii) $v(x + y) \geq \min\{v(x), v(y)\}$,   for all x, $y \in F$;
iii) $v(x) \geq v(y)$ implies $v(zx) \geq v(zy)$ for all x, y, $z \in F$.

Then $B_v = \{x \in F \mid v(x) \geq v(1)\}$ is a valuation ring of F. Conversely, $W_B = \{aB \mid 0 \neq a \in F\}$ with $aB \geq bB$ iff $aB \subseteq bB$ and $v_B(x) = xB$, $0 \neq x \in F$, $v_B(0) = \infty$, satisfy the conditions i), ii), iii).

If $V = B$ is a valuation ring in the commutative field $K = F$ then $W_V = G$ is an ordered group with $aVbV = abV$ as operation and iii) can be replaced by iii)$'$ $v(xy) = v(x)v(y)$ for all x, $y \in F$. We call G the group associated with V in this case. Conversely, Krull in [K] observed that for any ordered abelian group (G, P) with positive cone P the group ring $\mathbb{Q}[G]$ has a field K of quotients and a valuation ring

$$V = \mathbb{Q}[P]S^{-1} \quad \text{of } K \text{ exists with} \quad S = \{\Sigma a_g g \in \mathbb{Q}[P] \mid a_g \in \mathbb{Q}, \ g \in P, \ 0 \neq a_e\}$$

and G as associated group of values.

Commutative valuation rings occur in Algebraic Geometry and Number Theory and the following result is a particular rich source for examples: Let L be an extension field of K with K-valuation ring V. Then there exists an L-valuation ring V' with $V' \cap K = V$.

The ordered set W associated with a valued skew field (F, B) is an ordered group under multiplication of fractional principal right ideals as operation only if B is invariant, i.e. $dBd^{-1} = B$ for all $0 \neq d \in F$, which means that all one-sided ideals of B are two-sided.

In this case condition iii)′ holds and these are the valuation rings investigated by Schilling ([S]). For an arbitrary ordered group G valued skew fields (F, B) with G as associated value group exist, however they cannot in general be obtained as localizations of subrings of the group ring $\mathbb{Q}[G]$. Similarly, the above-mentioned extension result does not hold any longer. Only for $p = 2$ does the p-adic valuation subring $\mathbb{Z}_p$ of the center $\mathbb{Q}$ of the skew field H of quaternions over the rationals $\mathbb{Q}$ have an extension B in H, i.e. a valuation subring B of H with $B \cap \mathbb{Q} = \mathbb{Z}_p$, p is a prime number.

2. The order type of the chain of completely prime ideals $\neq (0)$ of B is called the rank of B where (F, B) is a valued skew field. It follows from Hölder's theorem that the value group associated with a commutative or, more general, invariant valuation ring B of rank 1, i.e. with $J = J(B)$ and (0) as its only completely prime ideals is a subgroup of $(\mathbb{R}, +)$, the group of real numbers under addition.

THEOREM 2.1. ([BBT] Cor. 6.3) *If B is a valuation ring with exactly two completely prime ideals J and (0), then one of the following cases occurs:*

α) *B is invariant;*

β) *B is nearly simple, i.e. no further two-sided ideals exist in B besides (0), J, B;*

γ) *There exists an additional non-completely prime ideal Q with $J \supset Q \supset (0)$, $\cap Q^n = (0)$, $J^2 = J$ and no other two-sided ideal between J and Q.*

PROBLEM 2.2: Do there exist valuation domains of type γ) in Theorem 2.1?

This problem has come up in various forms in [P], [0] and [BT]. Dubrovin in [D1] suggested an interesting construction, which contains a gap as is pointed out in [SM].

Nearly simple valuation domains do exist, Dubrovin and Mathiak constructed the first examples, and we will discuss one construction method. We say that a subset P of a group G is a generalized positive cone of P if $P \cup P^{-1} = G$, $P \cap P^{-1} = \{e\}$ and $PP \subseteq P$. Let R_0 be an integral domain and $\sigma : G \to \text{Aut}(R_0)$ be a group homomorphism from G into the automorphism group of R_0. We consider the skew group ring $R_1 = R_0[G, \sigma] = \{\Sigma gr_g \mid g \in G, r_g \in R_0\}$ with $rg = g\sigma_g(r)$. Then R_1 is an integral domain and contains the subring $R' = R_0[P, \sigma] = \{\Sigma gr_g \in R_1 \mid g \in P\}$ and $S = \{\Sigma gr_g \in R' \mid r_e \neq 0\}$ is a multiplicatively closed subset of R'.

THEOREM 2.3. *If R_1 is a right Ore domain with F as skew field of quotients then S is a right Ore subset of R' and $B = R'S^{-1}$ is a valuation domain of F.*

Here we say that R_1 is right Ore if $a \neq 0 \neq b$ in R_1 implies $aR_1 \cap bR_1 \neq (0)$. It is a difficult problem to decide when the group ring R_1 is right Ore. That S is a right Ore set in R' means that for $s \in S$ and $0 \neq r \in R'$ the set $sR \cap rS$ is non-empty and then the elements in S can be inverted. For the construction we consider the following:

Let K be an ordered (commutative) field, V an ordered K-vector space and let G be the group of affine K-linear transformations on V, i.e.

$$G = \{(a, v) \mid 0 < a \in K, \ v \in V\}$$

with

$$(a, v) \cdot (a', v') = (aa', av' + v)$$

as operation.

Since G is the semi-direct product of two torsion-free abelian groups $\{(1,v) \mid v \in V\} \cong (V,+)$ and $\{(a,0) \mid 0 < a \in K\} \cong (\{a \mid 0 < a \in K\}, \cdot)$, it follows that $L[G]$ is Ore for any field L.

We assume that V' is an ordered K-vector space that is a proper extension of the ordered K-vector space V and we choose $\alpha \in V'\backslash V$. Then α defines a Dedekind cut $C_\alpha = (U_\alpha, \mathcal{O}_\alpha)$ on V with $U_\alpha = \{v \in V \mid v < \alpha\}$ and $\mathcal{O}_\alpha = \{v \in V \mid v > \alpha\}$.

One can prove that $P_\alpha = \{(a,v) \in G \mid a\alpha + v \geq \alpha\}$ is a generalized positive cone of G and hence the valuation domain $B = B_\alpha$ exists using Theorem 2.3.

The next results (see [BS]) give conditions in terms of the cut C_α for B_α to be nearly simple or finally simple, i.e. for B_α to have a non-zero minimal two-sided ideal $I_{\min} \neq (0)$. This ideal is necessarily completely prime.

THEOREM 2.4. *The valuation domain B_α is nearly simple if and only if the following conditions hold:*
a) *For any pair of elements v, $0 \neq v' \in V$ there exists $a \in K$ with $v < av'$.*
b) *For $0 < d \in V$ there exists $r \in V$ with $r < \alpha < r + d$.*
c) *U_α has no largest and $\mathcal{O}_\alpha$ has no smallest element.*

THEOREM 2.5. *The valuation domain B_α is finally simple if and only if the following conditions hold:*
a) *There exists v' in V such that for any v in V there exists an element a in K with $v < av'$.*
b) *There exists r, s in V with $r < \alpha < s$.*

3. Let (F, B) be a valued skew field and $W = \{aB \mid 0 \neq a \in F\}$ be the totally ordered value set as before.

One considers ([M]) the group

$$G = \{\tilde{x} \mid 0 \neq x \in F\}$$

of automorphisms of W with $\tilde{x}(aB) = xaB$, $\widetilde{xy} = \tilde{x}\tilde{y}$ and $\tilde{x} \leq \tilde{y}$ if and only if $xaB \leq yaB$ for all aB in W defines a partial order on G.

Further the ring

$$R = \bigcap_{a \in F\backslash 0 = F^*} aBa^{-1} \qquad \text{is an invariant subring of } F.$$

It follows that $H(R) = \{aR \mid a \in F^*\}$ with $aRbR = abR$ is a group, it is partially ordered by $aR \leq bR$ if and only if $aR \supseteq bR$ and $\varphi(\tilde{x}) = xR$ defines an isomorphism from G to $H(R)$ as partially ordered groups.

LEMMA 3. *([BG3])* $H(R)$ *is lattice ordered if and only if $aR \cap BR \in H(R)$ for a, $b \in F^*$.*

COROLLARY 3.2. *If G is lattice ordered then F is the skew field of quotients $Q(R)$ of R.*

PROOF: Let $0 \neq a \in F$ and $aR \cap R = rR$ for some $0 \neq r \in R$. Hence $ar' = r$ and $a = rr'^{-1}$ for $r' \in R$.

REMARK: G is lattice ordered if F is finite dimensional over its center.

We are interested in the relationship between properties of R and properties of G. The ring R is called a Bezout domain if every finitely generated ideal is principal. The ring R is called distributive if $A \cap (B + C) = (A \cap B) + (A \cap C)$ for any ideals A, B, C of R.

PROPOSITION 3.3.   *Let R be distributive with $Q(R) = F$. Then G is lattice ordered if and only if R is Bezout.*

This result is similar to the result that says that a ring of algebraic integers is a UFD if and only if it is a principal ideal domain. Commutative distributive domains are just the Prüfer domains and the Dedekind domains are the noetherian Prüfer domains.

If B is an invariant valuation domain of F then $R = B$ and G is isomorphic to the associated ordered group of G. Hence R is distributive and G is lattice ordered in this case.

In [BG3] we construct examples for valued skew fields (F, B) such that R is not distributive and G is not lattice ordered in one case and R is distributive and G is not lattice ordered in the other case. We don't know whether the fourth possibility, R not distributive but G lattice ordered, can occur.

One essential step in the construction of an example of a valued skew field (F, B) with R distributive and G not lattice ordered is the realization of the following situation:

(K, V) is a valued field and there exists a group A of automorphisms of K such that
i) $R_0 = \cap \sigma(V)$, $\sigma \in A$, is distributive with $Q(R_0) = K$;
ii) R_0 is not Bezout.

To obtain such a K, let L be an algebraically closed field of characteristic $\neq 2, 3$.

$$K_0 = L(t) \quad \subset \quad L(t, \sqrt{t^3 + 1}) \quad \subset \quad K = L(t)(\{\sqrt{(t - \ell)^3 + 1} \mid \ell \in L\})$$

$$\begin{array}{ccc}
\cup & & \cup \\[4pt]
V_0 = L[t]_t & & V \\[4pt]
\cup & & \cup \\[4pt]
L[t] \quad \subset \quad S \quad \subset & & \displaystyle\bigcap_{\sigma \in A} \sigma(V) = R_0 \\[4pt]
\cup & & \\[4pt]
tL[t] \quad \subset \quad M_1, M_2 & &
\end{array}$$

Here, $K_0 = L(t)$ is the function field in one indeterminate t over L and K is an infinite Galois extension of K_0. The valuation ring $V_0 = L[t]_t$ is the t-adic valuation ring in K_0 and V an extension valuation ring of V_0 in K, i.e. $K_0 \cap V = V_0$. The group A is the group of automorphisms of K generated by $\mathrm{Gal}(K/K_0)$ and the set $\{\varphi_\ell \mid \ell \in L\}$ of automorphisms φ_ℓ of K that are extensions of the automorphisms $(\varphi_\ell)_0$ of K_0 that map t to $t - \ell$ and fix the elements of L.

One proves that R_0 is distributive as the integral closure in K of the distributive ring $L[t]$.

Let S be the integral closure of $L[t]$ in $L(t, \sqrt{t^3 + 1})$ and S is a Dedekind domain and one can show that $tS = M_1 M_2$, M_i maximal ideals in S and M_1 has order 3 in the class group of S.

If R_0 is Bezout, then $M_1 R_0 = a R_0$ is principal, $a \in R_0$ and 3 divides $[L(t, \sqrt{t^3 + 1}, a) : L(t, \sqrt{t^3 + 1})]$ which must in turn be a power of 2. This contradiction then shows that R_0 is not Bezout.

4. We consider the case in which the skew field $F = D$ is finite dimensional over its center K with V a valuation subring of K and $[D : K] = m^2$. As before, a valuation subring B of D with $B \cap K = V$ will be called an extension of V in D.

We have the following result ([BG1]):

THEOREM 4.1.   a) *There exists an extension B_i of V in K if and only if the set T of elements in D integral over V is a subring of D. In that case $T = \cap B_i$, B_i is an extension of V in K.*
b) *Any two extensions of V in D are conjugate in D.*
c) *The number of extensions of V in D is $\leq m$.*

Earlier results about invariant extensions were obtained by P.M. Cohn ([C]) and Wadsworth ([W1]). The extensions B_i considered in the last theorem are in general not invariant. However, the following group, associated with an extension B of V in D is useful to define a number that generalizes the ramification index in the commutative situation.

$G_B = \{ dB \mid 0 \neq d \in D, \, dB = Bd \}$ is an ordered group which contains the group G_V associated with V as a subgroup. Wadsworth in [W2] proved the following result:

THEOREM 4.2. $[D : K] = [B/J(B) : V/J(V)][G_B : G_V] n^2 \delta(B).$

Here, n is the number of extensions of V in D, $J(B)$ is the maximal ideal of B and $J(V)$ is the maximal ideal of V. The defect $\delta(B)$ is equal to 1 if $\text{char}(V/J(V)) = 0$ and is equal to p^a for some integer $a \geq 0$ if $\text{char}(V/J(V)) = p$, p a prime.
This last result is only a special case of the result proved in [W2] for Dubrovin valuation rings. (For Dubrovin valuation rings see also [D2], [BG2]). Finally, we mention the group of permutations induced by conjugation on the set of extensions B_i of V in the center K of D. Gräter in [G] showed that this group is solvable.

REFERENCES

[BBT] C. Bessenrodt, H.H. Brungs, G. Törner, *Right Chain Rings; Part 1*, Schriftenreihe des Fachbereichs Mathematik No. 181, Duisburg 1990.

[BG1] H.H. Brungs, J. Gräter, *Valuation rings in finite dimensional division algebras*, J. Algebra **120** (1989), 90–99.

[BG2] H.H. Brungs, J. Gräter, *Extensions of valuation rings in central simple algebras*, Trans. Amer. Math. Soc. **317** (1990), 287–302.

[BG3] H.H. Brungs, J. Gräter, *Value groups and distributivity*, Can. J. Math.

[BS] H.H. Brungs, M. Schröder, *Finally simple valuation rings*, preprint, 1991.

[BT] H.H. Brungs, G. Törner, *Chain rings and prime ideals*, Arch. Math. **27** (1976), 253–260.

[C] P.M. Cohn, *On extending valuations in division algebras*, Studia Sci. Math. Hungar. **16** (1981), 65–70.

[D1] N.I. Dubrovin, *An example of a nearly simple chain ring with nilpotent elements*, Mat. Sbornik **120** (1983), 441–447.

[D2] N.I. Dubrovin, *Noncommutative valuation rings in simple finite-dimensional algebras over a field*, Mat. USSR Sbornik **51** (1985), 493–505.

[G] J. Gräter, *A note on valued division algebras*, J. Algebra

[K] W. Krull, *Allgemeine Bewertungstheorie*, J. Reine Angew. Math. **167** (1932), 160–196.

[M] K. Mathiak, *Bewertungen nicht kommutativer Körper* , J. Algebra **48** (1977), 217–235.

[O] B.L. Osofsky, *Noncommutative rings whose cyclic modules have cyclic injective hulls*, Pacific J. Math. **25** (1968), 331–340.

[P] E.C. Posner, *Left valuation rings and simple radical rings*, Trans. Amer. Math. Soc. **107** (1963), 458–465.

[S] O.F.G. Schilling, *The Theory of Valuations*, Math Surveys 4, Amer. Math. Soc., Providence, R.I. 1950.

[SM] M. Schröder, *Über N.I. Dubrovin's Ansatz zur Konstruktion von nicht vollprimen Primidealen in Ketterringen*, Results in Math. **17** (1990), 296–306.

[W1] A.R. Wadsworth, *Extending valuations to finite dimensional division algebras*, Proc. Amer. Math. Soc. **98** (1986), 20–22.

[W2] A.R. Wadsworth, *Dubrovin valuation rings and Henselization*, Math. Ann. **283** (1989), 301–328.

Cotorsion-free Abelian Groups
Cotorsion as Modules Over Their Endomorphism Rings

Manfred Dugas

Department of Mathematics
Baylor University, Waco, TX 76798

Theodore G. Faticoni

Department of Mathematics
Fordham University, Bronx, NY 10458

1. INTRODUCTION

The abelian group G is *cotorsion-free* if $\text{Hom}_{\mathbf{Z}}(\hat{\mathbf{Z}}_p, G) = 0$ for each prime $p \in \mathbf{Z}$, where $\hat{\mathbf{Z}}_p$ denotes the (additive group of) the p-adic integers.

Throughout this note, the groups we consider are abelian, R denotes an associative ring with identity, and M denotes a left R-module such that M^+ is a cotorsion-free group and $\text{ann}_R(M) = 0$. We call M a *cotorsion-free left R-module*, and if R^+ is a cotorsion-free group then R is called a *cotorsion-free ring*.

In recent years, there have been a number of realization Theorems for cotorsion-free rings. (See e.g. the references in [6].) Each Theorem states that if R is a cotorsion-free ring then there is a group A and an isomorphism $\text{End}_{\mathbf{Z}}(A) \cong R$. By examining the proof of these theorems we find that A is constructed as an extension of a free left R-module by a free left $\mathbf{Q}R$-module (= flat left R-module). That is, there is an exact sequence

$$0 \longrightarrow R^{(\aleph)} \longrightarrow A \longrightarrow \mathbf{Q}C \longrightarrow 0$$

of left R-modules where $\aleph$ is a cardinal, C is a free left R-module, and $\text{End}_{\mathbf{Z}}(A) \cong R$. Then A is a faithfully flat left R-module, [1].

Furthermore, because A contains a copy of R, the finite topology on R is discrete so that A is a *self-small group*. i.e. For each cardinal c the natural map $\text{Hom}_{\mathbf{Z}}(A, A)^{(c)} \rightarrow \text{Hom}_{\mathbf{Z}}(A, A^{(c)})$ is an isomorphism, [4]. This seems to be quite a restriction on the structure of A, especially in view of the fact that if the ring R is choosen judiciously then A will exhibit a number of pathological direct sum decompositions.

In this paper, we use Shelah's Black Box and a technique that was introduced in [9] to prove a realization theorem that allows one to vary the structure of A as a left $\text{End}(A)$-module and as a group.

A Summary of Results

We require the following notations.

Given a set $\Pi \subset \mathbf{Z}$ of primes and a group X, we say that X is Π-*divisible* if $pX = X$ for each $p \in \Pi$, and X is Π-*reduced* if X does not contain a nonzero Π-divisible subgroup. The Π-*adic topology* on X is the linear topology whose open neighborhoods of 0 are of the form sX where s is a product of powers of elements of Π, and X is Π-*adically complete* if it is complete in its Π-adic topology. The subgroup Y of X is Π-*pure in* X if Y/X is p-torsion-free for each $p \in \Pi$.

We fix once and for all a cotorsion-free left R-module M such that $\text{ann}_R(M) = 0$, a set of primes $\Pi \subset \mathbf{Z}$ such that M is Π-reduced, and we assume that Π contains {primes $p \in \mathbf{Z} \mid pM = M$}. Furthermore, we let S denote the subring of $\mathbf{Q}$ generated by $\{\frac{1}{p} \mid p \in \Pi\}$.

Given a ring T and left T-module X we let

$$\Gamma(T, X) = \{\text{ann}_T(F) \mid F \subset X \text{ is finite}\}.$$

Then $\Gamma(T, X)$ is the base of open neighborhoods of 0 for a linear topology on T called the X-*topology on* T. (The open neighborhoods of $t \in T$ under the X-topology are of the form $t + I$ for some $I \in \Gamma(T, X)$.) The X-topology on T is Hausdorff iff $\text{ann}_T(X) = 0$.

We let $\Gamma(M) = \Gamma(R, M)$, and we let $\text{ann} = \text{ann}_R$. Observe that each $I \in \Gamma(SR, SM)$ satisfies $I = \text{ann}(F)$ for some finite $F \subset M$. Thus we refer to the SM-topology on SR as the M-topology.

Let $\widehat{SR}$ denote the completion of SR in the M-topology and let

$$\hat{\mathcal{O}}(M) = \{q \in \widehat{SR} \mid qM \subset M\} = \text{End}_{\mathbf{Z}}(M) \bigcap \widehat{SR}.$$

Then $\hat{\mathcal{O}}(M)$ is a subring of $\widehat{SR}$ and M is a left $\hat{\mathcal{O}}(M)$-module. Moreover, the M-topology on $\hat{\mathcal{O}}(M)$ is equal to the topology inherited by $\hat{\mathcal{O}}(M)$ as a subspace of $\widehat{SR}$. The main result of this paper is

Theorem 1.1 *Let M be a cotorsion-free left R-module such that $\text{ann}_R(M) = 0$. There is a cardinal $\aleph$ and an exact sequence*

1.2 $$0 \longrightarrow M^{(\aleph)} \longrightarrow A \longrightarrow SC \longrightarrow 0$$

of left R-modules such that

  (1) C is a direct sum of cyclic R-submodules of $M^{(\aleph)}$; and

  (2) There is a topological isomorphism $\hat{\mathcal{O}}(M) \cong \text{End}_{\mathbf{Z}}(A)$ of rings, where $\text{End}(A)$ is endowed with the finite topology and $\hat{\mathcal{O}}(M)$ is endowed with the M-topology.

Applications of Theorem 1.1 include the following.

Let $\Omega(R)$ denote the class of groups A for which there exists an isomorphism of rings $R \cong \text{End}(A)$. In 3.1 and 3.8 we show that if $n < \infty$ is the (weak) left global dimension of R and if $1 \leq k < n$ is an integer then there is a self-small $A_k \in \Omega(R)$ whose projective (flat) dimension as a left R-module is k. In particular, in 3.6 we characterize the rings R such that each $A \in \Omega(R)$ is an E-flat module.

Let Γ denote a base of open neighborhoods of 0 for a linear Hausdorff topology on R. Call Γ *a cotorsion-free topological base for R* if R/I is cotorsion-free for each $I \in \Gamma$. In 3.16 we characterize the cotorsion-free rings R such that R imbeds in $A^{(\aleph_0)}$ for each $A \in \Omega(R)$, and in 3.17 we characterize those cotorsion-free R such that each $A \in \Omega(R)$ is a self-small group.

2. THE BLACK BOX CONSTRUCTION

Throughout this section, M is a cotorsion-free left R-module such that $\operatorname{ann}(M) = 0$, and we fix a set of primes $\Pi \subset \mathbf{Z}$ such that M is Π-reduced, we assume that Π contains $\{$primes $p \in \mathbf{Z} \mid pM = M\}$, and we let S denote the subring of $\mathbf{Q}$ generated by $\{\frac{1}{p} \mid p \in \Pi\}$.

THE BLACK BOX

We begin by constructing the traps used in the Black Box.

Fix an infinite regular cardinal $\aleph$ and a cardinal $|M| \leq \kappa \leq \aleph$ such that $\aleph^\kappa = \aleph$. For each cardinal $a < \aleph$ and $k < \kappa$ let $(a,k) : M \to M(a,k)$ denote an isomorphism of left R-modules, and for each $m \in M$ let $m(a,k)$ denote the image of m under (a,k). We let group homomorphisms act on the left.

Let
$$B = \bigoplus_{\substack{a < \aleph \\ k < \kappa}} M(a,k),$$

and let $\hat{B}$ and $\hat{M}(a,k)$ denote the Π-adic completions of B and $M(a,k)$ respectively.

Each element $x \in \hat{B}$ has unique representation
$$x = \sum_{n < \omega} x_n(a_n, k_n)$$

as a convergent sum in the Π-adic topology on $\hat{B}$, where $0 \neq x_n \in \hat{M}$ and $(a_n, k_n) \neq (a_m, k_m)$ for $n \neq m$. The *support of x* is
$$[x] = \{(a_n, k_n) \mid n < \omega\},$$

the *norm of x* is
$$\|x\| = \sup\{a_n \mid n < \omega\},$$

and the *norm* of a subset $X \subset \hat{B}$ is
$$\|X\| = \sup\{\|x\| \mid x \in X\}.$$

Given a submodule $X \subset \hat{B}$ the *interior of X* is
$$X^\circ = \{x \in X \mid \|x\| < \|X\|\}.$$

A *canonical submodule* of B is any module P such that

1. The cardinality of P is κ;

2. If $0 \neq x \in P$ and $(a, k) \in [x]$ then $M(a, t) \subset P \ \forall \ t < \kappa$.

2.1 A *trap* is a triple (f, P, ϕ) such that

1. f is a monotonic increasing sequence of ordinals;

2. P is a canonical submodule of B such that

    (a) $M(f(n), k) \subset P$ for each $n < \omega$ and $k < \kappa$; and
    (b) $\|P\| = \sup\{f(n) \mid n < \omega\}$;

3. $\phi \in \text{Hom}_{\mathbf{Z}}(P, \hat{P})$ is such that $\|\phi(x)\| < \|P\|$ for each $x \in P$.

We are now ready to state the Black Box.

Theorem 2.2 [7, Theorem 2.6] *For some ordinal λ there exists a sequence of traps $\tau_\ell = (f_\ell, P_\ell, \phi_\ell)$ such that*

1. $\|P_\mu\| \leq \|P_\ell\|$ *for all $\mu < \ell < \lambda$;*

2. *If $\mu \neq \ell$ then $\text{Image } f_\mu \cap \text{Image } f_\ell$ is finite;*

3. *If $\mu + \kappa^\omega \leq \ell$ then $\bigoplus_{n<\omega} M(f_\ell(n), g(n)) \cap P_\mu \subset \bigoplus_{n<n_o} M(f_\ell(n), g(n))$ for some $n_o = n_o(\ell)$ and for each map $g : \omega \to \kappa$;*

4. *If X is a subset of $\hat{B}$, if $|X| \leq \kappa$, and if $\phi \in \text{End}_{\mathbf{Z}}(\hat{B})$ then there exists an $\ell < \lambda$ such that $X \subset \hat{P}_\ell$, $\|X\| < \|P_\ell\|$, and $\phi \lceil P_\ell = \phi_\ell$. ///*

SPECIAL ELEMENTS

Let $\ell \leq \lambda$. We call an element $y \in \hat{B}$ *special for* ℓ if y can be written as a convergent sum

2.3 $$y = b_o \pi + \sum_{n<\omega} y_n(f_\ell(n), g(n))$$

in $\hat{B}$ where

1. $\pi \in \hat{\mathbf{Z}}$, $b_o \in P_\ell^o$, $y_n \in M$;

2. $g : \omega \to \kappa$ is a map;

3. $\text{ann}(b_o) = \text{ann}(\{y_n \mid n < \omega\})$; and

4. For each $n < \omega$ the set $\{n' < \omega \mid \text{ann}(y_n) = \text{ann}(y_{n'})\}$ is infinite.

Lemma 2.4 *Let y be special for ℓ.*

1. *If $r_0 b_o \pi + r_1 \sum_{n<\omega} y_n(f_\ell(n), g(n)) = 0$ for some $r_0, r_1 \in R$ then $r_0 b_o \pi = 0$ and $r_1 y_n = 0$ for each $n < \omega$.*

2. $\text{ann}(y) = \text{ann}(b_o) \in \Gamma(M)$

3. *If $r \in R$ and $ry \neq 0$ then $\|ry\| = \|y\| = \|P_\ell\|$.*

Proof: (1) Let y be special for ℓ, let $r_0, r_1 \in R$, and consider the convergent sum

$$r_0 b_o \pi + r_1 \sum_{n < \omega} y_n(f_\ell(n), g(n)) = 0.$$

Given a finite set $F \subset \aleph \times \kappa$, let $\rho_F : \hat{B} \to \hat{B}$ be the canonical map whose kernel is $\oplus_{(a,k) \in F} M(a, k)$.

By 2.3.1 $[b_o]$ is finite, so

$$\begin{aligned} 0 &= \rho_{[b_o]} \left(r_0 b_o \pi + r_1 \sum_{n < \omega} y_n(f_\ell(n), g(n)) \right) \\ &= \sum_{(f_\ell(n), g(n)) \notin [b_o]} r_1 y_n(f_\ell(n), g(n)). \end{aligned}$$

Because f_ℓ is a monotonic increasing sequence, 2.1.1, the $r_1 y_n(f_\ell(n), g(n))$ are independent elements of B. Then $r_1 y_n(f_\ell(n), g(n)) = 0$ for all n such that $(f_\ell(n), g(n)) \notin [b_o]$. Furthermore, because $[b_o]$ is finite, 2.3.4 shows that if $(f_\ell(n), g(n)) \in [b_o]$ then there is an $n' < \omega$ such that $(f_\ell(n'), g(n')) \notin [b_o]$ and $\mathrm{ann}(y_n) = \mathrm{ann}(y_{n'})$. Hence $r_1 y_n(f_\ell(n), g(n)) = 0$ for each $n < \omega$, and thus $r_0 b_o \pi = 0$.

(2) follows from part 1 and 2.3.3.

(3) By 2.1.2(b) $\|P_\ell\| = \sup(\{ f_\ell(n) \mid n < \omega \}) = \|y\|$.

Let $r \in R$ and assume that $ry \neq 0$. Clearly $\|ry\| \leq \|y\|$. Fix $n < \omega$ such that $y_n(f_\ell(n), g(n)) \neq 0$ in the expression 2.3. Because $ry \neq 0$ there is an $n' < \omega$ such that $ry_{n'}(f_\ell(n'), g(n')) \neq 0$, and by 2.3.4 there is an $n < n'' < \omega$ such that $\mathrm{ann}(y_{n'}) = \mathrm{ann}(y_{n''})$. Then $(f_\ell(n''), g(n'')) \in [ry]$ and because f_ℓ is a monotonic increasing sequence, $f_\ell(n) < f_\ell(n'')$. Thus $\|y\| \leq \|ry\|$, and hence $\|y\| = \|ry\|$. ///

CONSTRUCT A

Now define a group A by transfinite induction as follows.

(I_0) $A_0 = B$;

       Given an ordinal $\ell < \lambda$

2.5 (I_ℓ) $A_\ell = \cup_{\mu < \ell} A_{\mu+1}$ for all $\ell < \lambda$

      (II_ℓ) Given A_ℓ then choose a special element y_ℓ for ℓ and let

          $A_{\ell+1}/A_\ell = S(Ry_\ell + A_\ell)/A_\ell \subset \hat{B}/A_\ell.$

Then $A = \bigcup_{\ell < \lambda} A_\ell$.

Lemma 2.6 *Let A_ℓ be as in 2.5.*

1. $SA_\ell \cap SRy_\ell = 0$.

2. A_ℓ is a Π-pure R-submodule of $A_{\ell+1}$ and of $\hat{B}$.

Proof: (1) Let $0 \neq x \in SA_\ell$ and let $r \in SR$ be such that $ry_\ell \neq 0$. From (I_ℓ) and (II_ℓ) it is clear that

$$SA_\ell = \sum_{\mu < \ell} SRy_\mu + SB. \tag{1}$$

Then $x = b + r_1 y_1 + \ldots + r_{n_x} y_{n_x}$ for some $b \in SB$, $r_i \in SR$, $\mu_i < \ell$, and some special elements

$$y_i = b_i \pi + \sum_{n < \omega} y_{n,i}(f_{\mu_i}(n), g_i(n))$$

for μ_i. If we let $[z]'$ denote the first coordinates of the pairs in $[z]$ then from 2.3 we have that

$$[r y_\ell]' \subset [y_\ell]' = \text{Image } f_\ell$$

and that

$$[x]' \subset [b]' \cup \left(\bigcup_{i=1}^{n_x} [b_i]' \right) \cup \left(\bigcup_{i=1}^{n_x} \text{Image } f_{\mu_i} \right).$$

Now $b, b_i \in B$ so that $[b]'$ and $[b_i]'$ are finite sets. Moreover, 2.2.2 states that Image $f_{\mu_i} \cap$ Image f_ℓ is finite for each $\mu_i < \ell$. Hence $[x]' \cap [r y_\ell]'$ is finite. However, 2.1.1 and 2.4.3 imply that $[r y_\ell]'$ is infinite. Thus $x \notin SR y_\ell$, and hence $SA_\ell \cap SR y_\ell = 0$.

(2) It readily follows from 2.5 that A_ℓ/B is an S-submodule of $\hat{B}/B$. Thus,

$$\hat{B}/A_\ell = \frac{\hat{B}/B}{A_\ell/B}$$

is an S-module. Because S-modules are Π-torsion-free, A_ℓ is Π-pure in $\hat{B}$, and hence A_ℓ is Π-pure in $A_{\ell+1}$. ///

Lemma 2.7   *1. A is a cotorsion-free Π-pure R-submodule of $\hat{B}$;*

2. $A/B \cong \bigoplus_{\ell < \lambda} SR \bar{y}_\ell$ where $\bar{y}$ denotes the image of y modulo B.

3. $\text{ann}(x) \in \Gamma(M)$ for each $x \in A$.

4. A is a left $\hat{\mathcal{O}}(M)$-module.

Proof: (1) The proof that A is cotorsion-free proceeds as in [7, page 94-95], and the Π-purity of A in $\hat{B}$ follows from 2.6.2.

(2) The independence of the y_ℓ follows immediately from 2.6.1. The rest follows from equation (1) in 2.6.

(3) We use induction on ℓ. Assume we have shown that $\text{ann}(x) \in \Gamma(M)$ for each $x \in A_\mu$ and each $\mu < \ell$. Given $x \in A_\ell$, there is an ordinal $\mu < \ell$ such that $x = a_\mu + r y_\mu \in SA_\mu \oplus SR y_\mu$. The sum $SA_\mu \oplus SR y_\mu$ is direct by 2.6.1, and by induction $\text{ann}(a_\mu) \in \Gamma(M)$. Furthermore, by 2.4.2 $\text{ann}(r y_\mu) \in \Gamma(M)$, so that $\text{ann}(x) = \text{ann}(a_\mu) \cap \text{ann}(r y_\mu) \in \Gamma(M)$.

(4) Let $x \in A$ and $q \in \hat{\mathcal{O}}(M)$. By part 3 $\text{ann}(x) \in \Gamma(M)$ and because $q \in \widehat{SR}$, there is an $r_x \in SR$ such that $q - r_x \in \text{ann}(x)$. Then $qx = r_x x$. There is a product $m \neq 0$ of powers of primes in Π such that $m r_x \in R$ and because A is a left R-module $m r_x x \in A$. But part 1 states that A is Π-pure in $\hat{B}$, so $qx = r_x x \in A$. Thus A is a left $\hat{\mathcal{O}}(M)$-module. ///

CONSTRUCTING TEST ELEMENTS

Let $\hat{B}[\Gamma] = \{x \in \hat{B} \mid Ix = 0 \text{ for some } I \in \Gamma(M)\}$.

We will refine the construction in 2.5 by including elements t_ℓ that will serve as test elements for endomorphisms of A.

Construct A as in 2.5 but include the following two steps.

2.8 Assume $\infty \notin \hat{B}$ and define $t_\ell \in \hat{B} \cup \{\infty\}$ as follows:

(III$_\ell$) $t_\mu \notin A_\ell$ for all $\mu \le \ell$.

(IV$_\ell$) (a) Either $t_\ell = \phi_\ell(y_\ell) \notin A_\ell$ or else

(b) $t_\ell = \infty$, and for all $w \in \hat{B}[\Gamma]$, we have $t_\mu \notin SA_\ell + SRw$ for each $\mu < \ell$ but $\phi_\ell(w) \in SA_\ell + SRw$.

A modified version of the proof of [7, Lemma 3.4] shows that this construction of t_ℓ can indeed be done. Observe that $t_\ell \notin A$ for each $\ell < \lambda$.

PROOF OF THEOREM 1.1

At this point, the argument proceeds as in [9, pages 7,8,9].

Lemma 2.9 $\Gamma(M) = \Gamma(R, B)$.

Proof: It is clear that $\Gamma(M) \subset \Gamma(R, B)$. Let $I \in \Gamma(R, B)$. There is a finite set $F \subset B$ such that $\text{ann}(F) = I$. Given $x \in F$ write

$$x = \sum_{n=1}^{n_x} x_n(f_x(n), g_x(n)) \tag{2}$$

where $x_n \in M$, f_x and g_x have finite domain $\{1, ..., n_x\}$, and $f_x(1) < ... < f_x(n_x)$. The terms in the sum (2) are independent, so $\text{ann}(x) = \text{ann}(\{x_n \mid 1 \le n \le n_x\}) \in \Gamma(M)$. Because $\Gamma(M)$ is closed under finite intersections $\text{ann}(F) = \bigcap_{x \in F} \text{ann}(x)$ is in $\Gamma(M)$. ///

Lemma 2.10 Let $\phi \in \text{End}(A)$ and $x \in B$. There is an $r_x \in SR$ such that $\phi(x) = r_x x$.

Proof: Given $\phi \in \text{End}(A)$ suppose to the contrary that there is an $x \in B$ such that $\phi(x) \ne rx$ for any $r \in SR$. Then $0 \ne \phi(x) \in A$, and by 2.5 there is an ordinal $\ell_o < \lambda$ such that $\phi(x) \in A_{\ell_o}$. By 2.2.4, with $X = A_{\ell_o}$, there is an ordinal $\ell_o < \ell < \lambda$ such that $A_{\ell_o} \subset \hat{P}_\ell$, $\|A_{\ell_o}\| < \|P_\ell\|$, and $\phi\lceil P_\ell = \phi_\ell$.

An obvious contradiction occurs if ℓ satisfies condition (IV$_\ell$)(a), ($y_\ell \in A$ but $\phi(y_\ell) = \phi_\ell(y_\ell) \notin A$), so ℓ must satisfy (IV$_\ell$)(b).

Construct a special element $w = b_o + \sum_{n < \omega} w_n(f_\ell(n), g(n))$ for ℓ such that $\text{ann}(w) = \text{ann}(x)$. (e.g. Write x as in equation (2), let $x = b_o$, and let $w_n = n! x_k$ iff $n \equiv k(\text{mod } n_x)$.) By 2.9 $\text{ann}(x) \in \Gamma(M)$, so $w \in \hat{B}[\Gamma]$. Then (IV$_\ell$)(b) states that $\phi(w) = \phi_\ell(w) \in SA_\ell + SRw$. (Here we identify $\phi : A \to A$ with its unique lifting $\phi : \hat{B} \to \hat{B}$.) There is an $r \in SR$ such that

$$\phi_\ell(w) \equiv rw(\text{mod } SA_\ell). \tag{3}$$

Next let $y = w + x$. Since $\text{ann}(x)y = 0$, $y \in \hat{B}[\Gamma]$, and then $\phi_\ell(y) \in SA_\ell + SRy$ by (IV$_\ell$)(b). There is an $s \in SR$ such that

$$\phi_\ell(y) \equiv s(w + x)(\text{mod } SA_\ell). \tag{4}$$

Combining equations (3) and (4) reveals that

$$\phi_\ell(x) = \phi_\ell(y - w) = (s - r)w + (sx + z)$$

for some $z \in SA_\ell$. By 2.6.1 and our choice of ℓ

$$\phi_\ell(x) - (sx + z) = (s - r)w \in SA_\ell \cap SRw = 0 \tag{5}$$

so $(s - r)w = 0$.

Now, by our choice of x, $(\phi_\ell - s)(x) \neq 0$, and by 2.7.1 A_ℓ is cotorsion-free, so there is a $\pi \in \hat{\mathbf{Z}}$ such that $(\phi_\ell - s)(x)\pi \notin A_\ell$. Inasmuch as $A_\ell = SA_\ell \cap \hat{B}$, 2.6.2, $(\phi_\ell - s)(x)\pi \notin SA_\ell$.

Finally, let $u = w + x\pi$. As above there is a $t \in SR$ such that

$$\phi_\ell(u) \equiv t(w + x\pi)(\bmod SA_\ell).$$

Combining our choices of r, s, t, and π yields the equation

$$\begin{aligned}
(\phi_\ell - s)(x\pi) &= (\phi_\ell - s)(u - w) \\
&= (t - r)w + (t - s)x\pi + z'
\end{aligned} \tag{6}$$

for some $z' \in SA_\ell$. In a manner similar to 2.6.1 one shows that $(SA_\ell + SA_\ell\pi) \cap SRw = 0$. Then $(t - r)w = 0$ as in equation (5). Since we have shown that $(s - r)w = 0$, $(t - s) = (t - r) - (s - r) \in \text{ann}(w) = \text{ann}(x)$. Substituting $(t - s)x = 0$ into (6) then shows that $(\phi_\ell - s)(x\pi) = z' \in SA_\ell$.

This contradiction to our choice of π proves that for each $x \in B$ there is an $r_x \in SR$ such that $\phi(x) = r_x x$. ///

Lemma 2.11 Let $\phi \in \text{End}(A)$ and let $F \subset B$ be a finite set. There is an $r_F \in SR$ such that $\phi(x) = r_F x$ for each $x \in F$.

Proof: Let $F \subset B$ be a finite set. Given $x \in F$ write x as in equation (2) in 2.9. Let $E = \{x_n \mid x \in F\}$. Then $E \subset M$ is a finite set. Let

$$\{(a_y, k_y) \in \aleph \times \kappa \mid y \in E\}$$

be a finite set indexed by E, and let

$$y_F = \sum_{y \in E} y(a_y, k_y).$$

Notice that $y_F \in B$.

Several applications of 2.10 will yield r_y, $r_F \in SR$ such that

$$\begin{aligned}
\phi(y(a_y, k_y)) &= r_y y(a_y, k_y) \\
\phi(y_F) &= r_F y_F,
\end{aligned}$$

whence

$$\sum_{y \in E} r_F y(a_y, k_y) = \sum_{y \in E} r_y y(a_y, k_y). \tag{7}$$

Because the sum $\oplus_{y \in E} M(a_y, k_y) \subset B$ is direct we can compare the terms in equation (7) to show that

$$\phi(y(a_y, k_y)) = r_y y(a_y, k_y) = r_F y(a_y, k_y) \tag{8}$$

for each $y \in E$.

Finally, let $x \in F$ and write x as in equation (2). Because $x_n \in E$ and because the pairs (a_y, k_y) were arbitrarily chosen, equation (8) shows that

$$\phi(x_n(f_x(n), g_x(n))) = r_F x_n(f_x(n), g_x(n))$$

for each n. Then $\phi(x) = r_F x$, and hence $\phi(x) = r_F x$ for each $x \in F$. This completes the proof. ///

Lemma 2.12 *Let A be as constructed in 2.5. There is an isomorphism $\hat{\mathcal{O}}(M) \cong \mathrm{End}(A)$.*

Proof: By 2.7.4 A is an $\hat{\mathcal{O}}(M)$-submodule of $\hat{B}$. Because $\mathrm{ann}(A) \subset \mathrm{ann}(M) = 0$ there is an imbedding $\sigma : \hat{\mathcal{O}}(M) \to \mathrm{End}(A)$ given by sending $q \in \hat{\mathcal{O}}(M)$ to left multiplication by q. It remains to show that σ is a surjection.

Let $\phi \in \mathrm{End}(A)$. By 2.11 for each finite subset $F \subset B$ there exists an $r_F \in SR$ such that $(\phi - r_F)(F) = 0$. Clearly $\{r_F \mid F \subset B\}$ is a Cauchy net under the B-topology on SR. Because $\Gamma(M) = \Gamma(R, B)$, 2.9, the net $\{r_F \mid F \subset B\}$ has a limit $\hat{r}$ in $\widehat{SR}$. Then for each finite set $F \subset B$, $(r_F - \hat{r})(F) = 0$. Therefore $\hat{r} \in \hat{\mathcal{O}}(M)$ and $(\phi - \hat{r})(B) = 0$. Because A/B is Π-divisible and because A is Π-reduced $\phi - \hat{r} = 0$. Thus σ is a surjection, and the proof is complete. ///

Lemma 2.13 *Let τ denote the topology on $\hat{\mathcal{O}}(M)$ induced by the isomorphism given in 2.12 and the finite topology on $\mathrm{End}(A)$. Then τ is equivalent to the M-topology on $\hat{\mathcal{O}}(M)$.*

Proof: Proceed as in [9, Lemma 2.11]. ///

Proof of Theorem 1.1: Part 2 follows from 2.12 and 2.13. By 2.7.2, $A/B = SC$ where C is the direct sum of cylic modules of the form Ry_ℓ for some $\ell < \lambda$. By 2.7.3 $\mathrm{ann}(y_\ell) \in \Gamma(M)$, so that $\mathrm{ann}(y_\ell) = \mathrm{ann}(F)$ for some finite set $F \subset M$. Let $F = \{x_1, \ldots, x_n\}$ and let $x = \sum_{k=1}^{n} x_k(k, k) \in B$. Then $\mathrm{ann}(x) = \mathrm{ann}(F) = \mathrm{ann}(y_\ell)$ and we conclude that $Ry_\ell \cong Rx \subset B$, as required by part 1. ///

The group A is *self-small* if for each cardinal c the natural imbedding $\mathrm{Hom}(A, A)^{(c)} \longrightarrow \mathrm{Hom}(A, A^{(c)})$ is an isomorphism.

Corollary 2.14 *Let R be a cotorsion-free ring, let M be a cotorsion-free left R-module, and let $\Pi \subset \mathbf{Z}$ be a set of primes that contains $\{$primes $p \in \mathbf{Z} \mid pM = M\}$. Assume that M is Π-reduced, and let S be the subring of $\mathbf{Q}$ generated by $\{\frac{1}{p} \mid p \in \Pi\}$. There is a cardinal $\aleph$ and an exact sequence*

2.15 $$0 \longrightarrow (M \oplus R)^{(\aleph)} \longrightarrow A \longrightarrow SC \longrightarrow 0$$

of left R-modules such that

1. *A is a self-small group;*

2. *$R \cong \mathrm{End}(A)$; and*

3. *C is a free left R-module.*

Proof: Consider the cotorsion-free left R-module $M \oplus R$. Because R is a direct summand of $M \oplus R$, the $M \oplus R$-topology on R is discrete and $\hat{\mathcal{O}}(M \oplus R) = R$. Then 1.1 states that there is a cardinal $\aleph$ and an exact sequence 2.15 of left R-modules such that $C \cong \oplus_y Ry$ for some special elements y, 2.7.2, and $R \cong \text{End}(A)$. Because the isomorphism is topological, $\text{End}(A)$ is discrete in the finite topology, so that A is a self-small group, [4]. This proves parts 1 and 2.

Because $1 \in R \subset M \oplus R$, the special elements y that we choose in 2.5 can be selected so that $\text{ann}(y) = 0$. (Construct y in 2.3 and 2.5 so that $b_o = 1$. Then $\text{ann}(y) = \text{ann}(b_o) = 0$ by 2.4.2.) Then C is a free left R-module, as required by part 3. ///

3. APPLICATIONS

This section examines some of the consequences of Theorem 1.1. Fix a cotorsion-free ring R, and let $\Omega(R)$ denote the class of groups A for which there is a ring isomorphism $R \cong \text{End}(A)$. In this section any application of Theorem 1.1 will assume that $S = \mathbf{Q}$.

FLAT DIMENSIONS

The flat dimension, flatdim.(X), of the left R-module X is the least integer m such that $\text{Tor}_R^{m+1}(\cdot, X) = 0$, provided this minimum exists. Otherwise we write flatdim.$(X) = \infty$. The *weak left global dimension of R* is the supremum of the flat dimensions of the left R-modules. All flat dimensions in this section are calculated as left R-modules. See [2, 12] for details concerning flat dimension and weak global dimension.

We will prove the following result.

Theorem 3.1 *Let R be a cotorsion-free ring and let n be the weak left global dimension of R.*

1. *If $n < \infty$ then for each integer $0 \leq k < n$ there is a group $A_k \in \Omega(R)$ such that* flatdim.$(A_k) = k$.

2. *If $n = \infty$ then there is a group $A_\infty \in \Omega(R)$ such that* flatdim.$(A_\infty) = \infty$.

Lemma 3.2 *Let R, M, A, and C be as in 2.14. Identify $R = \text{End}(A)$.*

1. $\mathbf{Q}C$ *is a flat left R-module.*

2. $R/I \otimes_R A \neq 0$ *for each proper right ideal $I \subset R$.*

3. $\text{Tor}_R^k(\cdot, A) \cong \text{Tor}_R^k(\cdot, M^{(\aleph)})$ *for each integer $1 \leq k$.*

Proof: (1) Observe that $\mathbf{Q}C$ is the direct union of the free (= flat) left R-modules $(n!)^{-1}C$. Then $\mathbf{Q}C$ is a flat left R-module.

(2) Let $I \subset R$ be a proper right ideal. By part 1 $\text{Tor}_R^1(R/I, \mathbf{Q}C) = 0$, so an application of $R/I \otimes_R \cdot$ to the exact sequence 2.15 yields the exact sequence

$$0 \to R/I \otimes_R (M \oplus R)^{(\aleph)} \longrightarrow R/I \otimes_R A \longrightarrow R/I \otimes_R \mathbf{Q}C \to 0$$

of abelian groups. Inasmuch as $R/I \otimes_R R \neq 0$ is a direct summand of the left hand term in this sequence, $R/I \otimes_R A \neq 0$.

(3) Let $1 \leq k$ and apply $\cdot \otimes_R$ to the exact sequence 2.15 to produce the long exact sequence

$$\operatorname{Tor}_R^{k+1}(\cdot, \mathbf{Q}C) \longrightarrow \operatorname{Tor}_R^k(\cdot, (M \oplus R)^{(\aleph)}) \longrightarrow \operatorname{Tor}_R^k(\cdot, A) \longrightarrow \operatorname{Tor}_R^k(\cdot, \mathbf{Q}C)$$

of abelian groups. By part 1 $\operatorname{Tor}_R^k(\cdot, \mathbf{Q}C) = 0$ for each $1 \leq k$, so

$$\operatorname{Tor}_R^k(\cdot, M^{(\aleph)}) \cong \operatorname{Tor}_R^k(\cdot, (M \oplus R)^{(\aleph)}) \cong \operatorname{Tor}_R^k(\cdot, A)$$

for each $1 \leq k$. ///

Corollary 3.3 *Let R, M, and A be as in 2.14.*
Then flatdim.(A) = flatdim.(M). ///

As a consequence of 3.3, in order to find a group $A_k \in \Omega(R)$ such that flatdim.$(A_k) = k$, it suffices to find a cotorsion-free left R-module M_k of flat dimension k. Of course R is a cotorsion-free left R-module of flat dimension 0.

Lemma 3.4 *Let R be a cotorsion-free ring.*

1. *If there is a left R-module of flat dimension $0 < n < \infty$ then for each integer $0 \leq k \leq n - 1$ there is a cotorsion-free left R-module M_k such that* flatdim.$(M_k) = k$.

2. *If there is a left R-module of infinite flat dimension then there is a cotorsion-free left R-module M_∞ such that* flatdim.$(M_\infty) = \infty$.

Proof: 1. Suppose M is a left R-module such that $0 < n = $ flatdim.$(M) < \infty$. Write $M = P_n/M_{n-1}$ where P_n is a free left R-module. Then $M_{n-1} \subset P_n$ is a cotorsion-free left R-module, and the usual dimension shifting argument, (see e.g. [12, page 169, Theorem 2(3)]), shows that flatdim.$(M_{n-1}) = n - 1$. A simple induction on $n - 1$ then produces for each $0 \leq k \leq n - 1$ a cotorsion-free R-module M_k such that flatdim.$(M_k) = k$.
2. If M has infinite flat dimension then write $M = P/M_\infty$ where P is a (cotorsion-free) projective right R-module. As above flatdim.$(M_\infty) = \infty$. ///

Proof of Theorem 3.1: 1. If $0 \leq k < n < \infty$ is given then use 3.4.1 or the comment preceeding it to produce a cotorsion-free left R-module M_k of flat dimension k. Then use 2.14 to construct $A_k \in \Omega(R)$. By 3.3, flatdim.$(A_k) = k$.
2. If $n = \infty$ then use 3.4.2 to produce a cotorsion-free left R-module M_∞ of infinite flat dimension. Proceed as above to construct $A_\infty \in \Omega(R)$ such that flatdim.$(A_\infty) = \infty$. ///

The group A is *(faithfully) E-flat* if it is (faithfully) flat as a left End(A)-module.

Corollary 3.5 *Let R, M, A, and C be as in 2.14.*

1. *A is a (faithfully) E-flat module iff M is a (faithfully) flat left R-module.*

2. *[1, Theorem 2.8] Each cotorsion-free ring R is the group endomorphism ring of a self-small faithfully E-flat group A.*

Proof: Part 1 follows immediately from 3.2.2 and 3.3, while part 2 follows immediately from part 1 if we let $M = R$ in 3.1.1. ///

In [10, Corollary 3.11] it is shown that the reduced torsion-free finite rank ring R is hereditary iff each finite rank $A \in \Omega(R)$ is an E-flat group. We extend this result to cotorsion-free rings.

Theorem 3.6 *The following are equivalent for a cotorsion-free ring R.*

1. *Each $A \in \Omega(R)$ is an E-flat group.*

2. *Each self-small $A \in \Omega(R)$ is an E-flat group.*

3. *Each finitely generated cotorsion-free left R-module is flat.*

Proof: $(1) \Rightarrow (2)$ is clear.

$(2) \Rightarrow (3)$ Assume each self-small $A \in \Omega(R)$ is an E-flat group, and let M be a finitely generated cotorsion-free left R-module. By 2.14 there is an exact sequence 2.15 such that A is a self-small group, and there is a isomorphism $R \cong \text{End}(A)$. By hypothesis A is an E-flat group, so by 3.3 M is a flat left R-module.

$(3) \Rightarrow (1)$ Assume each finitely generated cotorsion-free left R-module is flat, and let $A \in \Omega(R)$. Inasmuch as each finitely generated left R-submodule of A is cotorsion-free, A is the direct union of (finitely generated) flat left R-modules. Then A is a flat left R-module, and because $R \cong \text{End}(A)$, A is an E-flat group. This completes the proof. ///

Corollary 3.7 *Let R be a left Noetherian cotorsion-free ring. If each $A \in \Omega(R)$ is an E-flat group then R is a left hereditary ring.*

Proof: Assume that each $A \in \Omega(R)$ is an E-flat group, and let $I \subset R$ be a left ideal. Because R is left Noetherian cotorsion-free ring, I is a finitely generated cotorsion-free left R-module. By 3.6 I is a flat left R-module, and because finitely generated flat modules are projective over a left Noetherian ring, [2], I is a projective left R-module. Thus R is a left hereditary ring. ///

PROJECTIVE DIMENSIONS

We maintain the assumptions on R and S made at the beginning of this section.

The *projective dimension*, proj.dim.(X), of the left R-module X is the least integer m such that $\text{Ext}_R^{m+1}(X, \cdot) = 0$, provided this minimum exists. Otherwise, we write proj.dim.$(X) = \infty$. Throughout this section proj.dim. is calculated over R.

The *left global dimension of R*, l.gl.dim.(R), is the supremum of the projective dimensions of left R-modules. See [2] and [12] for details.

Angaud-Gaur [3] shows that if k is a positive integer then there is a torsion-free group A of finite rank whose projective dimension as a left $\text{End}(A)$-module is k. Vinsonhaler and Wickless [13] show that for each integer $k > 0$ there is a completely decomposable group of finite rank whose projective dimension as a left $\text{End}(A)$-module is k. We will prove the following.

Theorem 3.8 *Let R be a cotorsion-free ring and let $n = $ l.gl.dim.(R).*

1. *If $n < \infty$ then for each integer $1 \le k \le n - 1$ there is a group $A_k \in \Omega(R)$ such that* proj.dim.$(A_k) = k$.

2. *If $n = \infty$ then there is a group $A_\infty \in \Omega(R)$ such that* proj.dim.$(A_\infty) = \infty$.

Lemma 3.9 *Let R, M, A, and C be as in 2.14.*

1. *A is not a projective left R-module.*

2. proj.dim.$(\mathbf{Q}C) = 1$.

Proof: (1) Because $M \oplus R \subset A$ we may view $\mathrm{Hom}_R(A, R)$ as a subgroup of $\mathrm{End}(A) = R$. We claim $\mathrm{Hom}_R(A, R) = 0$. Let $\phi \in \mathrm{Hom}_R(A, R)$. Because $R \cong \mathrm{End}(A)$ there is an $r \in R$ such that $\phi(x) = rx$ for each $x \in A$. Thus $rA \subset R$. But then $r\mathbf{Q}C = 0$. Because C is a free left R-module $r = 0$, which proves the claim. Since the dual of a nonzero projective module is nonzero, A is not a projective left R-module.

(2) $\mathbf{Q}C$ is the directed union of the chain $\{(q!)^{-1}C | 0 < q \in \mathbf{Z}\}$ of free left R-modules. Fix $0 < q \in \mathbf{Z}$. Because $0 \to C \xrightarrow{q} C \to C/qC \to 0$ is a projective resolution of C/qC and because the cotorsion R-module C/qC is a not a projective R-module, proj.dim.$(C/qC) = 1$. Then proj.dim.$(\mathbf{Q}C) \le 1$ by [8, page 374, 18.18]. Since R is cotorsion-free $\mathbf{Q}C$ is not a projective R-module, so proj.dim.$(\mathbf{Q}C) = 1$. ///

Apply $\mathrm{Hom}_R(\ , \cdot)$ to the exact sequence 2.15 to obtain the long exact sequence

$$\mathrm{Ext}_R^m(\mathbf{Q}C, \cdot) \longrightarrow \mathrm{Ext}_R^m(A, \cdot) \longrightarrow \mathrm{Ext}_R^m((M \oplus R)^{(\aleph)}, \cdot) \longrightarrow \mathrm{Ext}_R^{m+1}(\mathbf{Q}C, \cdot)$$

of abelian groups. By 3.9.2 $\mathrm{Ext}_R^m(\mathbf{Q}C \cdot) = 0$ for all $m \ge 2$, so that

$$\mathrm{Ext}_R^m(A, \cdot) \cong \mathrm{Ext}_R^m((M \oplus R)^{(\aleph)}, \cdot) \cong \mathrm{Ext}_R^m(M^{(\aleph)}, \cdot) \tag{9}$$

for all $m \ge 2$.

Lemma 3.10 *Let R, M, and A be as in 2.14.*

1. *If* proj.dim.$(M) \ge 2$ *then* proj.dim.$(A) =$ proj.dim.(M).

2. *If* proj.dim.$(M) \le 1$ *then* proj.dim.$(A) = 1$.

Proof: (1) Suppose proj.dim.$(M) = k \ge 2$. Then by setting $m = k \ge 2$ in equation (9) we have $\mathrm{Ext}_R^k(A, \cdot) \ne 0$. By setting $m = k + 1$ in equation (9) we have $\mathrm{Ext}_R^{k+1}(A, \cdot) = 0$. Hence proj.dim.$(A) = k$.

(2) Suppose proj.dim.$(M) \le 1$. By setting $m = 2$ in equation (9) we have $\mathrm{Ext}_R^2(A, \cdot) = 0$, so that proj.dim.$(A) \le 1$. By 3.9.1 proj.dim.$(A) \ne 0$, so proj.dim.$(A) = 1$. ///

As a consequence of 3.10, in order to find a group $A_k \in \Omega(R)$ such that proj.dim.$(A_k) = k$ it suffices to find a cotorsion-free left R-module M_k of projective dimension k. Of course R is a cotorsion-free left R-module of projective dimension 0, so by 3.10.2 there is a cotorsion-free left R-module of projective dimension 1

Lemma 3.11 *Let R be a cotorsion-free ring.*

1. *If there exists a left R-module M of projective dimension $1 < n < \infty$ then for each integer $1 \leq k \leq n - 1$ there exists a cotorsion-free left R-module M_k such that proj.dim.$(M_k) = k$.*

2. *If l.gl.dim.$(R) = \infty$ then there is a cotorsion-free left R-module M_∞ such that proj.dim.$(M_\infty) = \infty$.*

Proof: (1) Let M be a left R-module of projective dimension $1 < n < \infty$. Write $M = P_n/M_{n-1}$ where P_n is a free left R-module, and $M_{n-1} \subset P_n$. Then M_{n-1} is a cotorsion-free left R-module, and the usual dimension shifting argument shows that proj.dim.$(M_{n-1}) = n - 1$. A simple induction on n shows that for each $1 \leq k \leq n - 1$ there is a cotorsion-free left R-submodule M_k such that proj.dim.$(M_k) = k$.

(2) Suppose l.gl.dim.$(R) = \infty$. For each integer $k > 0$ there is a left R-module L_k such that $k \leq$ proj.dim.(L_k). Then $L = \oplus_{k>0} L_k$ has infinite projective dimension. Write $L = P/M_\infty$ where P is a free left R-module and $M_\infty \subset P$. As above proj.dim.$(M_\infty) =$ proj.dim.$(L) = \infty$. ///

Proof of Theorem 3.8: Let $n = $ l.gl.dim.(R).

1. Assume $n < \infty$. There is a left R-module M of projective dimension n. By 3.11.1 and the remarks preceeding it, for each integer $1 \leq k \leq n - 1$ there is a cotorsion-free left R-module M_k such that proj.dim.$(M_k) = k$. Now apply 2.14, 3.10.1, and .2 to produce a group $A_k \in \Omega(R)$ such that proj.dim.$(A_k) = k$.

2. Assume $n = \infty$. By 3.11.2 there is a cotorsion-free left R-module M_∞ of infinite projective dimension. Apply 2.14 and 3.10.1 to M_∞ to produce a group $A_\infty \in \Omega(R)$ such that proj.dim.$(A_\infty) = \infty$. ///

We do not know if it is possible in general to construct a group A_n whose projective dimension over R is $n = $ l.gl.dim.(R).

COTORSION-FREE FULL TOPOLOGIES

Recall $\Gamma(T, X)$ from the Introduction.

It is natural to ask which linear Hausdorff topologies on a cotorsion-free ring R are of the form $\Gamma(M)$ for some cotorsion-free module M. It is also natural to ask for conditions on M and R under which $R \cong \hat{\mathcal{O}}(M)$.

3.12 Let R be a cotorsion-free ring, and let Γ be a set of left ideals of R. Call Γ a *(cotorsion-free) topological base* for R if Γ is a base of neighborhoods of 0 for some linear Hausdorff topology on R, (and R/I is a cotorsion-free group for each $I \in \Gamma$.)

Call Γ *full* if Γ contains $(I : r) = \{s \in R \mid sr \in I\}$ for each $r \in R$ and $I \in \Gamma$. It is readily shown that $(I : r) = \text{ann}(r + I)$ where $r + I \in R/I$.

It is clear that a topological base Γ for R generates the discrete topology iff $0 \in \Gamma$. Also, $\Gamma(M)$ is a cotorsion-free full topological base for R if M is a cotorsion-free group.

Lemma 3.13 *The following are equivalent for a set of left ideals Γ of R.*

1. Γ *is a (cotorsion-free) full topological base for R.*

2. *There is a (cotorsion-free) left R-module M such that $\Gamma(M) = \Gamma$.*

Proof: (2) $\Rightarrow$ (1) follows from the above comments.

(1) $\Rightarrow$ (2) Assume Γ is a (cotorsion-free) full topological base and let

3.14 $$M = \oplus_{I \in \Gamma} R/I.$$

Observe that M is cotorsion-free if Γ is a cotorsion-free topological base for R, 3.12. We claim that $\Gamma(M) = \Gamma$. Clearly $I = \text{ann}(1 + I)$ for each $I \in \Gamma$, so that $\Gamma \subset \Gamma(M)$. Next, let $F \subset M$ be a finite set. For each $x \in F$ we can write

$$x = \sum_{i=1}^{n_x} (r_{i,x} + I_{i,x})$$

for some $n_x \in \mathbf{Z}$, $r_{i,x} \in R$, and $I_{i,x} \in \Gamma$. Because the terms of this sum are independent elements of M and because Γ is closed under finite intersections, 3.12,

$$\text{ann}(F) = \bigcap_{x \in F} \text{ann}(x) = \bigcap_{x \in F} \bigcap_{i=1}^{n_x} \text{ann}(r_{i,x} + I_{i,x}) = \bigcap_{x \in F} \bigcap_{i=1}^{n_x} (I_{i,x} : r_{i,x}) \in \Gamma.$$

Thus $\Gamma = \Gamma(M)$. ///

Let Γ be a topological base for R. In an abuse of terminology, we will say that R is *complete in* Γ if R is complete in the linear Hausdorff topology generated by Γ.

Corollary 3.15 *Let Γ be a cotorsion-free full topological base for R and let M be the module in 3.14. Then R is complete in Γ iff $R = \hat{\mathcal{O}}(M)$.*

Proof: Assume R is complete in Γ, let M be the module constructed in 3.14, and let $q \in \widehat{SR}$ be such that $qM \subset M$. For each $I \in \Gamma$ there is an $r_I \in R$ such that $q + I = r_I + I \in M$. Then $\{r_I\}_{I \in \Gamma}$ forms a Cauchy net in R, and because R is complete the net converges to some $\hat{r} \in R$. Thus $q - \hat{r} = (q - r_I) - (\hat{r} - r_I) \in I$ for each $I \in \Gamma$. Because Γ generates a Hausdorff topology, 3.12, $q = \hat{r}$, and therefore $R = \hat{\mathcal{O}}(M)$.

Conversely, let $R = \hat{\mathcal{O}}(M)$. By 3.13.2 $\Gamma = \Gamma(M)$. By 1.1 there is a cotorsion-free group A and a topological isomorphism $(R, \Gamma) = (\hat{\mathcal{O}}(M), \Gamma) \cong \text{End}(A)$ where $\text{End}(A)$ is endowed with the finite topology. It is well known, (see e.g. [11]), that $\text{End}(A)$ is a complete Hausdorff space in the finite topology. Then (R, Γ) is a complete space, which completes the proof. ///

SELF-SMALL MODULES

We use cotorsion-free full topological bases to determine when each $A \in \Omega(R)$ is a self-small group.

Theorem 3.16 *The following are equivalent for a cotorsion-free ring R.*

1. *If R is complete in a cotorsion-free full topological base Γ then $0 \in \Gamma$.*

2. *The finite topology on $\text{End}(A)$ is discrete for each $A \in \Omega(R)$.*

3. *R imbeds in $A^{(\aleph_0)}$ for each $A \in \Omega(R)$.*

Proof: The equivalence of parts 3 and 2 is straight forward.

$(1) \Rightarrow (2)$ Let $A \in \Omega(R)$ and assume part 1. Then $\Gamma(R, A) = \Gamma$ is a cotorsion-free full topological base for R and there is a topological isomorphism $(R, \Gamma) \cong \text{End}(A)$. Because $\text{End}(A)$ is complete in the finite topology, R is complete in Γ. Part 1 implies that $0 \in \Gamma$, so the finite topology on $\text{End}(A)$ is discrete.

$(2) \Rightarrow (1)$ Suppose R is complete in a cotorsion-free full topological base Γ, and let M be the module given in 3.14. Then M is a cotorsion-free left R-module, $\Gamma = \Gamma(M)$ is the M-topology on R, 3.13.2, and $R = \hat{\mathcal{O}}(M)$, 3.15. By 1.1 there is a group A and a topological isomorphism $(R, \Gamma) \cong \text{End}(A)$. By part 2 the finite topology on $\text{End}(A)$ is discrete, and because the isomorphism is topological Γ is discrete. Hence $0 \in \Gamma$, as required by part 1. ///

The set Γ of left ideals of R satisfies the *null dcc* (null descending chain condition) if each chain $I_1 \supset I_2 \supset \cdots$ in Γ such that $\bigcap_{n < \omega} I_n = 0$ is eventually constant. If $\Gamma \subset \text{End}(A)$ is the set of annihilators of finite subsets of A then [4, Proposition 1.1] states that A is a self-small group iff Γ satisfies the null dcc.

Theorem 3.17 *The following are equivalent for a cotorsion-free ring R.*

1. *If R is complete in a cotorsion-free full topological base Γ then Γ satisfies the null dcc.*

2. *Each $A \in \Omega(R)$ is a self-small group.*

Proof: $(1) \Rightarrow (2)$ Let $A \in \Omega(R)$. As above R is complete in the cotorsion-free full topological base $\Gamma(R, A) = \Gamma$. Part 1 states that Γ satisfies the null dcc, so [4, Proposition 1.1] shows that A is a self-small group.

$(2) \Rightarrow (1)$ Let R be complete in the cotorison-free full topological base Γ, and let M be the module in 3.14. By 3.13.2 $\Gamma = \Gamma(M)$, and by 1.1 there is a group A and a topological isomorphism $(R, \Gamma) \cong \text{End}(A)$. By part 2 A is a self-small group, and because $\Gamma = \Gamma(M)$, each $I \in \Gamma$ is of the form $\text{ann}(F)$ for some finite set $F \subset M \subset A$. Then [4, Proposition 1.1] states that Γ satisfies the null dcc. This completes the proof. ///

4. ACKNOWLEDGEMENTS

The authors would like to thank the organizers for the opportunity to present this paper at the Conference on Torsion-free Abelian Groups held at the University of the Netherlands Antilles, Willemstad, Curaçao from August 18 to 25, 1991. The second author would like to thank Professor Laszlo Fuchs for suggesting the problem of determining which positive integers are the projective dimension of an abelian group over its endomorphism ring. The application in this paper to that problem is influenced by several conversations with Professor Fuchs.

References

[1] U. Albrecht, *Locally A-projective abelian groups and generalizations*, Pac. J. Math. **141**, No. 2, (1990), 209-228

[2] F.W. Anderson; K.R. Fuller, "Rings and Categories of Modules", Graduate texts in Mathematics **13**, Springer-Verlag, New York-Berlin, (1974)

[3] H.M.K. Angad-Gaur, *The homological dimension of a torsion-free group of finite rank as a module over its ring of endomorphisms*, Rend. Sem. Mat. Univ. Padova **57**, (1977), 299-309.

[4] D.M. Arnold; C.E. Murley, *Abelian groups A such that* Hom(A, ·) *preserves direct sums of copies of A*, Pac. J. Math. **56**, (1), (1975), 7-20

[5] A.L.S. Corner, *Every countable torsion-free ring is an endomorphism ring*, Proc. London Math. Soc. **13**, (1963), 23-33

[6] A.L.S. Corner; R. Gobel, *Prescribing endomorphism algebras*, Proc. London Math. Soc. **50**, (3), (1985), 447-494

[7] M. Dugas; A. Mader; C. Vinsonhaler, *Large E-rings Exist*, J. Algebra **108**, 1, (1987), 88-101.

[8] C. Faith, "Algebra: Rings, Modules, and Categories I", Graduate Texts in Math **190**, Springer- Verlag, New York- Berlin, (1973)

[9] T.G. Faticoni, *Examples of torsion-free groups torsion as modules over their endomorphism rings*, to appear in *Comm. Algebra*.

[10] T.G. Faticoni; P.H. Goeters, *Examples of torsion-free groups flat as modules over their endomorphism rings*, Comm. Algebra **19**, (1), (1991), 1-27.

[11] L. Fuchs, "Infinite Abelian Groups I, II", Academic Press, New York-London, (1969, 1970).

[12] I. Kaplansky, "Fields and Rings", The University of Chicago Press, (1972).

[13] C. Vinsonhaler, W. Wickless, *Homological dimension of a completely decomposable abelian group*, preprint.

Near Isomorphism Invariants for a Class of Almost Completely Decomposable Groups

Manfred Dugas* and Ed Oxford

Department of Mathematics, Baylor University
Waco, Texas 76798

INTRODUCTION

All groups in this paper are torsion-free abelian groups of finite rank and our undefined notations are standard as in [FI/II]. In particular, if X is a subset of a group G, then X_* denotes the pure subgroup of G generated by X. If $g \in G$, then $\|g\|$ is the type of g in G and $|g|_p$ is the p-height of g in G. Note that if g is an integer, then $|g|_p$ is the exponent of the highest power of p that divides g. A group G is almost completely decomposable (a.c.d.) if G is a finite extension of a subgroup C where C is completely decomposable (c.d.), i.e. C is a (finite) direct sum of subgroups of the group of rationals, $\mathbb{Q}$. Special classes of these groups have been studied by several authors, e.g. [BU1], [KM], [BU2], [LA1]. We will study some a.c.d. groups up to near isomorphism [LA2]. Two groups G and H are nearly isomorphic iff $\bigoplus_n G \cong \bigoplus_n H$ for some natural number n iff for each m there is a subgroup H' of G, $H' \cong H$ and $|G/H'|$ is relatively prime to m, cf. [LA1]. We write $G \approx_n H$ if G is nearly isomorphic to H. We restrict ourselves to the following class of groups:

Let $T = (\tau_1, ..., \tau_n)$ be an ordered n-tuple of pairwise incomparable types. A group G is in the class $\mathscr{C}(T)$ iff rk(G) =

*Research partially supported by NSF grant DMS-8900350

$n = |T|$, and $G(\tau_i) = \{g \in G: \|g\| \geq \tau_i\}$ has rank 1, and $G/\sum_{i=1}^{n} G(\tau_i)$ is finite. Since $rk(G) = n$, we conclude $\sum_{i=1}^{n} G(\tau_i) = \bigoplus_{i=1}^{n} G(\tau_i)$. Note that the groups investigated in [BU] are special cases of groups in $\mathscr{C}(T)$. First we prove

<u>Theorem</u> 0.1. Let G, $H \in \mathscr{C}(T)$. For each prime p let G_p be the subgroup of G such that $G_p/\bigoplus_{i=1}^{n} G(\tau_i)$ is the p-primary part of $G/\bigoplus_{i=1}^{n} G(\tau_i)$. Similarly define H_p. Then H is nearly isomorphic to G if and only if G_p is nearly isomorphic to H_p for all primes p.

This allows us to restrict our attention to the groups in $\mathscr{C}(T,p)$ which is the class of all group G in $\mathscr{C}(T)$ such that $G/\bigoplus_{i=1}^{n} G(\tau_i)$ is a finite p-group. We will construct an example to show that Theorem 0.1 is false for "near isomorphism" replaced by "isomorphism". This may indicate that "near isomorphism" is a very useful notion for a.c.d. groups.

A major tool of our investigation are some invariants which may prove useful in other settings as well. For $G \in \mathscr{C}(T,p)$ and $E \subseteq T$, $\tau \in E$ we define

$$\bar{\mu}_{\tau,E}(G) = \frac{\left(G/\left(\sum_{\sigma \in T-E} G(\sigma)\right)_*\right)(\tau)}{\left(G(\tau) + \left(\sum_{\sigma \in T-E} G(\sigma)\right)_*\right) / \left(\sum_{\sigma \in T-E} G(\sigma)\right)_*}$$

A trivial but important observation is that $\bar{\mu}_{\tau,E}(G) \cong \bar{\mu}_{\tau,E}(H)$ whenever $G \cong H$. For G's in $\mathscr{C}(T,p)$ it is easy to see that $\bar{\mu}_{\tau,E}(G)$ is a cyclic p-group. Moreover, if $G_i \in \mathscr{C}(T,p)$, $1 \leq i \leq k$, then $\bar{\mu}_{\tau,E}(\bigoplus_{i=1}^{k} G_i) \cong \bigoplus_{i=1}^{k} (\bar{\mu}_{\tau,E}(G_i))$. This shows that $\bar{\mu}_{\tau,E}(G) \cong \bar{\mu}_{\tau,E}(G')$ if G, $G' \in \mathscr{C}(T,p)$ and G and G' are nearly isomorphic. For all $G \in \mathscr{C}(T,p)$, let $\mu_{\tau,E}(G)$ denote the exponent of the cyclic p-group $\bar{\mu}_{\tau,E}(G)$ for all $\tau \in E \subseteq T$. The following is crucial for our investigation:

<u>Theorem</u> 0.2. For groups G in $\mathscr{C}(T,p)$, $\mu_{\tau,E}(G)$, $(\tau \in E \subseteq T)$, are near isomorphism invariants.

Let us explain what the groups $\bar{\mu}_{\tau, E}(G)$, $G \in \mathscr{C}(T,p)$ "really" are. Let $p^e = |G/\bigoplus_{i=1}^{n} G(\tau_i)|$. Then $\bigoplus_{i=1}^{n} G(\tau_i) \subseteq G \subseteq \bigoplus_{i=1}^{n} p^{-e} G(\tau_i) \subseteq \bigoplus_{i=1}^{n} \mathbb{Q}$. Now let $\pi_E: \bigoplus_{i=1}^{n} p^{-e} G(\tau_i) \rightarrow \bigoplus_{\tau_i \in E} p^{-e} G(\tau_i)$ be the natural projection. Then

$$\bar{\mu}_{\tau, E}(G) = [\pi_E(G) \cap p^{-e} G(\tau)]/G(\tau) \subseteq p^{-e} G(\tau)/G(\tau) \cong \mathbb{Z}/p^e\mathbb{Z}$$ if $pG(\tau) \neq G(\tau)$ and $= 0$ otherwise. Now we are ready to define the subclass of $\mathscr{C}(T,p)$ which will concern us most: $\mathscr{C}(T,p,e,r)$ is the class of all groups $G \in \mathscr{C}(T,p)$ such that $G/\bigoplus_{i=1}^{n} G(\tau_i)$ is a direct sum of r copies of $\mathbb{Z}(p^e)$, the cyclic group of order p^e. We call the groups $G \in \mathscr{C}(T,p,e,r)$ *uniform* and observe that if $H \in \mathscr{C}(T,p)$ is nearly isomorphic to G, then $H \in \mathscr{C}(T,p,e,r)$. We are now ready to introduce our last set of invariants to pin down the groups in $\mathscr{C}(T,p,e,r)$ up to near isomorphism. Let $G \in \mathscr{C}(T,p,e,r)$ and recall that T is an *ordered* n-tuple of types. Since $G/\bigoplus_{i=1}^{n} G(\tau_i) \cong \bigoplus_{r} \mathbb{Z}(p^e)$ there are $a_i \in \bigoplus_{i=1}^{n} G(\tau_i)$, $1 \leq i \leq r$ and $G = \bigoplus_{i=1}^{n} G(\tau_i) + \sum_{i=1}^{r} p^{-e} a_i \mathbb{Z}$. Let $a_i = (\alpha_{ij})_{j=1}^{n}$, $\alpha_{ij} \in G(\tau_j)$, $1 \leq j \leq n$. W.l.o.g. we may assume $\alpha_{ij} \in \mathbb{Z}$. Let $M = (\alpha_{ij})$ be the r×n-matrix with entries α_{ij}. We identify M with $(\alpha_{ij} + p^e \mathbb{Z})$, i.e. we view M as a matrix over $\mathbb{Z}/p^e\mathbb{Z}$. Like in elementary linear algebra, we may compute the reduced row echelon form F of M and we'll see that we obtain r columns with only O's except a single 1. Let g be the set of τ_i's such that the i'th column is one of those columns. With the help of the $\mu_{\tau, E}$'s we show that g is a near isomorphism invariant. Let $\mathscr{P} = T-g$. If i is the column label of one of the r canonical columns in F with 1 in the j'th row, we label that row by τ_i. Abusing notation, we let $F = [I_r | (\alpha_{ij})]$, I_r the r×r-identity matrix, $\tau_i \in g$, $\tau_j \in \mathscr{P}$ and identify i and τ_i.

We find it useful to borrow some language from incidence geometry; see [De] or any other book on (finite) geometry. In general, a geometry is a triplet $\mathscr{G} = (\mathscr{P}, g, I)$ where $\mathscr{P}, g$ are disjoint sets and $I \subseteq \mathscr{P} \times g$ is a relation. The elements of $\mathscr{P}$ are called points and the elements of g are called lines and I is the incidence relation. Instead of $(P,h) \in I$ we write P I h and P is called incident with (or lying on) the line h.

We define relations I_ε, $\varepsilon = 0,1,\ldots,e-1$ by $P\ I_\varepsilon\ h$ iff $|\alpha_{hP}|_p = \varepsilon$, i.e. the p-height of α_{hP} is ε. Moreover, we define $I = \bigcup_{\varepsilon=0}^{e-1} I_\varepsilon$. This gives rise to a geometric structure: We call the elements in $\mathscr{P}$ "points", the elements in g "lines" and P is ε-incident with h iff $P\ I_\varepsilon\ h$. Thus, for each $G \in \mathscr{C}(T,p,e,r)$ we obtain a geometry $\mathscr{G} = \mathscr{G}(G) = (\mathscr{P},g,I)$ with $T = \mathscr{P}\ \dot{\cup}\ g$ and prove

Theorem 0.3: Let G, $\overline{G} \in \mathscr{C}(T,p,e,r)$ be nearly isomorphic. Then $g = g(G) = g(\overline{G})$ and $\mathscr{P} = \mathscr{P}(G) = \mathscr{P}(\overline{G})$ and $\mathscr{G}(G) = \mathscr{G}(\overline{G})$. Moreover $P\ I_\varepsilon\ h$ $\Leftrightarrow P\ \overline{I}_\varepsilon\ h$ for all $P \in \mathscr{P}$ and $h \in g$.

Thus we obtain a geometry as a near isomorphism invariant. Actually, this geometry is coded in the $\mu_{\tau,E}$'s and is just a particular way to read the $\mu_{\tau,E}$'s. On the other hand we need the geometric notation to define our final invariants. Moreover, we will show that G is indecomposable iff $\mathscr{G}(G)$ is connected, i.e. for any two points P,Q in $\mathscr{G}(G) = \mathscr{G}$ there is a path from P to Q. A circuit Δ (or loop) in $\mathscr{G}$ is a finite sequence of points and lines P_i, h_i, such that P_i, $P_{i+1}\ I\ h_i$, $0 \le i \le n$ and all P_i's and h_i's are distinct, except $P_{n+1} = P_0$. Then define $\delta(\Delta)$ to be the largest ε such that $P_i\ I_\varepsilon\ h_{i-1}$ or $P_i\ I_\varepsilon\ h_i$. We define $\alpha_{hP}^\perp = \alpha_{hP}p^{-|\alpha_{h,P}|_p}$ for $P\ I\ h$ and $\alpha_{h,P}^{\perp*}$ is the inverse of $\alpha_{hP}^\perp$ modulo p^e. Then we define

$$\rho(\Delta) = (\alpha_{h_0P_0}^\perp\ \alpha_{h_0P_1}^{\perp*})\ (\alpha_{h_1P_1}^\perp\ \alpha_{h_1P_2}^{\perp*})\ \ldots\ (\alpha_{h_nP_n}^\perp\ \alpha_{h_nP_0}^{\perp*})\ \bmod\ p^{e-\delta(\Delta)}.$$

It turns out that the $\rho(\Delta)$'s, Δ a loop in $\mathscr{G}$, are near isomorphism invariants of G! We obtain the following result

Theorem 0.4. Let T be an ordered set of pairwise incomparable primes, p a prime and r,e natural numbers. If G,H are in $\mathscr{C}(T,p,e,r)$, the G is nearly isomorphic to H if and only if

(1) $\mu_{\tau,E}(G) = \mu_{\tau,E}(H)$ for all $\tau\in E \subseteq T$. (Note that (1) implies that the groups G and H induce the same geometry $\mathscr{G}$.)

(2) All loop invariants $\rho(\Delta)$ defined by the labellings induced by H and G are equal for each loop Δ of $\mathscr{G}$.

An immediate consequence is:

Corollary 0.5 (a) For the groups $G \in \mathscr{C}(T,2,1,r)$, the invariants

$\mu_{\tau,E}(G)$, $\tau \in E \subseteq T$, are a complete set of near isomorphism invariants, i.e. if G, $\overline{G} \in \mathscr{C}(T,2,1,r)$, then G is nearly isomorphic to $\overline{G}$ if and only if $\mu_{\tau,E}(G) = \mu_{\tau,E}(\overline{G})$ for all $\tau \in E \subseteq T$.

(b) If $\mathscr{G} = \mathscr{G}(G)$ has no loops, i.e. $\mathscr{G}$ is a tree, and G, $\overline{G} \in \mathscr{C}(T,p,e,r)$, then G is nearly isomorphic to $\overline{G}$ if and only if $\mu_{\tau,E}(G) = \mu_{\tau,E}(\overline{G})$ for all $\tau \in E \subseteq T$.

We treated T as an *ordered* n-tuple of types since this allowed us to compute *the* row echelon form of the matrix M. Without this—rather artificial—condition, each group $G \in \mathscr{C}(T,p,e,r)$ gives rise to many different geometries $\mathscr{G}(G,g) = \mathscr{G}(G) = (\mathscr{P},g,I)$, $T = \mathscr{P} \cup g$.

(If G is nearly isomorphic to $\overline{G}$ then $g \subseteq T$ is the set of lines of a geometry derived from G iff g is also the set of lines of one of the geometries of $\overline{G}$ and $\mathscr{G}(G,g) = \mathscr{G}(\overline{G},g)$.)

If one such geometry is a tree, it still may happen that another geometry is *not* a tree. What are the loop invariants $\rho(\Delta)$ in that case?

Another open question is: Is there a more algebraic or functorial way to describe the $\rho(\Delta)$?

We conclude this paper by presenting some numerical algorithms and study groups G of rank 3, $r = 2$, that are *not* uniform. We find a complete set of near isomorphism invariants, which—for the time being—discourage us to pursue this general case in higher rank.

I.   NOTATIONS AND A FIRST REDUCTION

Throughout this paper, let $n \in \mathbf{N}$ and $T = \{\tau_1, \tau_2, ..., \tau_n\}$ a set of *pairwise incomparable* types, $R_i \subseteq \mathbb{Q}$ of type τ_i. We write $G \approx_n \overline{G}$ if G, $\overline{G}$ are nearly isomorphic.

Let G be a torsion-free group of rank n such that $G(\tau_i)$ has rank 1, $G(\tau_i) = \{g \in G \mid \|g\| \geq \tau_i\}$, and $F = G/(\sum_{i=1}^{n} G(\tau_i))$ is finite. Note that since rk $G = n$ and rk $G(\tau_i) = 1$ we have

$$C = \sum_{i=1}^{n} G(\tau_i) = \bigoplus_{i=1}^{n} G(\tau_i).$$

Thus G is almost completely decomposable. It is easy to see that n, $\{\tau_1, \ldots, \tau_n\}$, F are near-isomorphism invariants of G. Now let F = $\underset{p}{\oplus}$ F_p be the decomposition of F into the p-primary components of F and $G_p \subseteq G$ with $G_p/C = F_p$. With these notations in mind, we have the following crucial

<u>Theorem 1.1</u>. Let G, $\overline{G}$ be as above. Then $G \approx_n \overline{G}$ if and only if $G_p \approx_n \overline{G}_p$ for each prime p.

Proof: Suppose $G \approx_n \overline{G}$. Then G and $\overline{G}$ are quasi-isomorphic which implies that G, $\overline{G}$ have the same rank and typeset $\{\tau_1, \ldots, \tau_n\}$. Since $G^{(k)} \cong \overline{G}^{(k)}$ for some $k \in \mathbf{N}$, we infer $C^{(k)} \cong \overline{C}^{(k)}$ and $(G/C)^{(k)}$ is isomorphic to $(\overline{G}/\overline{C})^{(k)}$. Since these factor groups are finite we may cancel the exponent and conclude $G/C \cong \overline{G}/\overline{C}$ where $C = \overset{n}{\underset{i=1}{\oplus}} G(\tau_i)$, $\overline{C} = \overset{n}{\underset{i=1}{\oplus}} \overline{G}(\tau_i)$. Since $C^{(k)}$ ($\overline{C}^{(k)}$) are fully invariant, we conclude $G_p^{(k)} \cong \overline{G}_p^{(k)}$ and $G_p \approx_n \overline{G}_p$ for all primes p. Now assume that $G_p \approx_n \overline{G}_p$ for all p and let $m = |G/C| = |\overline{G}/\overline{C}| = \underset{p \in P_m}{\Pi} p^{e_p}$.

Since $G_p \approx_n \overline{G}_p$, there exists an embedding $f_p : G_p \to \overline{G}_p$ such that $|\overline{G}_p/f_p(G_p)|$ is relatively prime to m. Thus

$(\overline{G}_p/f_p(G_p))/(f_p(G_p) + \overline{C}/f_p(G_p)) \cong \overline{G}_p/(f_p(G) + \overline{C}) \cong$

$(\overline{G}_p/\overline{C})/((f_p(G_p)+\overline{C})/\overline{C})$ is a p-group of order relatively prime to p. Therefore $\overline{G}_p = f_p(G) + \overline{C}$ and f_p induces an onto homomorphism from G_p/C into $\overline{G}_p/\overline{C}$. Since any such map is 1-1, we conclude $\overline{C} \cap f_p(G_p) = f_p(C)$ and consequently $\overline{G}_p/f_p(G_p) \cong \overline{C}/f_p(C)$. Since $G(\tau_i) \cong \overline{G}(\tau_i)$ for all $1 \leq i \leq n$, we may assume $G(\tau_i) = \overline{G}(\tau_i)$, i.e. $C = \overline{C}$. Thus $f_p = (f_{p,i})_{i=1,\ldots,n}$ and $f_{p,i} \in \text{End}(G(\tau_i))$. We apply the Chinese Remainder Theorem for these rings and obtain $g_i \in \text{End}(G(\tau_i))$ with $g_i \equiv f_{p,i} \mod p_i^{e_{p_i}} \text{End}(G(\tau_i))$, $1 \leq i \leq n$. We set $g = (g_i)_{i=1,\ldots,n}$. Note that g and f_p induce the same bijective map $\theta : G_p/C \to \overline{G}_p/\overline{C}$. This implies $g(G_p) + \overline{C} = \overline{G}_p$ and $g(G_p) \cap \overline{C} =$

$g(C)$. Moreover, as above, $\overline{G}_p/g(G_p) \cong \overline{C}/g(C) = \overset{n}{\underset{i=1}{\oplus}} \overline{G}(\tau_i)/g_i G(\tau_i)$.
Note that $g_i = f_{p,i} + p^e p_{y_{p,i}}$ for all $p \in P_m$. If $p \in P_m$ and p/g_i,
then $p/f_{p,i}$, a contradiction. Thus $(p,g_i) = 1$ for all $p \in P_m$ which
implies that $\overline{G}_p/g(G_p)$ is generated by elements of order relatively
prime to m. Since $\overline{G}_p + g(G)/g(G)$ is an epimorphic image of
$\overline{G}_p/g(G_p)$, this group has the same property. Since $\overline{G}/g(G) =$
$\underset{p \in P_m}{\sum} \overline{G}_p + g(G)/g(G)$ we conclude that $|\overline{G}/g(G)|$ is relatively prime to
m. Thus $G \approx_n \overline{G}$.

We use a counterexample to show that $G_p \cong \overline{G}_p$ for all $p \in P_m$
doesn't imply $G \cong \overline{G}$: Let
$$G = (\mathbb{Q}^{(7)} \oplus \mathbb{Q}^{(11)}) + (1/13, 1/13)\mathbb{Z} + (1/17,1/17)\mathbb{Z}$$
and
$$\overline{G} = (\mathbb{Q}^{(7)} \oplus \mathbb{Q}^{(11)}) + (1/13, 1/13)\mathbb{Z} + (10/17,1/17)\mathbb{Z}.$$
Note that $G_{17} \cong \overline{G}_{17}$ and $G_{13} = \overline{G}_{13}$. To show that $G_{17} \cong \overline{G}_{17}$, let $U(p)$
be the group of units modulo p. Then $U(13) = \langle 2 \rangle$, $U(17) = \langle 3 \rangle$ and
$11 = 3^7$ mod 17 and $7 = 3^{11}$ mod 17. Thus $U(17) = \langle 11 \rangle = \langle 7 \rangle$.
Moreover $10 = 3^3$ mod 17. Let ε, $\delta \in \mathbb{Z}$ be solutions of $11\varepsilon \equiv 7\delta+3$
modulo 16. Then $7^\varepsilon \equiv 11^\delta 10$ mod 17 and $(1,1)(7^\varepsilon,11^\delta) = (7^\varepsilon,11^\delta) =$
$(11^\delta 10,11^\delta) = (10,1)11^\delta$ mod 17. Thus $\alpha = (7^\varepsilon,11^\delta)$ may be viewed as
an automorphism of $\mathbb{Q}^{(7)} \oplus \mathbb{Q}^{(11)}$ with $\alpha(G_{17}) = \overline{G}_{17}$. This shows
$G_{17} \cong \overline{G}_{17}$. Now suppose $G \cong \overline{G}$. Then there are ε, $\delta \in \mathbb{Z}$ and
$\alpha = (\pm 7^\varepsilon, \pm 11^\delta): G \to \overline{G}$ is an antomorphism of C. Then
$$\pm 7^\varepsilon \equiv \pm 11^\delta \text{ mod 13 and}$$
$$\pm 7^\varepsilon \equiv \pm 11^\delta 10 \text{ mod 17.}$$
Since $7 \equiv 2^{11}$ mod 13 and $11 \equiv 2^7$ mod 13 and $7 \equiv 3^{11}$ mod 17 and $11 \equiv$
3^7 mod 17 we infer
$$\pm 2^{11\varepsilon} \equiv \pm 2^{7\delta} \text{ mod 13 and}$$
$$\pm 3^{11\varepsilon} \equiv \pm 3^{7\delta} 3^3 \text{ mod 17. Moreover } -1 \equiv 2^6 \text{ mod 13 and}$$
$-1 \equiv 3^8$ mod 17. Thus we have 2 cases.

<u>Case 1</u>: $11\varepsilon \equiv 7\delta$ mod 12 and
          $11\varepsilon \equiv 7\delta + 3$ mod 16 or
<u>Case 1</u>: $11\varepsilon \equiv 7\delta +6$ mod 12 and
          $11\varepsilon \equiv 7\delta + 11$ mod 16 in case a negative sign shows up.

Since $11 \cdot 11 = 121 \equiv 1 \bmod 12$ and $11 \cdot 3 = 33 \equiv 1 \bmod 16$ we obtain

$\varepsilon \equiv 5\delta \bmod 12$ and $\varepsilon \equiv 5\delta + 9 \bmod 16$

or $\varepsilon \equiv 5\delta + 6 \bmod 12$ and $\varepsilon \equiv 5\delta + 1 + 9 \bmod 16$.

Since $\gcd(12,16) = 4$ we infer $9 \equiv 0 \bmod 4$ or $6 \equiv 1 \bmod 4$. Both are absurd, which is the desired contradiction that shows G is *not isomorphic to* $\overline{G}$.

Since we want to study almost completely decomposable groups up to near-isomorphism, we may restrict ourselves to the case of groups G with G/C a p-group. We will restrict our scope even more:
Let $T = \{\tau_1, \ldots, \tau_n\}$ be a set of n incomparable types and

$\mathscr{C}(T,p,e,r)$ the class of groups G such that

(1) $G(\tau)$ has rank 1 for all $\tau \in T$

(2) $\text{rk } G = n$

(3) $G / \bigoplus_{i=1}^{n} G(\tau_i) \cong (\mathbb{Z}(p^e))^{(r)}$, i.e. a direct sum of r copies of $\mathbb{Z}(p^e)$.

We call such groups G *uniform*. Note that if $\tau_i(p) = \infty$ for some $1 \leq i \leq n$, then $G(\tau_i)$ is a summand of G. Thus w.l.o.g. we will assume $\tau_i(p) = 0$ for all $1 \leq i \leq n$.

1.1   Some geometric notions

Let $\mathscr{G} = (\mathscr{P}, g, I)$ be a geometric structure, i.e. $\mathscr{P}$ and g are non empty, disjoint sets and $I \subseteq \mathscr{P} \times g$. The elements of $\mathscr{P}$ are called points, the elements of g are called lines, and most of the time we write $P \, I \, h$ instead of $(P,h) \in I$. The elements of I are called flags. We refer to [De] for basic definitions and terminology relating to (incidence) geometry.

Definition 1.1. Let e be a natural number. The geometric structure $\mathscr{G} = (\mathscr{P}, g, I)$ is called an e-geometry if
(i)   I is the disjoint union of distinct subsets $I_0, I_1, \ldots, I_{e-1}$, which are part of the structure of $\mathscr{G} = (\mathscr{P}, g; I_0, I_1, \ldots, I_{e-1})$.
(ii) For all $h \in g$ there is $P \in \mathscr{P}$ such that $P \, I_0 \, h$.

A _path_ Π from a point P to a point Q is a sequence of flags $P_i I \, h_i$ and $P_{i+1} \, I \, h_i$, $0 \leq i \leq n$, where $P = P_0$ and $Q = P_{n+1}$. We write $\Pi = (P_i, h_i)_{0 \leq i \leq n}$. The *defect* $\delta(\Pi)$ of the path Π is $\delta(\Pi) = \sup\{\varepsilon_i, \overline{\varepsilon}_i : P_i \, I_{\varepsilon_i} \, h_i, \, P_{i+1} \, I_{\overline{\varepsilon}_i} \, h_i, \, 0 \leq i \leq n\}$. A path Π is called *reduced* if

$h_i \neq h_j$ for all $i \neq j$ and $P_i \neq P_j$ for all $i \neq j$, $(i,j) \neq (0,n+1)$. The path Π is a *loop* (or circuit), if $P_o = P_{n+1}$. We call two points P,Q *connected* and write $P \sim_I Q$ if there is a path Π with $P = P_o$ and $Q = P_{n+1}$. Note that $\sim_I$ is an equivalence relation and the equivalence classes of $\sim_I$ are called the connected components of $\mathcal{G}$. We call $\mathcal{G}$ connected if there is only one connected component.

The e-geometry $\mathcal{G}$ is p-labelled by the map α (p a prime) if $\alpha: I \rightarrow \mathbf{Z}$ has the property that $|\alpha(P,h)|_p = m$, $0 \leq m < e-1$, whenever $P \ I_m \ h$, i.e. $(P,h) \in I_m$. For the sake of convenience we write $\alpha_{h,P} = \alpha(P,h)$. For $x \in \mathbf{Z}$ we write $x = p^m x^\perp$ with $(x^\perp, p) = 1$ and if $|x|_p = 0$ we define x^* such that $xx^* \equiv 1 \bmod p^e$.

If $\Pi = (P_i, \ h_i)_{0 \leq i \leq n}$ is a path in $\mathcal{G}$, $\mathcal{G}$ labelled by α, we define $\rho_\alpha(\Pi) = \prod_{i=0}^{n} \alpha_{h_i P_i}^\perp (\alpha_{h_i P_{i+1}}^\perp)^* \in \mathbf{Z}$. These numbers will play a prominent role in our later investigation.

Suppose $\overline{\alpha}$ is another labelling of g. We call α *compatible* to $\overline{\alpha}$ if there are $f_P \in \mathbf{Z}-p\mathbf{Z}$, $P \in \mathcal{P}$ and $f_h \in \mathbf{Z}-p\mathbf{Z}$, $h \in g$ such that $\alpha_{hP} \ f_P \cong \overline{\alpha}_{hP} \ f_h \bmod p^e$ for all flags $P \ I \ h$.

Note that $\alpha_{hP} \ f_P \cong \overline{\alpha}_{hP} \ f_h \bmod p^e$ implies $\alpha_{hP} \ f_h^* \cong \overline{\alpha}_{hP} \ f_h^* \bmod p^e$. It is easy to see that "compatibility" of (p-) labellings α is an equivalence relation on (p-)labellings. The next theorem is of crucial importance in the later sections:

<u>Theorem 1.2</u>. Two labellings α, $\overline{\alpha}$ of $\mathcal{G}$ are compatible if and only if $\rho_\alpha(\Pi) \equiv \rho_{\overline{\alpha}}(\Pi) \bmod p^{e-\delta(\Pi)}$ for all loops Π of $\mathcal{G}$.

Proof: Suppose α is compatible to $\overline{\alpha}$ and let $\Pi = (P_i; h_i)_{0 \leq i \leq n}$ be a loop with $P_o = P_{n+1}$ and $\delta(\Pi) = m$. This implies

$\alpha_{h_i P_i} \ f_{P_i} \equiv \overline{\alpha}_{h_i P_i} \ f_{h_i} \bmod p^e$ and

$|\alpha_{h_i P_i}|_p = |\overline{\alpha}_{h_i P_i}|_p \leq m$.   Therefore

$f_{h_i} \equiv \alpha_{h_i P_i}^\perp \ \overline{\alpha}_{h_i P_i}^{\perp *} \ f_{P_i} \bmod p^{e-m}$ and

$f_{h_i} \equiv \alpha_{h_i P_{i+1}}^\perp \ \overline{\alpha}_{h_i P_{i+1}}^{\perp *} \ f_{P_{i+1}} \bmod p^{e-m}$ which implies

$f_{P_{i+1}} \equiv \alpha_{h_i P_{i+1}}^{\perp *} \overline{\alpha}_{h_i P_{i+1}}^{\perp} \alpha_{h_i P_i}^{\perp} \overline{\alpha}_{h_i P_i}^{*} f_{P_i}$. An obvious induction yields

$$f_{P_{n+1}} \equiv \prod_{i=0}^{n} (\alpha_{h_i P_i}^{\perp} \alpha_{h_i P_{i+1}}^{\perp *}) \; (\prod_{i=0}^{n} (\overline{\alpha}_{h_i P_i}^{\perp} \overline{\alpha}_{h_i P_{i+1}}^{\perp *}))^{*} \; f_{P_0} \bmod p^{e-m}$$

i.e. $f_{P_{n+1}} = \rho_\alpha(\Pi) \; (\rho_{\overline{\alpha}}(\Pi))^{*} f_{P_0} \bmod p^{e-m}$ since $P_0 = P_{n+1}$ and $|f_P|_p = 0$ for all $P \in \mathscr{P}$ we obtain $\rho_\alpha(\Pi) \equiv \rho_{\overline{\alpha}}(\Pi))^{*} \bmod p^{e-m}$,
$m = \delta(\Pi)$, for all loops Π of $\mathscr{G}$.

To show the converse, let $X_1^{(0)}, \ldots, X_{t_o}^{(0)}$ be the connected components of $\mathscr{P} = \mathscr{P}^{(0)}$ w/t the incidence relation I_0. We pick $P_j^{(1)} \in X_j^{(0)}$, $1 \le j \le t_o$ and let $\mathscr{P}^{(1)} = \{\mathscr{P}_j^{(1)}: 1 \le j \le t_o\}$. Note that $\mathscr{P}^{(1)} \subseteq \mathscr{P}^{(0)}$ is totally disconnected w/t I_0. Let $\mathscr{P}^{(1)} = \bigcup_{j=1}^{t_i} X_j^{(1)}$ be the decomposition of $\mathscr{P}^{(1)}$ into connected components w/t I_1. Again, pick $P_j^{(2)} \in X_j^{(1)}$, $1 \le j \le t_1$ and let $\mathscr{P}^{(2)} = \{P_j^{(2)} | 1 \le j \le t_1\}$. Then $\mathscr{P}^{(2)}$ is totally disconnected w/t $I_0 \cup I_1$. We proceed like this by induction and define $\mathscr{P}^{(k)}$, $0 \le k \le e$, where $\mathscr{P}^{(e)}$ is totally disconnected w/t $I = \bigcup_{j=0}^{e-1} I_j$. By induction we define $f_P \in \mathbb{Z}\text{-}p\mathbb{Z}$ for $P \in \mathscr{P}^{(k)}$, $k=1, 1-1, \ldots, 0$. For $P \in \mathscr{P}^{(e)}$, we set $f_P = 1$. The set $\mathscr{P}^{(e-1)}$ has a partition into connected components $X_j^{(e-1)}$ with $\mathscr{P}^{(e)}$ a complete system of representatives. If $Q \in X_j^{(e-1)}$ then there is a uniquely determined element $P_j^{(e)} \in \mathscr{P}^{(e)} \cap X_j^{(e-1)}$. Thus there is a path Π from $P_j^{(e)}$ to Q of defect $e-1$ and each path from $P_j^{(e)}$ to Q has defect $e-1$. Define $f_Q = \rho_\alpha(\Pi) \; (\rho_{\overline{\alpha}}(\Pi))^{*} f_{P_j^{(e)}}$.

Note that if Π' is another path from $P_j^{(e)}$ to Q, we let Π'^{-1} be the corresponding path from Q to $P_j^{(e)}$ and note that $\Delta_\alpha(\Pi^{-1}) = \Delta_\alpha(\Pi)^{*}$. Moreover, $\Pi \cup \Pi^{-1}$ is a loop. This implies that $f_Q \bmod p^{e-(e-1)}$ is well defined. This concludes the definition of f_Q, $Q \in \mathscr{P}^{(e-1)}$. Suppose f_Q, $Q \in \mathscr{P}^{(i+1)}$, is already defined. Then $\mathscr{P}^{(i+1)}$ is totally disconnected w/t $I_0 \cup \ldots \cup I_i$ and each $Q \in \mathscr{P}^{(i)}$ is connected to some unique $P_j^{(i+1)} \in \mathscr{P}^{(i+1)}$. Thus there is a path of defect i from $P_j^{(i+1)}$ to Q and we set $f_Q = \rho_\alpha(\Pi) \; \rho_{\overline{\alpha}}(\Pi)^{*} f_{P_j(i+1)}$ and note that f_Q is well defined mod p^{e-i}. This defines f_P for all $P \in \mathscr{P}$. Let h be a line. By (ii) there is some $P = P(h) \in \mathscr{P}$ with $P \; I_0 \; h$. We define $f_h = \alpha_{h,P} \overline{\alpha}_{h,P}^{*} f_P$. Now suppose $Q \; I_m \; h$, $P \in X_j^{(0)}$. Then $P_j^{(1)}$ is connected to Q via $I_0 \cup \ldots \cup I_m$. Let δ be minimal such that $P_j^{(1)}$ and Q are connected

via $I_0U...UI_\delta$. Then there exists a path Π from some $P_j^{(\delta+1)}$ to P and a path $\tilde{\Pi}$ from $P_j^{(\delta+1)}$ to Q with $\delta(\Pi) = \delta = \delta(\tilde{\Pi})$ and
$f_P \equiv \rho_\alpha(\Pi) \; \rho_{\overline{\alpha}}(\Pi)^* \; f_{P_j}^{(\delta+1)}) \bmod p^{e-\delta}$ and

$$f_Q \equiv \rho_\alpha(\tilde{\Pi}) \; \rho_{\overline{\alpha}}(\tilde{\Pi})^* \; f_{P_j}^{(\delta+1)} \bmod p^{e-\delta}. \quad \text{Now}$$

$$\alpha^\perp_{hQ} \, f_Q \equiv \alpha^\perp_{hQ} \, \rho_\alpha(\tilde{\Pi}) \; \rho_{\overline{\alpha}}(\tilde{\Pi})^* \; f_{P_j}^{(\delta+1)}) \equiv$$

$$\equiv \alpha^\perp_{hQ} \, \rho_\alpha(\tilde{\Pi}) \; \rho_{\overline{\alpha}}(\tilde{\Pi})^* \; \rho_\alpha(\Pi)^* \; \rho_{\overline{\alpha}}(\Pi) \; f_P$$

$$\equiv \overline{\alpha}^\perp_{hQ} \, \overline{\alpha}^{\perp *}_{hQ} \, \alpha^{\perp *}_{hP} \, \overline{\alpha}^\perp_{hp} \, \alpha^\perp_{hQ} \, \rho_{\overline{\alpha}}(\tilde{\Pi}) \; \rho_{\overline{\alpha}}(\tilde{\Pi})^* \; \rho_{\overline{\alpha}}(\Pi) \; \rho_{\overline{\alpha}}(\tilde{\Pi})^* \; f_h$$

$$\equiv K \, \overline{\alpha}^\perp_{hQ} \, f_h \bmod p^{e-\delta} \text{ where}$$

$$K = [\alpha^\perp_{hQ} \, \alpha^{\perp *}_{hp} \, \rho_\alpha(\tilde{\Pi}) \; \rho_\alpha(\Pi)^*] \; [\overline{\alpha}^\perp_{hQ} \, \overline{\alpha}^{\perp *}_{hp} \, \rho_{\overline{\alpha}}(\Pi)^* \; \rho_{\overline{\alpha}}(\tilde{\Pi})]^*.$$

Consider the following loop $\hat{\Pi}$ with defect m: Start with Q, go to P via h; take Π^{-1} to $P_j^{(\delta+1)}$ and then back to Q on $\tilde{\Pi}$. Notice that $K = \rho_\alpha(\hat{\Pi}) \; \rho_{\overline{\alpha}}(\hat{\Pi})^*$. Thus $K \equiv 1 \bmod p^{e-m}$, $m = \delta(\hat{\Pi})$. This implies $\alpha^\perp_{hQ} \, f_Q \equiv \overline{\alpha}^\perp_{hQ} \, f_1 \bmod p^{e-m}$ and $\alpha_{hQ} f_Q \equiv \overline{\alpha}_{hQ} f_h \bmod p^e$ follows. This shows that α is compatible to $\overline{\alpha}$.

<u>Lemma 1.3</u>. Let α, $\overline{\alpha}$ be p-labellings of the e-geometry $\mathcal{G}$ such that $\rho_\alpha(\Pi) \equiv \rho_{\overline{\alpha}}(\Pi) \bmod p^{e-\delta(\Pi)}$ for all reduced loops Π. Then $\rho_\alpha(\Pi) \equiv \rho_{\overline{\alpha}}(\Pi) \bmod p^{e-\delta(\Pi)}$ for all loops Π.

<u>Proof</u>: Let $\Pi = (P_i; h_i)_{0 \le i \le n}$ and assume $P_i = P_j$ for $(i,j) \ne (0, n+1)$. W.l.o.g. we may assume $i = 0$ and $0 < j \le n$. Now consider the subloops $\Pi^{(0)}$, $\Pi^{(1)}$ of Π where $\Pi^{(0)} = (P_i; h_i)_{0 \le i \le j-1}$ and $\Pi^{(1)} = (P_{j+i}; h_{j+i})_{0 \le i \le n-j}$. Then $\rho_\alpha(\Pi) = \rho_\alpha(\Pi^{(0)}) \; \rho_\alpha(\Pi^{(1)})$ and $\delta(\Pi) = \max\{\delta(\Pi^{(0)}), \delta(\Pi^{(1)})\}$. By induction hypothesis, $\rho_\alpha(\Pi^{(\varepsilon)}) \equiv \rho_{\overline{\alpha}}(\Pi^{(\varepsilon)}) \bmod p^{e-\delta(\Pi(\varepsilon))}$ for $\varepsilon = 0;1$ and $\rho_\alpha(\Pi) \equiv \rho_{\overline{\alpha}}(\Pi) \bmod p^{e-\delta(\Pi)}$ follows.
Now suppose $h_i = h_j$ for $i \ne j$. We may assume $i = 0$ and $0 \le j \le n$. Again, we consider two subloops of Π. Let $\Pi^{(0)}$ be the loop starting with P_1, follow Π to P_j I $h_j = h_0$. Then goes back to P_1 on $h_0 = h_j$. Let $\Pi^{(1)}$ be the loop starting at P_0, goes on h_0 to P_{j+1}, then follows Π back to P_0.
Then $\rho_\alpha(\Pi^{(0)}) = (\alpha^\perp_{h_1P_1} \, \alpha^{\perp *}_{h_1P_2}) \cdots (\alpha^\perp_{h_{j-1}P_{j-1}} \, \alpha^{\perp *}_{h_{j-1}P_j}) \cdot (\alpha^\perp_{h_0P_j} \, \alpha^{\perp *}_{h_0P_1})$

$$= \prod_{i=1}^{j-1} \alpha^{\perp}_{h_i P_i} \alpha^{\perp *}_{h_i P_{i+1}} \cdot (\alpha^{\perp}_{h_o P_j} \alpha^{\perp *}_{h_o P_1}).$$

Furthermore

$$\rho_\alpha(\Pi^{(1)}) = \alpha^{\perp}_{h_o P_o} \alpha^{\perp *}_{h_o P_{j+1}} = \prod_{v=1}^{n-j} (\alpha^{\perp}_{h_{j+v} P_{j+v}} \alpha^{\perp *}_{h_{j+v} P_{j+v+1}}) \cdot$$

and

$$\rho_\alpha(\Pi^{(0)}) \; \rho_\alpha(\Pi^{(1)}) = \prod_{i=1}^{j-1} \alpha^{\perp}_{h_i P_i} \alpha^{\perp *}_{h_i P_{i+1}} \prod_{i=j+1}^{n} \alpha^{\perp}_{h_i P_i} \alpha^{\perp *}_{h_i P_{i+1}} \alpha^{\perp}_{h_o P_j} \alpha^{\perp *}_{h_o P_1}$$

$$= [\rho_\alpha(\Pi) \; \alpha^{\perp *}_{h_o P_o} \alpha^{\perp}_{h_o P_j} \alpha^{\perp *}_{h_j P_j} \alpha^{\perp}_{h_j P_{j+1}}] \; (\alpha^{\perp}_{h_o P_j} \alpha^{\perp *}_{h_o P_1}) (\alpha^{\perp}_{h_o P_o} \alpha^{\perp *}_{h_o P_{j+1}}) = \rho_\alpha(\Pi)$$

since $h_j = h_o$. We use an induction hypothesis to show
$\rho_\alpha(\Pi) = \rho_{\bar\alpha}(\Pi') \bmod p^{e-\delta(\Pi')}$.

<u>Corollary 1.4.</u>   The two labellings α and $\bar\alpha$ are compatible iff
$\rho_\alpha(\Pi) \equiv \rho_{\bar\alpha}(\Pi) \bmod p^{e-\delta(\Pi)}$ for every reduced loop Π.

II. UNIFORM ALMOST COMPLETELY DECOMPOSABLE GROUPS

In all what follows, let G be an almost completely
decomposable group, i.e. G has a completely decomposable subgroup
C with G/C finite. $\tau(G)$ denotes the typeset of G. Let p be a prime
integer and T a set of n pairwise incomparable primes. We define a
class $\mathscr{C} = \mathscr{C}(T,p)$ of groups:

   $G \in \mathscr{C}(T,p)$ if and only if

   (a) $rk(G) = n$ and G is torsion-free

   (b) $T \subseteq \tau(G)$ is the set of maximal types in $\tau(G)$

   (c) $G/\sum_{\tau \in T} G(\tau) \cong (\mathbb{Z}(p^e))^r$ for some e, $r \in \mathbb{N}$.

Note that $\sum_{\tau \in T} G(\tau) = \bigoplus_{\tau \in T} G(\tau)$ and $G(\tau)$ has rank 1. $\mathscr{C}(T,p,e,r)$ is the
obvious subclass of $\mathscr{C}(T,p)$. It is easy to see that if $G \approx_n \bar G$ and
$G \in \mathscr{C}(T,p)$, then $\bar G \in \mathscr{C}(T,p)$ and if $G \in \mathscr{C}(T,p,e,r)$, then
$\bar G \in \mathscr{C}(T,p,e,r)$. We refer to groups $G \in \mathscr{C}(T,p)$ as p-uniform groups
of rank n and typeset T. (Recall that we <u>always</u> assume that the
elements in T are pairwise incomparable.)
If τ is a type, let τ_q be the entry of τ at the prime q. If $G \in$
$\mathscr{C}(T,p)$, then $\oplus\{G(\tau) : \tau \in T, \tau_p = \infty\}$ is a direct summand of G. Thus,
w.l.o.g. we always assume $\tau_p = 0$ for all $\tau \in T$. If $E \subseteq T$ then $E^c = T - E$.

Definition 2.1. Let $E \subseteq T$ and $\tau \in E$. Then

$$\mu_{\tau, E}(G) = \frac{\left(G/\left(\sum_{\sigma \in E^c} G(\sigma)\right)_*\right)(\tau)}{\left(G(\tau) + \left(\sum_{\sigma \in E^c} G(\sigma)\right)_*\right)/\left(\sum_{\tau \in E^c} G(\sigma)\right)_*}.$$

Note that $\bigoplus_{\sigma \in T} G(\sigma) \subseteq G \subseteq \bigoplus_{\sigma \in T} p^{-e}G(\sigma)$ for all $G \in \mathscr{C}(T,p,e,r)$ and for $E \subseteq T$, let $\Pi_E: G \to \bigoplus_{\sigma \in E} p^{-e}G(\sigma)$ be the natural projection. It is easy to see that there is a natural isomorphism

$\Pi_E(G) \cong G/\left(\sum_{\sigma \in E^c} G(\sigma)\right)_*$. Thus $\mu_{\tau, E}(G)$ is a subgroup of $\mathbb{Z}(p^e)$ for all $G \in \mathscr{C}(T,p,e,r)$. Since $\mu_{\tau, E}$ commutes with arbitrary direct sums, $\mu_{\tau, E}(G)$ is a near isomorphism invariant of G, i.e. if $G \approx_n \bar{G}$, then $\mu_{\tau, E}(G) = \mu_{\tau, E}(\bar{G})$ for all τ, E. We assume that $1 = 1_\tau \in G(\tau)$ and $1_\tau \notin pG(\tau)$.

We will now study representing matrices of G and their row-echelon forms. Recall that $G/C \cong (\mathbb{Z}(p^e))^r$, $C = \bigoplus_{\tau \in T} G(\tau)$.

We pick $a_i \in C$, $1 \le i \le r$ such that $G/C = \bigoplus_{i=1}^r <p^{-e}a_i + C>$, $a_i = (e_\tau \alpha_{i\tau})_{\tau \in T}$, where $e_\tau \in G(\tau) - pG(\tau)$. W.l.o.g. we may assume $\alpha_{i\tau} \in \mathbb{Z}$ for all i, τ. The matrix $M = (\alpha_{i\tau})$ as well as $\bar{M} = (\alpha_{i\tau} + p^e\mathbb{Z})$ is called a representing matrix of G. (Sometimes we will identify M and $\bar{M}$, but this will cause no confusion).

Definition 2.2 Let M be a matrix over $\mathbb{Z}$. A p-elementary row operation is

   (a)   any permutation of rows
   (b)   adding a multiple of one row to another row
   (c)   multiplying one row by some $c \in \mathbb{Z} - p\mathbb{Z}$.

We call M p-row equivalent to M' if one obtains M' by applying a (finite) sequence of p-elementary row operations to M. Note that in that case M' is again a representing matrix for G.

Definition 2.3. A $r \times n$-matrix $M = (\alpha_{ij})$ over the integer is in p^e-row echelon form if

   (a)   For each i, $1 \le i \le r$ there is i^*, $1 \le i^* \le n$ such that
         $\alpha_{ii^*} \equiv 1 \mod p^e$ and
   (b) $\alpha_{ij} \equiv 0 \mod p^e$ for $1 \le j < i^*$ and

$\alpha_{ji*} \equiv 0 \mod p^e$ for $j \neq i^*$

(c)   If $i < j$, then $i^* < j^*$.

The set $\{i^*: 1 \leq i \leq r\}$ is the set of column indices where 1's occur and is called the pivot set of the p^e-row echelon form matrix M.

<u>Lemma 2.4</u>.   Let G, $\tilde{G} \in \mathcal{G}(T,p,e,r)$ and M($\tilde{M}$) representing matrices for G($\tilde{G}$).   Then M and $\tilde{M}$ can be transformed into p^e-row echelon form using elementary p-row operations. If $G \approx_n \tilde{G}$, then there are p^e-row echelon forms for M, $\tilde{M}$ with the same pivot set.

<u>Proof</u>:   Let $T = \{\tau_1, \tau_2, \ldots, \tau_n\}$ and recall that the τ_i's are pairwise incomparable.   Let $M = (\alpha_{ij})$, $1 \leq i \leq r$, $1 \leq j \leq n$, $\alpha_{ij} \in \mathbb{Z}$ be a representing matrix for G.   We will use elementary row operations to compute a sequence $M = M^{(0)}, M^{(1)}, \ldots, M^{(k)}$ such that $M^{(k)}$ is in row echelon form. By abuse of, and ease for, notation, we will call the ij'th entry of $M^{(m)}$ always α_{ij}.   We begin our procedure by considering $s_1 = \inf \{|\alpha_{i1}|_p: 1 \leq i \leq r\}$. Ad hoc convention: if $x \in \mathbb{Z}$, $|x|_p \geq e$, we set $|x|_p = e$.   Note that $e-s_1 = \mu_{\tau_1, \{\tau_1\}}(G)$, i.e. s_1 is a near isomorphism invariant.   If $s_1 \neq 0$ we do nothing and set $M^{(1)} = M$ and proceed to the second column of M.   If $s_1 = 0$, we find i, $1 \leq i \leq r$ such that $|\alpha_{i1}|_p = 0$. Now we exchange first and i-th row and via (c) obtain $\alpha_{11} \equiv 1 \mod p^e$ in $M^{(1)}$.   Moreover we can do a few row operations and have $\alpha_{i1} \equiv 0 \mod p^e$ for all $i > 1$.   Here we say that 1^* is defined and we set $1^* = 1$.   This finishes our work for the first column. Notice that our action was determined by $\mu_{\tau_1, \{\tau_1\}}(G)$.   By induction, let us assume that we proceeded to the k'th column and let $D_k = \{i | i^*$ defined, $1 \leq i < k\}$ and assume $i^* < j^*$ for $i < j$ and that the i^*'s are the elements introduced in Definition 2.3.

Case I: $D_k = \emptyset$.   We compute $s_k = \inf \{|\alpha_{ik}|_p : 1 \leq i \leq r\}$ and observe $e-s_k = \mu_{\tau_k, \{\tau_k\}}(G)$.   If $s_k > 0$, we do nothing and proceed to the next column.   (Note that if $|D_n| < r$, then we get a contradiction to $G/C \cong (\mathbb{Z}(p^e))^r!$)

If $s_k = 0$, proceed as at the beginning and conclude $k^* = 1$.

Case II. $D_k \neq \emptyset$.   Now consider $\mu_{\tau_k, D_k \cup \{\tau_k\}}(G) = m_k$.   A quick look at the matrix $M^{(k-1)}$ shows that for $s_k = \inf \{|\alpha_{ik}|_p : i > |D_k|\}$ we have

$e - s_k = m_k$. If $s_k = 0$, we set $k^* = |D_k| + 1$ and suitable row operations will yield $\alpha_{k^*k} \equiv 1 \bmod p^e$ and $\alpha_{ik} \equiv 0 \bmod p^e$ for $i \neq k^*$. We set $D_{k+1} = D_k \cup \{k\}$ and proceed. If $s_k > 0$ we may be lazy and go to the next column. The set $D = D_n$ is the pivot set of the row echelon matrix that this procedure yields after at most n steps.

Since the "μ-invariants" determined D, any group $\tilde{G} \approx_n G$ will have the same pivot set D as long as we label $T = \{\tau_1, \ldots, \tau_n\}$ for $\tilde{G}$ the same way as we did for G.

A little elementary linear algebra shows the following

<u>Corollary 2.5</u>. Let $G \in \mathscr{C}(T,p,e,r)$ with representing matrix $M = (\alpha_{i\tau})$, $0 \leq i \leq r$, $\tau \in T$ in row echelon form with pivot set D. We may identify $i = d \in D$ whenever $\alpha_{id} \equiv 1 \bmod p^e$, i.e. In this way M decomposes into an $r \times r$-identity matrix $I_{D,D}$ and $M^{\#} = (\alpha_{\tau\sigma})$, $\tau \in D$, $\sigma \in D^c = T - D$. Then $\min\{e, |\alpha_{\sigma,\tau}|_p\} = \mu_{\sigma,D \cup \{\tau\}}(G)$ for all $\sigma \in D$, $\tau \in D^c$, and are near isomorphism invariants for G.

Let M, D be as in 2.5. We use these data to define a labelled e-geometry $\mathscr{G} = \mathscr{G}(G,D)$. Let $\mathscr{P} = D^c$ and $\mathit{g} = D$. We define $P \ I_m \ h$ iff $|\alpha_{hP}|_p = m < e$. If $|\alpha_{hP}| \geq e$, then $P \not{I} h$. Finally, if $P \ I \ h$ in $\mathscr{G}$, α_{hP} is the label of the flag (P,h).

An immediate consequence of 2.5 and our definition of $\mathscr{G}$ is the

<u>Corollary 2.6</u>. If $G, \overline{G} \in \mathscr{C}(T,p,e,r)$, D is a pivot set for G and $G \approx_n \tilde{G}$, then $\mathscr{G}(G,D) = \mathscr{G}(\overline{G},D)$.

<u>Theorem 2.7</u>. Let $G \in \mathscr{C}(T,p,e,r)$ and D a pivot set for G. Then $\mathscr{G}(G,D)$ is connected if and only if G is indecomposable. Thus if $G \approx_n \overline{G}$ and G indecomposable, then $\overline{G}$ is indecomposable.

<u>Proof</u>: Note that $\text{End}(G) \subseteq \prod_{\tau \in T} \text{End}(G(\tau))$. Thus any idempotent $f \in \text{End}(G)$ is a vector $f = (f_\tau)_{\tau \in T}$ with $f_\tau \in \{0,1\}$. Define $P \in \mathscr{P}^{(\varepsilon)}$ iff $P \in \mathscr{P}$ and $f_P = \varepsilon$ and $h \in \mathit{g}^{(\varepsilon)}$ iff $f_h = \varepsilon$, $\varepsilon = 0,1$. Note that $G = C + \sum_{h \in \mathit{g}} a_h \ \mathbf{Z}$, where $a_h = (\alpha_{h\tau})$, $a_{hh} = 1$ and $a_{h\tau} = 0$ for $h \neq \tau \in \mathit{g}$ we have $a_h f \equiv a_h f_h \bmod p^e$ for all $h \in \mathit{g}$. Thus if $P \in \mathscr{P}$ we have $\alpha_{hP} f_P \equiv \alpha_{hP} f_h \bmod p^e$ which implies: If $P \ I \ h$, $P \in \mathscr{P}^{(\varepsilon)}$,

h $\in g^{(\delta)}$ then $\varepsilon = \delta$. Thus $\mathscr{P}^{(0)}$, $\mathscr{P}^{(1)}$ are not connected. The other direction is easy and left to the reader.

Lemma 2.8. Let G, $\overline{G} \in \mathscr{C}(T,p,e,r)$. Then $G \approx_n \overline{G}$ if and only if there exists $f = (f_\tau)_{\tau \in T} : G \to \overline{G}$ with $f_\tau \in \text{Hom}(G(\tau), \overline{G}(\tau)) - \text{pHom}(G(\tau), \overline{G}(\tau))$.

Proof: Suppose $G \approx_n \overline{G}$. Then there is an embedding $f = (f_\tau)_{\tau \in T} : G \to \overline{G}$ such that $|\overline{G}/G|$ is relatively prime to p. Since $f_\tau \neq 0$ for all τ, f preserves types i.e. $G(\tau)f = \overline{G}(\tau) \cap Gf$. Therefore $\overline{G}(\tau)/\overline{G}(\tau)f = \overline{G}(\tau)/\overline{G}(\tau) \cap Gf = (\overline{G}(\tau)+Gf)/Gf \subseteq \overline{G}/Gf$. Thus $|\overline{G}(\tau)/G(\tau)f_\tau|$ is relatively prime to p.

Thus $f_\tau \notin \text{pHom}(G(\tau), \overline{G}(\tau))$.

For the other direction, let $f = (f_\tau)_{\tau \in T} : G \to \overline{G}$ with $f_\tau \in \text{Hom}(G(\tau), \overline{G}(\tau)) - \text{pHom}(G(\tau), \overline{G}(\tau))$. Let $e_\tau \in G(\tau) - pG(\tau)$ and $\overline{e}_\tau = f_\tau e_\tau$. Note that if $\overline{e}_\tau = px_\tau$, $x_\tau \in \overline{G}(\tau)$ then $p^{-1}f_\tau : G(\tau) \to \overline{G}(\tau)$, a contradiction. Thus p divides neither $\overline{e}_\tau$ nor $|\overline{C}/f(C)|$, $C = \bigoplus_\tau G(\tau)$, $\overline{C} = \bigoplus_\tau \overline{G}(\tau)$ and $f(C) \subseteq \overline{C}$, $\overline{C} \cap f(G) = f(C)$. Moreover

$$G/C \cong \overline{G}/\overline{C} \supseteq \frac{f(G)+\overline{C}}{\overline{C}} \cong f(G)/\overline{C} \cap f(G) \cong f(G)/f(C) \cong G/C.$$

Since G/C finite, $f(G)+\overline{C} = \overline{G}$ follows.

Thus $\overline{G}/f(G) = \overline{C}+f(G)/f(G) \cong \overline{C}/\overline{C} \cap f(G) = \overline{C}f(C)$ and $|\overline{C}/f(C)| = \prod_{\tau \in T} \overline{G}(\tau)/f_\tau(G(\tau))$ is relatively prime to p. By [LA2, Thm 11], $G \approx_n \overline{G}$.

Corollary 2.9. If $G \approx_n \overline{G}$, then for every pivot set D of G we have $\mathscr{G}(G,D) = \mathscr{G}(\overline{G},D)$ and $P \, I_\varepsilon \, h \Leftrightarrow P \, \overline{I}_\varepsilon \, h$ for all points P and lines h. If $\mathscr{G}(G,D)$ is a tree, then $\mathscr{G}(G,D) = \mathscr{G}(\overline{G},D)$ implies $G \approx_n \overline{G}$.

Proof: Suppose $G \approx_n \overline{G}$. By Lemma 2.8, there exists

$f_\tau : G(\tau) \to \overline{G}(\tau)$, not divisible by p and $f = (f_\tau)_{\tau \in T} : G \to \overline{G}$ is an embedding.

By Lemma 2.4, every pivot set D of G is also a pivot set for $\overline{G}$. Let $M(\overline{M})$ be the representing matrix of $G(\overline{G})$ with pivot set D, $M = (\alpha_{hP})$, $\overline{M} = (\overline{\alpha}_{hP})$. Then $\alpha_{hP}f_P \equiv \overline{\alpha}_{hP}f_h$ mod p^e for all P I h. Since p doesn't divide f_P, we conclude $|\alpha_{hP}|_p = |\overline{\alpha}_{hP}|_p$. This implies $\mathcal{G}(G,D) = \mathcal{G}(\overline{G},D)$.

If $\mathcal{G}(G,D)$ is a tree then any labellings α, $\overline{\alpha}$ are compatible by 1.2. The latter means that there exist $f_\tau \in \mathbf{Z}-p\mathbf{Z}$ and $f = (f_\tau) : G \to \overline{G}$ is an embedding. By 2.8, $G \approx_n \overline{G}$.

<u>Theorem 2.10</u>. Let G, G' $\in \mathscr{C}(T,p,e,r)$ and D a pivot set for G. Then $G \approx_n G'$ if and only if D is also a pivot set for G', $\mathcal{G}(G,D) = \mathcal{G}(G',D)$ and $\delta_\alpha(\Pi) \equiv \delta_{\alpha'}(\Pi)$ mod $p^{e-\delta(\Pi)}$ for each reduced loop Π of $\mathcal{G}(G,D) = \mathcal{G}(G',D)$.
<u>Proof</u>: Theorem 1.2, Corollary 2.6, Lemma 2.8.

<u>Corollary 2.11</u>: For G $\in \mathscr{C}(T,2,1,r)$, the invariants $\mu_{\tau,E}$, $\tau \in E \subseteq T$, form a complete set of near isomorphism invariants.

<u>Proof</u>: All loop invariants $\delta(\Delta)$ are equal to 1.
<u>Example</u> ([Bu, p. 1489]).
Let $|T| = 4$, p odd prime, $\alpha \in \mathbf{Z}(p)$. Let
$M_\alpha = \begin{pmatrix} 1 & 0 & 1 & 1 \\ 0 & 1 & 1 & \alpha \end{pmatrix} \begin{matrix} l_1 \\ l_2 \end{matrix}$ be the representing matrix of
$\qquad\qquad\quad P_1\ P_2$
$G_\alpha \in \mathscr{C}(T,p,1,2)$. Then $\mathcal{G} = \mathcal{G}(G_\alpha)$ is the loop

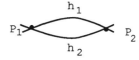

if $0 \not\equiv \alpha$ mod p and $\mathcal{G}$ is $\underline{\quad P_1 \qquad P_2 \diagup \qquad}\ h_1$ if $0 \equiv \alpha$ mod$_p$.
$\qquad\qquad\qquad\qquad\qquad\qquad\qquad h_2$

This shows that G_α is not nearly isomorphic to G_0 for $\alpha \not\equiv 0$ mod p. Let $\Delta = (P_1,P_2,h_1,P_2,P_1,h_2)$. Then $\delta(\Delta) = 0$ and $\delta(\Delta) = 11^*\alpha1^* = \alpha$ mod p. This shows that $G_\alpha \approx_n G_\beta$ iff $\alpha \equiv \beta$ mod p. It is routine to verify that $\mu_{\tau,E}(M_\alpha) = \mu_{\tau,E}(M_\beta)$ for all $\tau \in E \subseteq T$ if only $\alpha \not\equiv 0$

mod ρ and $\beta \not\equiv 0$ mod p. This shows that the $\mu_{\tau, E}$'s are *not* a complete set of near isomorphism invariants for $\mathscr{C}(T,p,1,2)$.

III. NON-UNIFORM RANK 3 GROUPS IN $\mathscr{C}(T,p)$.

Let $T = (\tau_1, \tau_2, \tau_3)$, G, $\overline{G} \in \mathscr{C}(T,p)$ with
$\overline{G}/\bigoplus_{i=1}^{3} \overline{G}(\tau_i) \cong \bigoplus_{i=1}^{3} G(\tau_i) \cong \mathbf{Z}(p^{e-d}) \oplus \mathbf{Z}(p^e)$, $0 \leq d < e$.
Let $G = (\bigoplus_{i=1}^{3} G(\tau_i)) + a_1 p^{-e}\mathbf{Z} + a_2\, p^{-e}\mathbf{Z}$.

We assume that $G \approx_n \overline{G}$. Let i be minimal such that $\mu_{\tau_i, \{\tau_i\}}(G) = \mu_{\tau_i, \{\tau_i\}}(\overline{G}) = e$. Then $i < 3$ and there exists $i < j \leq 3$ such that $\mu_{\tau_j, \{\tau_i, \tau_j\}}(G) = \mu_{\tau_j, \{\tau_i, \tau_j\}}(\overline{G}) = d$. W.l.o.g. we may assume $i = 1$, $j = 2$. Then some elementary row operations give us representing matrices $M = \begin{pmatrix} 1 & \alpha_{11} & \alpha_{12} \\ 0 & p^d & p^d\alpha_{22} \end{pmatrix}$ and

$\overline{M} = \begin{pmatrix} 1 & \overline{\alpha}_{11} & \overline{\alpha}_{12} \\ 0 & p^d & p^d\overline{\alpha}_{22} \end{pmatrix}$ modulo p^e of G, $\overline{G}$.

We compute the μ-invariants:
$\mu_{\tau_i, \{\tau_i\}} = e$, $\mu_{\tau_2, \{\tau_2\}} = e - \min\{d, |\alpha_{11}|_p\}$
$\mu_{\tau_3, \{\tau_3\}} = e - \min\{d, |\alpha_{12}|_p, d\}$,
$\mu_{\tau_1, \{\tau_1, \tau_2\}} = \min\{e - d + |\alpha_{11}|_p, e\}$,
$\mu_{\tau_2, \{\tau_1, \tau_2\}} = e - d$
$\mu_{\tau_1, \{\tau_1, \tau_3\}} = \min\{e, e - d + |\alpha_{12}|_p\}$
$\mu_{\tau_3, \{\tau_1, \tau_3\}} = e - d$ (Note that p doesn't divide α_{22}),

$\mu_{\tau_2, \{\tau_2, \tau_3\}} = e$, since $\Delta = \text{def} \begin{pmatrix} \alpha_{11} & \alpha_{12} \\ 1 & \alpha_{22} \end{pmatrix}$ is relatively prime to p.
Finally, $\mu_{\tau_3, \{\tau_2, \tau_3\}} = e$. Moreover $\mu_{\tau_1, \{\tau_1, \tau_2, \tau_3\}} = 0$ since the $G(\tau_i)$'s are pure in G. Now suppose $f = (f_1, f_2, f_3): G \to \overline{G}$ is an embedding with $(p, f_i) = 1$ for all i. A little elementary linear algebra and Kramer's rule show:

$$p^d/\Delta^\star \det \begin{pmatrix} \overline{\alpha}_{11} & \overline{\alpha}_{12} \\ \overline{\alpha}_{11f2} & \overline{\alpha}_{12f3} \end{pmatrix} = p^d z_2, \quad z_1 = \Delta^\star \det \begin{pmatrix} \overline{\alpha}_{11}f_2 & \overline{\alpha}_{12}f_3 \\ 1 & \overline{\alpha}_{22} \end{pmatrix} = f_1.$$

Moreover $f_2 \bar{\alpha}_{22} \equiv f_3 \alpha_{22} \mod p^{e-d}$. Thus $f_3 \equiv f_2 \bar{\alpha}_{22} \alpha_{22}^* + p^{e-d}x$ and

$$\bar{\Delta}f_1 = \det \begin{vmatrix} \alpha_{11} \ f_2, & f_2 \ \alpha_{12} \ \bar{\alpha}_{22} \ \alpha_{22}^* + p^{e-d}x \ \alpha_{12} \\ \\ 1 & \bar{\alpha}_{22} \end{vmatrix}$$

$$= f_2 \det \begin{vmatrix} \alpha_{11} \ \alpha_{12} \ \bar{\alpha}_{22} \ \alpha_{22}^* \\ \\ 1 & \bar{\alpha}_{22} \end{vmatrix} + p^{e-d}x \ \alpha_{12}$$

$$= f_2 \ \bar{\alpha}_{22} \ \alpha_{22}^* \ \Delta + p^{e-d}x \ \alpha_{12}$$

Thus we need that

$$p^d \ / \ \begin{vmatrix} \bar{\alpha}_{11} & \bar{\alpha}_{12} \\ \\ \alpha_{11}f_2 & \alpha_{12}f_3 \end{vmatrix} = \begin{vmatrix} \bar{\alpha}_{11} & \bar{\alpha}_{12} \\ \\ \alpha_{11}f_2 & \alpha_{12}f_2 \ \bar{\alpha}_{22} \ \alpha_{22}^* + \alpha_{12} \ p^{e-d}x \end{vmatrix}$$

$$= f_2 \begin{vmatrix} \bar{\alpha}_{11} & \bar{\alpha}_{12} \\ \\ \alpha_{11} & \alpha_{12} \ \bar{\alpha}_{22} \ \alpha_{22}^* \end{vmatrix} + \bar{\alpha}_{11} \ \alpha_{12} \ p^{e-d}x$$

Note that

$$\begin{vmatrix} \bar{\alpha}_{11} & \bar{\alpha}_{12} \\ \\ \alpha_{11} & \alpha_{12} \ \bar{\alpha}_{22} \ \alpha_{22}^* \end{vmatrix} =$$

$$= \alpha_{22}^* \begin{vmatrix} \bar{\alpha}_{11} & \bar{\alpha}_{12} \ \alpha_{22} \\ \\ \alpha_{11} & \alpha_{12} \ \bar{\alpha}_{22} \end{vmatrix} = \alpha_{22}^* \ (\alpha_{12} \ \bar{\Delta} - \bar{\alpha}_{12} \ \Delta) \ .$$

For being able to solve for x, we need

$$\min\{d, |\alpha_{12}\bar{\Delta} - \bar{\alpha}_{12} \ \Delta|_p\} \geq \min\{|\bar{\alpha}_{11}|_p + |\alpha_{12}|_p + e-d, d\}.$$

Since Δ, $\bar{\Delta}$ are p-adic units, we obtain

$$|\alpha_{12}\bar{\Delta} - \bar{\alpha}_{12} \ \Delta|_p = |\alpha_{12}\Delta^* - \bar{\alpha}_{12} \ \bar{\Delta}^*|_p \text{ and}$$

$$\min\{d, |\bar{\alpha}_{11}|_p + |\bar{\alpha}_{12}|_p + e-d\} =$$
$$\min\{d, \ e-d + (e-\mu_{\tau_3, \{\tau_3\}}(G)) + (e-\mu_{\tau_2, \{\tau_2\}}(G))\}$$
$$= \gamma(G) \text{ is a near isomorphism invariant and}$$

$\alpha_{12} \Delta^* \equiv \overline{\alpha}_{12} \overline{\Delta}^* \bmod p^{\gamma(G)}$.

This proves the following

<u>Theorem</u>: Let G, $\overline{G}$ be as above. Then $G \approx_n \overline{G}$ iff $\mu_{\tau, E}(G) = \mu_{\tau, E}(\overline{G})$

for all $\tau \in E \subseteq T$ and $\alpha_{12} \Delta^* \equiv \overline{\alpha}_{12} \overline{\Delta}^* \bmod p^{\gamma(G)}$.

This implies that $\alpha_{12} \Delta^* \bmod p^{\gamma(G)}$ is a near isomorphism invariant

for these groups as well as $\gamma(G)$, i.e. $\gamma(G) = \gamma(\overline{G})$ if $G \approx_n \overline{G}$.

<u>Question</u>: What is the "group theoretic meaning" of $\gamma(G)$ and

$\alpha_{12} \Delta^*$? Recall that Δ^* is the inverse of $\Delta \bmod p^e$.

<u>Example</u>: Let G_α be the group defined by d = 3, e = 4 and $\alpha \not\equiv 1$

mod p. $M = \begin{pmatrix} 1 & 1 & \alpha \\ 0 & p^3 & p^3 \end{pmatrix}$. Then $\gamma(G_\alpha) = \min\{3, |\alpha|_p + 1\} =$

$= 1$ if p divides α and $\alpha_{12} \Delta^* = \alpha(1-\alpha)^*$. Note that

$\alpha(1-\alpha)^* \equiv \beta(1-\beta)^* \bmod p$ iff $\alpha \equiv \beta \bmod p$. Thus

$G_\alpha \approx_n G_\beta$ iff $\alpha \equiv \beta \bmod p$ for p-adic units α, β. If

$|\alpha|_p = |\beta|_p = 1$, then the same holds iff $\alpha \equiv \beta \bmod p^2$.

4.   <u>An algorithm</u>

Let $T = \{\tau_1, ..., \tau_n\}$ be any set of n types and $R_i \subseteq \mathbb{Q}$ of type τ_i. Any

group G in $\mathscr{C}(T,p)$ can be written as $G = \bigoplus_{i=1}^{n} e_i R_i + \sum_{i=1}^{m} p^{-e} a_i \mathbb{Z}$

where $a_i = (\alpha_{ij})_{j=1,...,n} \in \mathbb{Z}^n$. On the other hand, any such

$G \subseteq \bigoplus_{i=1}^{n} e_i \mathbb{Q}$ is a.c.d. and $G(\tau_i) \supseteq e_i R_i$. In general we can't

expect $e_i R_i = G(\tau_i)$ to hold. We have $e_i R_i = G(\tau_i)$ iff $e_i R_i$ is pure

in G. Since $G/\bigoplus_{i=1}^{n} R_i$ is a p-group, p-purity of $e_i R_i$ implies

purity. We will give an algorithm that yields an isomorphic copy

of G such that the $e_i R_i$'s are pure in that copy of G.

We set $C = \bigoplus_{i=1}^{n} e_i R_i$ and $\hat{\Pi}_1: \bigoplus_{i=1}^{n} e_i \mathbb{Q} \to \bigoplus_{i=2}^{n} e_i \mathbb{Q}$ is the natural

projection with kernel $e_1 \mathbb{Q}$. Suppose $e_n q \in G$ for some $q \in \mathbb{Q}$ and

let $A = (\alpha_{ij})$ be the m × n-matrix over $\mathbb{Z}$ where $a_i = (a_{ij})_{j=1,...,n}$.

Since $e_n q \in G$, there are $x_i \in \mathbb{Z}$, $i \le i \le m$ such that $(x_1, ..., x_m) p^{-e} A$

$= e_n q + c$ for some $c \in C$. Thus $(x_1, ..., x_m) A \equiv e_n q p^e \bmod p^e C$. We

set $\overline{q} = q p^e$ and pick i_o, $i \le i_o \le m$ such that

$|\alpha_{i_o 1}|_p = \min\{|\alpha_{i1}|_p : 1 \le i \le m\}$. We do p-elementary row operations to

change A into $\overline{A} = (\overline{\alpha}_{ij})$ such that $\overline{\alpha}_{i1} \equiv 0 \bmod p^e \mathbb{Z}$ for $i \ne i_o$.

Note that still $G = C + \sum_{i=1}^{m} p^{-e}\bar{\alpha}_i \mathbb{Z}$ where $\bar{\alpha}_i = (\bar{\alpha}_{ij})_{1\leq j\leq n}$. Abusing

notation we set $A = \bar{A}$. This implies that for

$(x_1,\ldots,x_m)A \equiv e_n\bar{q}$ mod p^e C that $x_{i_0} \in p^{e-|\alpha_{i_0 1}|}p\mathbb{Z}$, i.e.

$x_{i_0} = \tilde{x}_{i_0} p^{e-|\alpha_{i_0 1}|}p$. Let $A^{(1)}$ be the $m\times(n-1)$ matrix obtained from

A by deleting the first column and multiplying the i_0'th row by

$p^{e-|\alpha_{i_0 1}|}p$. Then $(x_1,\ldots,x_m)A \equiv e_n\bar{q}$ mod $p^e C$ iff $(x_1,\ldots, x_{i_0-1}, \bar{x}_{i_0},$

$x_{i_0+1},\ldots,x_m)A^{(1)} \equiv e_n\bar{q}$ mod $p^e C$ where $C^{(1)} = \hat{\Pi}_1(C)$. We now consider

the group $G^{(1)} = \hat{\Pi}_1(C) + \sum_{i=1}^{m} p^{-e} a_i^{(1)} \mathbb{Z}$ where $a_i^{(1)}$ is the i'th row

of $A^{(1)}$ and repeat the step described above. After $n-1$ many steps

we obtain an $m\times 1$-matrix $A^{(n-1)}$ such that there are $x_i \in \mathbb{Z}$ with

$(x_1,\ldots,x_m)A \equiv e_n\bar{q}$ mod $p^e C$ iff there are $y_i \in \mathbb{Z}$ with

$(y_1,\ldots,y_m)A^{(n-1)} \equiv e_n\bar{q}$ mod $e_n p^e R_n$. This implies that,

since $A^{(n-1)}$ is $m \times 1$, that $|e_n|_p^G = e-\min\{|\alpha_j^{(n-1)}|_p: 1\leq j\leq m\}$ where

$A^{(n-1)} = (\alpha_1^{(n-1)},\ldots,\alpha_m^{(n-1)})^t$.

   This gives rise to the following, easy to implement, algorithm for computing $|e_n|_p^G$:

(0)   Start with $A = A^{(0)} = (\alpha_{ij})$ modulo $p^e\mathbb{Z}$, an $m\times n$-matrix

(1)   Pick i_0 with $|\alpha_{i_0,1}| = \min\{|\alpha_{i1}|_p:1\leq i\leq m\} < e$ and add multiples

of the i_0'th row to the others forcing the $(i,1)$-entries $i \neq i_0$ to

be 0 mod $p^e\mathbb{Z}$.

(2)   Multiply the i_0'th row of A by $p^{e-|\alpha_{i_0,1}|}p$ and delete the 1'st

column of A. Call this new $m\times(n-1)$ matrix B.

(3)   If B is $m\times 1$-matrix, go to (4). Otherwise set $A:= B$ and go

back to (0)

(4)   Here $A:= B$ is an $m\times 1$-matrix and

   $|e_n|_p^G = e-\min\{e, \min\{|\alpha_i| : 1\leq i\leq m\}\}$ (This algorithm works for

any finite set T of types!) We may compute the p-heights of all

e_i's by a suitable permutation of the columns of A.

Let $\delta_i = |e_i|_p^G$. Then

$$G \cong G' = \bigoplus_{i=1}^{n} e_i p^{\delta_i} R_i + \sum_{i=1}^{m} p^{-e}(\alpha_{ij} p^{\delta_j})_{j=1}^{n} \mathbb{Z}.$$ Let $R_i^1 = p^{\delta_i}R_i \cong R_i$.

Then $R_i^1 = G(\tau_i)$ is pure in G. By renaming we assume

$$G = \bigoplus_{i=1}^{n} e_i R_i + \sum_{i=1}^{m} p^{-e}\alpha_i\mathbb{Z}$$ and R_i is pure in G. Again, we consider

$A = (\alpha_{ij})$, $a_i = (a_{ij})_j$. We may perform some p-elementary row operations and - after a permutation of rows, mod $p^e\mathbb{Z}$, A has the form

$$A = \left(\begin{array}{c|cc} (\alpha_{ij}) & p^{\delta_1} & 0 \\ & \star & p^{\delta_r} \\ \hline 0 & 0 & \end{array} \right) \} r \qquad \text{where}$$

$$\underbrace{}_{n-r} \qquad \underbrace{}_{r}$$

$e > \delta_1 \geq \delta_2 \geq \delta_3 \geq ... \geq \delta_r$. This tells us that--from now on we assume that the τ_i's are pairwise incomparable--the structure of

$$G / \bigoplus_{i=1}^{n} G(\tau_i) = \bigoplus_{j=1}^{r} \mathbb{Z}(p^{e-\delta_j}).$$

If $\delta_i = \delta_j$ for all $1 \leq i, j \leq r$, then G is uniform i.e.
$G \in \mathscr{C}(T, p, e-\delta_r, r)$ and we may proceed with more p-elementary row operations and obtain

$$A = \left(\begin{array}{ccc} \alpha_{ij} & 1 & 0 \\ & 0 & 1 \\ 0 & & 0 \end{array} \right) \mod p^{e-\delta_r}$$

This allows us now to describe $\mathscr{G}(G)$ and all of its near isomorphism invariants as described in the earlier sections.

REFERENCES

[BU1]   R. Burkhardt, On a Special class of almost completely decomposable torsion free groups I, Abelian Groups and Modules, CISM Courses and Lectures No. 287, Proceedings of the Udine Conference, Udine, April 9-14, 1984, p. 141-150.

[BU2]   R. Burkhardt, Elementary abelian extensions of finite rigid systems, Comm. Algebra, 11(13) (1983) 1473-1499.

[De]   P. Dembowski, Finite Geometries, Springer, New York, Heidelberg, Berlin, 1968

[FI/II]   L. Fuchs, Infinite Abelian Groups, Vols. I and II, Academic Press, New York, 1970 and 1973.

[KM]   K. J. Krapf and O. Mutzbauer, Classification of almost completely decomposable groups, Abelian groups and Modules, CISM Courses and Lectures No 287, Proceedings of the Udine Conference, Udine, April 9-14, 1983, p. 151-162.

[LA1]   E. L. Lady, Almost completely decomposable torsion-free abelian groups, Proc. Amer. Math. Soc. 45 (1974), 41-47.

[LA2]   E. L. Lady, Nearly isomorphic torsion free abelian groups, J. of Algebra 35 (1975), 235-238.

Butler Quotients of Torsion-free Abelian Groups Modulo Prebalanced Subgroups

L. FUCHS*

Department of Mathematics, Tulane University, New Orleans, LA 70118

and

C. METELLI

Dipartimento di Matematica, Università di Napoli
Napoli, Italy

A torsion-free group B of finite rank is called a *Butler group* if it satisfies either of the following two conditions: 1. B is a pure subgroup of a completely decomposable group (of finite rank); 2. B is an epimorphic image of a completely decomposable group of finite rank. The equivalence of these conditions has been proved by Butler [4]. The same class of groups was discovered independently by Bican [1].

The observation that Butler's conditions on B are equivalent to the condition that $\text{Bext}(B,T) = 0$ for all torsion groups T (where Bext denotes the group of all balanced extensions) led to a natural generalization to groups of countable rank by Bican-Salce [3]. For groups of infinite rank, two important generalizations have been investigated. Call a torsion-free group B (of any rank)

    1. a B_1-*group* if $\text{Bext}^1(B,T) = 0$ for all torsion groups T;

    2. a B_2-*group* if there is a continuous well-ordered ascending chain of pure subgroups,

$$(1) \qquad 0 = B_0 < B_1 < ... < B_\alpha < ... < B_\tau = B = \cup B_\alpha$$

with finite rank (equivalently, rank 1) factors such that, for each $\alpha < \tau$, $B_{\alpha+1} = B_\alpha + G_\alpha$ for some finite rank Butler group G_α.

In case the factors are of rank 1, the last condition on the B_α can also be phrased by saying that B_α is prebalanced in $B_{\alpha+1}$. Following Fuchs-Viljoen [9], a pure subgroup A of a torsion-free group G is called *prebalanced* if for each rank 1 pure subgroup C/A

---

* This work was partially supported by NSF.

of G/A there is a finite rank Butler group B such that $C = A + B$. The concept of prebalancedness is more fully developed in Fuchs-Metelli [8].

It is not difficult to see that all B_2-groups are B_1-groups. The converse has been established under the assumption of the Continuum Hypothesis for groups of cardinalities not exceeding $\aleph_\omega$ (see Dugas-Hill-Rangaswamy [6] and the literature quoted there). In what follows, a B_2-group B of any rank will be called a *Butler group*.

This note is motivated by the following consideration.

Let A be a pure subgroup of a torsion-free group G. If A is balanced in G and G/A is completely decomposable, then $G = A \oplus B$ for a completely decomposable subgroup B of G. If we weaken the hypotheses, and we only assume that A is prebalanced in G and G/A is a Butler group, then obviously A need not be a summand of G. Can anything be said in this situation? What we wish to prove here is that an analogous conclusion can be drawn, namely, that $G = A + B$ holds for a suitable Butler subgroup B of G. In other words, a quotient which is a Butler group modulo a prebalanced subgroup can 'almost' be lifted to a Butler subgroup.

If G/A is of finite rank, the claim is an easy consequence of the definitions (see (1.1)). If, however, G/A is of infinite rank, then a more sophisticated argument is required which is based on a strengthened form of the Torsion Extension Property of finite rank Butler groups.

1. Preliminaries

We start with a simple lemma. It is a straightforward consequence of the definitions.

LEMMA 1.1. *Let G be a torsion-free group, and H a prebalanced subgroup of G. G/H is a finite rank Butler group if and only if $G = H + B$ with B a finite rank Butler subgroup of G.*

PROOF. Sufficiency is clear, since $B/(H \cap B)$ is, as a torsion-free quotient of B, a Butler group. Conversely, let $G/H = G_1/H + ... + G_n/H$ with G_i/H of rank 1. As H is prebalanced in each G_i, $G_i = H + B_i$ with B_i a finite rank Butler group. Then $B = B_1 + ... + B_n$ is a finite rank Butler group such that $G = H + B$. $\square$

In the next two lemmas, $I = \{1, 2, ..., t\} \subset \mathbf{N}$; i, j will denote elements of I.

LEMMA 1.2. *Let $X = \bigoplus_{i \in I} S_i x_i$ be a completely decomposable group where $\mathbf{Z} \leq S_i \leq \mathbf{Q}$ and $S_i \wedge S_j = \mathbf{Z}$ whenever $i, j \in I$ are distinct. The kernel of the epimorphism $\rho : X \to R = \sum_i S_i \leq \mathbf{Q}$ given by $\rho(x_i) = 1$ $(i \in I)$ is free.*

PROOF. Evidently, $g = \sum_{i \in I} s_i x_i \in \operatorname{Ker} \rho$ $(s_i \in S_i)$ if and only if $\sum_{i \in I} s_i = 0$. To show that here all the s_i must be integers, write $s_i = m_i/n_i$ with $(m_i, n_i) = 1$ and $n_i > 0$. In view of the hypothesis on the S_i, the integers n_i are pairwise coprime. Setting $n = \prod_{i \in I} n_i = n_i k_i$, we have $\sum_{i \in I} m_i k_i = 0$. For every i, n_i divides all k_j for $j \neq i$; thus $n_i \mid m_i$, and therefore $n_i = 1$. We conclude that $\operatorname{Ker} \rho$ is contained in the subgroup generated by the x_i. $\square$

The next result is a far reaching generalization of (1.1) for factor groups G/H of rank 1.

LEMMA 1.3. *Let H be a prebalanced subgroup of the torsion-free group G such that G/H is of rank 1. There is a finite rank Butler subgroup B of G such that $G = H + B$ and $H \cap B$ is a free group.*

PROOF. Write (irredundantly) $G = H + \sum_{i \in I} \langle b_i \rangle_*$; thus no $b_i \in H$. Then we can find nonzero integers $n_i (i \in I)$ such that $n_i b_i + H = n_j b_j + H$ holds for all i, j. Replacing the b_i's by suitable integral multiples we will have $b_i + H = b_j + H$ for all i, j. Set $\langle b_i \rangle_* = R_i b_i$ with $\mathbf{Z} \leq R_i \leq \mathbf{Q}$; then $G/H = R(b' + H)$, where $R = \sum_i R_i$ and b' is any of the b_i. Thus $b = \sum_{i \in I} r_i b_i \in H$ ($r_i \in R_i$ for each i) if and only if

$$\sum_{i \in I} r_i b_i + H = \sum_{i \in I} (r_i b_i + H) = \sum_{i \in I} r_i (b_i + H) = (\sum_{i \in I} r_i)(b' + H) = 0, \text{ that is, if and only if}$$

$$\sum_{i \in I} r_i = 0.$$

Now we define for each i a subgroup S_i of R_i as follows. Consider a prime p. If α is the p-height of 1 in R, then evidently it is the p-height of 1 in one of the R_i. Fix an index $i(p)$ for which this happens; for all indices $i \neq i(p)$ set the p-height of 1 in S_i to be 0, while for $i = i(p)$ the p-height of 1 in S_i should be the same α as in R_i. If this is done for each prime p, then $S_i \wedge S_j = \mathbf{Z}$ for all $i \neq j$, and $\sum_i S_i = \sum_i R_i = R$. Let $Y = \bigoplus_i S_i x_i \leq X = \bigoplus_i R_i x_i$ and $\rho : Y \to \sum_i S_i$ with $\rho(x_i) = 1$ ($i \in I$). From (1.1) we conclude that $\operatorname{Ker} \rho$ is free. If B denotes the Butler subgroup $\sum_{i \in I} S_i b_i$ of G, then $H \cap B$ is isomorphic to a subgroup of $\operatorname{Ker} \rho$, so it is free. Moreover, $B/(H \cap B) = R(b' + H \cap B)$ yields $(B + H)/H = R(b' + H) = G/H$, hence $H + B = G$. $\square$

A slightly stronger form of (1.3) is needed later on.

PROPOSITION 1.4. *Let G be a torsion-free group, H a prebalanced subgroup of corank 1, and A' a subgroup of H. There is a finite rank Butler subgroup B of G such that $H + B = G$ and $B \cap A$ is a free subgroup of A' where A is the purification of A' in G.*

PROOF. In view of (1.3) we have $G = H + B'$ with B' a finite rank Butler group such that $H \cap B'$ is a free group. With this choice of B', $B' \cap A$ is a free group. If $B' \cap A = 0$, there is nothing to prove. If this is not the case, then $B' \cap A'$ has a finite index in $B' \cap A$. Hence $B'/(B' \cap A')$ splits into a direct sum $B/(B' \cap A') \oplus (B' \cap A)/(B' \cap A')$ for some subgroup B of B' which - as a subgroup of finite index in B' - has to be a Butler group. This B satisfies $H + B = G$ and $B \cap A = B \cap B' \cap A = B' \cap A' \leq A'$; in particular, $B \cap A$ is free. $\square$

2. A stronger form of TEP for finite rank Butler groups

Torsion Extension Property (TEP) was discovered in an equivalent form by Procházka [10]. Bican [2] verified that in finite rank Butler groups all pure subgroups share this property. TEP plays an important role in the study of B_1-groups of infinite ranks (see Dugas-Rangaswamy [5] and Dugas-Hill-Rangaswamy [6]). Recall that a pure subgroup A of a torsion-free group G is said to be a *TEP-subgroup* of G, or to *have* TEP in G, if every homomorphism $A \to T$ (T any torsion group) extends to a homomorphism $G \to T$. This is easily seen to be equivalent to the following: for every subgroup A' of A such that A/A' is torsion, A/A' is a summand of G/A'.

If G is a finite rank Butler group, A' a subgroup of G, and $A = (A')_*$ (the purification of A' in G), then A being a TEP-subgroup of G, we have $G/A' = B/A' \oplus A/A'$ for some B with $A' \leq B \leq G$. It is natural to raise the question: Can B be chosen to be a Butler group? Note that for B to be Butler it is necessary to assume that A' is a Butler group, since $B \cap A = A'$ - as a pure subgroup of the finite rank Butler group B - is again a Butler group. Our theorem (2.1) will show that if A' is itself a Butler group, then B can indeed be chosen to be a Butler group.

Note that we cannot expect each such B to be Butler. A counterexample is given by $G = \mathbf{Q}_p \oplus \mathbf{Q}_p$ ($\mathbf{Q}_p$ = rationals with denominators powers of a prime p), B the (non Butler) indecomposable Pontryagin group of rank 2 (see e.g. [7, p.125, Example 5]), $A = \mathbf{Q}_p \leq G$ (any pure rank 1 subgroup), and $A' = B \cap A$.

THEOREM 2.1. *Let G be a Butler group of finite rank, and $A' \leq G$, $A = (A')_*$. If A' is a Butler group, then there is a Butler group B such that $A' \leq B \leq G$ and $G/A' = B/A' \oplus A/A'$. In this case, $B_* = G$.*

PROOF. Note that $A' \leq B$ implies $A \leq B_*$; thus if $G = B + A$, then $B_* = G$.

First assume G/A is of rank 1. Because of (1.4), we can find a Butler subgroup B' of G such that $B' + A = G$ and $B' \cap A$ is a free subgroup of A'. Then $B = B' + A'$ is a Butler group which satisfies both $B + A = G$ and $B \cap A = (C \cap A) + A' = A'$.

Turning to the general case, let $A < G_1 < \ldots < G_n = G$ ($n > 1$) be a chain of pure subgroups of G with rank one factors. What we have already proved can be applied to G_1 to conclude that there is a Butler group $B_1 \leq G_1$ such that $A' \leq B_1$, $G_1/A' = B_1/A' \oplus A/A'$, and $(B_1)_* = G_1$. Next we repeat this argument with G_2, B_1 playing the roles of G, A' to argue that for some Butler group $B_2 \leq G_2$, we have $G_2/B_1 = B_2/B_1 \oplus G_1/B_1$, and $(B_2)_* = G_2$. Note that B_2 satisfies the requirements of the theorem for G_2, since $G_2 = B_2 + G_1 = B_2 + B_1 + A = B_2 + A$ and $B_2 \cap A = B_2 \cap G_1 \cap A = B_1 \cap A = A'$. Thus proceeding, we arrive at a Butler group B_n such that $G = B_n + A$ and $B_n \cap A = A'$. □

Observe that, in the notation of (2.1), A/A' is the torsion subgroup of G/A', thus B/A' is a torsion-free group.

3. Prebalanced subgroups with Butler quotients

We are ready to prove the mentioned generalization of the necessity part of (1.1). The main difficulty in the proof is that one has to assure that in the construction of B the stepwise extensions are pure extensions. This can be done with the aid of (2.1).

THEOREM 3.1. *Let G be a torsion-free group, A a prebalanced subgroup of G such that G/A is a Butler group. Then $G = A + B$ holds with B a Butler subgroup of G (rk B is finite if rk G/A is finite, and rk B = rk G/A if the latter is infinite).*

Moreover, if A' is a subgroup of G such that $A = (A')_*$, then B can be chosen so as to satisfy $B \cap A \leq A'$.

PROOF. If G/A is a Butler group, then definition implies that there is a continuous well-ordered ascending chain $A = H_0 < H_1 < \ldots < H_\alpha < \ldots$ ($\alpha < \tau$ for some ordinal τ) of pure subgroups of G whose union is G, where each H_α is prebalanced in G and $H_{\alpha+1}/H_\alpha$ is of rank 1 for each α. We will build the subgroup B inductively as the union of a continuous well-ordered ascending chain $0 = B_0 < B_1 < \ldots < B_\alpha < \ldots$ of Butler subgroups such that, for each $\alpha < \tau$,

(i) $B_{\alpha+1}/B_\alpha$ is torsion-free;

(ii) $B_{\alpha+1} = B_\alpha + C_{\alpha+1}$ where $C_{\alpha+1}$ is a finite rank Butler group;

(iii) $H_\alpha = A + B_\alpha$;

(iv) $B_\alpha \cap A \leq A'$.

If α is a limit ordinal, B_α is defined as the union of its predecessors.

If B_α has been defined for some $\alpha < \tau$, then $B_{\alpha+1}$ will be defined as follows. An appeal to (1.4) yields $H_{\alpha+1} = H_\alpha + G_{\alpha+1}$ with $G_{\alpha+1}$ a finite rank Butler subgroup of G such that $G_{\alpha+1} \cap H_\alpha$ is a free subgroup of $H'_\alpha = A' + B_\alpha$. Here H_α is the purification of H'_α. Consider the finite rank Butler group $G_{\alpha+1}$ and its subgroup $L' = B_\alpha \cap G_{\alpha+1}$ which is itself free as a subgroup of $H_\alpha \cap G_{\alpha+1}$. We can apply (2.1) to $G_{\alpha+1}$, L' in order to obtain a finite rank Butler group C satisfying $G_{\alpha+1}/L' = C/L' \oplus L/L'$ where L denotes the pure closure of L' in $G_{\alpha+1}$. If we set $B_{\alpha+1} = B_\alpha + C$, then $B_{\alpha+1}/B_\alpha = (B_\alpha + C)/B_\alpha \cong C/(B_\alpha \cap C) = C/L'$ is torsion-free (cf. remark after (2.1)), so (i) holds. (ii) is obviously satisfied. (iii) for $\alpha+1$ follows from $A + B_{\alpha+1} = A + B_\alpha + C = H_\alpha + C = H_\alpha + L + C = H_\alpha + G_{\alpha+1} = H_{\alpha+1}$. Finally, in order to verify (iv) observe that

$$H_\alpha \cap B_{\alpha+1} = H_\alpha \cap (B_\alpha + C) = B_\alpha + (H_\alpha \cap C) \leq B_\alpha + (H_\alpha \cap G_{\alpha+1}) \leq B_\alpha + A',$$

so that $A \cap B_\alpha \leq A'$ implies $A \cap B_{\alpha+1} = A \cap H_\alpha \cap B_{\alpha+1} = A \cap (A' + B_\alpha) = A' + (A \cap B_\alpha) \leq A'$, establishing (iv).

Conditions (i) and (ii) guarantee that the union B of the groups B_α is a Butler group, while (iii) and (iv) ensure that $G = A + B$ and $B \cap A \leq A'$ hold. $\square$

It is worth while pointing out that in (3.1) the subgroup A' can be chosen to be a free subgroup of A, in which case we have that the intersection $B \cap A$ is free.

Finally, we wish to give a generalization of the sufficiency part of (1.1).

THEOREM 3.2. *The following are equivalent for a prebalanced subgroup A of the torsion-free group G:*

(i) *G/A is a Butler group;*

(ii) *$G = A + B$ for a suitable Butler subgroup B which is the union of a well-ordered ascending chain $0 = B_0 < B_1 < \ldots < B_\alpha < \ldots < B_\tau = B$ of prebalanced subgroups with finite rank factors such that for each $\alpha < \tau$, $A + B_\alpha$ is pure in G.*

PROOF. (i) $\Longrightarrow$ (ii) In view of (3.1) it suffices to verify that B has a chain of the indicated type. Let $0 = B_0 < B_1 < \ldots < B_\alpha < \ldots$ $(\alpha < \tau)$ be the continuous well-ordered ascending chain with union B obtained in the proof of (3.1). Then the subgroups $H_\alpha = A + B_\alpha$ are pure in G.

(ii) $\Longrightarrow$ (i) Let $0 = B_0 < B_1 < \ldots < B_\alpha < \ldots$ $(\alpha < \tau)$ be a continuous well-ordered ascending chain of pure subgroups with rank 1 factors and union B such that $B_{\alpha+1} = B_\alpha + G_\alpha$ for some finite rank Butler group G_α, and, moreover, the subgroups $H_\alpha = A + B_\alpha$ are pure in G. In this way, we obtain a continuous chain of pure subgroups H_α from A to G where $H_{\alpha+1} = H_\alpha + G_\alpha$ for all $\alpha < \tau$; therefore, H_α is prebalanced in $H_{\alpha+1}$. Hence G/A is a Butler group, indeed. $\square$

Let us draw a corollary of independent interest. It gives a sufficient condition for a quotient G/A of a completely decomposable group G modulo a prebalanced subgroup A to be a Butler group.

COROLLARY 3.4. *Let $G = \bigoplus G_i$ be a completely decomposable torsion-free group, G_i of rank 1, and A a prebalanced subgroup of G. The quotient G/A is a Butler* group if the G_i admit a well-ordering G_α $(\alpha < \tau)$ such that in the ascending chain $A_\alpha = A + \bigoplus_{\beta < \alpha} G_\beta$ $(\alpha < \tau)$ there are no infinite intervals of non-pure subgroups A_α. $\square$

REFERENCES

[1] L. Bican, Splitting in abelian groups, Czech. Math. J. 28 (1978), 356-364.

[2] ———— , Purely finitely generated abelian groups, Comment. Math. Univ. Carolin. 21 (1980), 209-218.

[3] L. Bican and L. Salce, Butler groups of infinite rank, Abelian Group Theory, Lecture Notes in Math. 1006 (Springer, 1983), 171-189.

[4] M. C. R. Butler, A class of torsion-free abelian groups of finite rank, Proc. London Math. Soc. 15 (1965), 680-698.

[5] M. Dugas and K. M. Rangaswamy, Infinite rank Butler groups, Trans. Amer. Math. Soc. 305 (1988), 129-142.

[6] M. Dugas, P. Hill and K. M. Rangaswamy, Infinite rank Butler groups, II, Trans. Amer. Math. Soc. 320 (1990), 643-664.

[7] L. Fuchs, Infinite Abelian Groups , vol. 2 (Academic Press, 1973).

[8] L. Fuchs and C. Metelli, Countable Butler groups, Manuscripta Math. (to appear).

[9] L. Fuchs and G. Viljoen, Note on the extensions of Butler groups, Bull. Austral. Math. Soc. 41 (1990), 117-122.

[10] L. Procházka, Über die Spaltbarkeit der Faktorgruppen torsionsfreier abelscher Gruppen endlichen Ranges, Czech. Math. J. 11 (1961), 521-557.

Torsion-free Abelian Groups with Precobalanced Finite Rank Pure Subgroups

ANTHONY J. GIOVANNITTI

The University of Southern Mississippi
Hattiesburg, MS USA 39406–5045

§1: **Introduction.** One of the best understood classes of torsion-free abelian groups is that of direct sums of subgroups of the rationals called the completely decomposable abelian groups. In 1965, M. C. R. Butler [Bu] introduced the investigation of pure subgroups of completely decomposable abelian groups of finite rank, which have come to be known as *Butler groups*. Bican in [B] showed that in the class of finite rank torsion-free abelian groups, such a group B is a Butler group if and only if (*) whenever $0 \to T \to G \to B \to 0$ is balanced exact and T is a torsion abelian group, then the sequence splits. Bican and Salce [BS] use (*) to define the class of Butler groups of arbitrary rank. There have been many authors investigating this class and more tractable subclasses. An interesting result in the countable rank case is that these are the groups in which all their pure subgroups of finite rank are in the class defined by Butler. This class is called the class of *finitely Butler groups*. Bican and Salce [BS] gave an example to show that there is an uncountable finitely Butler group that does not satisfy (*), while Arnold [A] has constructed an example of a countable finitely Butler group that is not a pure subgroup of a completely decomposable abelian group. Recently, Fuchs and Metelli [FM] have shown that in the countable case that the finitely Butler groups are also defined by the property that all their pure finite rank subgroups are prebalanced in the sense of Fuchs and Viljoen [FV]. Rangaswamy and the author [GR] have dualized this property which they call *precobalanced*. Thus it is natural to investigate torsion-free abelian groups in which all their pure finite rank subgroups are precobalanced. We let $\mathfrak{R}^*$ be the collection of such groups. We are able to derive the following:

Theorem 2: For a torsion-free group A, the following are equivalent:

    (1) $A \in \mathfrak{R}^*$;

    (2)   Every pure rank one subgroup of A is precobalanced;

    (3)   Every pure finite rank subgroup of A is a Butler group and is precobalanced.

and

Theorem 3: The class $\mathfrak{R}^*$ is closed with respect to pure subgroups and direct sums.

Thus we see that pure subgroups of completely decomposable groups are contained in $\mathfrak{R}^*$ and that $\mathfrak{R}^*$ is contained in the class of finitely Butler groups. For finite rank groups, the three classes agree. In the uncountable case the three classes are distinct, while for countable torsion-free groups there are finitely Butler groups that are not in $\mathfrak{R}^*$, but we have:

Theorem 4: A countable torsion-free group is in $\mathfrak{R}^*$ if and only if it is a pure subgroup of a completely decomposable group.

Throughout, we will use the term *group* to mean abelian group, $\mathbb{Q}$ will denote the group of rationals, and $\mathbb{Z}$ the group of integers. Let $\mathfrak{R}$ be the class of finite rank Butler groups, that is, pure subgroups of finite rank completely decomposable groups (equivalently, torsion–free homomorphic images of finite rank completely decomposable groups [Bu]), and C the class of finite rank completely decomposable groups. We say a subgroup K of a group B is a *corank–1* subgroup if B/K is isomorphic to a subgroup of $\mathbb{Q}$ (i. e., B/K is torsion–free rank–1). For an element b of B, we let $|b|_B$ denote the height sequence of b in B. For a prime p, we let $|b|_B^p$ denote the pth–component of $|b|_B$. For height sequences $h_1, \cdots, h_n$ indexed by the primes, we let $\bigwedge_{i=1}^n h_i$ be the sequence with p^{th}–component equal to $\inf\{h_1^p, \cdots, h_n^p\}$. The basic properties of height sequences can be found in Fuchs [F].

The following definition was introduced in [GR] and was shown to dualize the notion of prebalanced subgroups as defined in [FV].

Definition 1: A subgroup A of a group B is said to be *precobalanced* in B if for any corank–1 subgroup K of A there are corank–1 subgroups $K_1, \cdots, K_n$ of B satisfying

(i) $K = \bigcap_{i=1}^{n} (K_i \cap A)$

and

(ii) $|a + K|_{A/K} = \bigwedge_{i=1}^{n} |a + K_i|_{B/K_i}$ for each $a \in A$.

Since $A/(K_i \cap A) \cong (A + K_i)/K_i$ is a subgroup of B/K_i , either $K_i \cap A = A$ or it is corank–1 in A . So, if it is not A , then $K_i \cap A = K$. Thus (i) can be replaced by

(i′) $K = K_i \cap A$ for $1 \leq i \leq n$.

We say that the exact sequence $0 \to A \xrightarrow{\alpha} B \xrightarrow{\beta} C \to 0$ is *precobalanced* if the image of α is a precobalanced subgroup of B . Such sequences have a nice property with respect to maps from A to groups in $\mathfrak{R}$.

Definition 2. Let $\mathcal{I}$ be a nonempty class of groups. For a map $f : A \to G$ and an exact sequence $0 \to A \xrightarrow{\alpha} B \xrightarrow{\beta} C \to 0$, we call the triple $(\alpha′, H, f′)$ an $\mathcal{I}$*–semi–extension* of f with respect to α if

(a)   the diagram

$$\begin{array}{ccc} A & \xrightarrow{\alpha} & B \\ {\scriptstyle f}\downarrow & & \downarrow{\scriptstyle f'} \\ G & \xrightarrow{\alpha'} & H \end{array}$$

        commutes,

(b) $H \in \mathcal{I}$, and

(c) $\alpha′$ is a pure monomorphism.

We list two equivalent conditions for precobalanced exactness. The proofs are to be found in [GR] .

Theorem 1: For an exact sequence of torsion-free groups $E : 0 \to A \xrightarrow{\alpha} B \longrightarrow C \to 0$ the following are equivalent:

1) E is precobalanced;

2) every map $f : A \to X$ with $X \in \mathfrak{R}$ has a C-semi-extension with respect to α ;

3) every map $f : A \to X$ with $X \in \mathfrak{R}$ has a $\mathfrak{R}$-semi-extension with respect to α .

If $B \in \mathfrak{R}$, then $A \in \mathfrak{R}$. Then for any $f:A \to X$ with $X \in \mathfrak{R}$, the pushout of f and α will give an $\mathfrak{R}$-semi-extension. Thus we have:

Corollary 1: Every pure subgroup of a finite rank Butler group is precobalanced.

As we shall see later, the property that every pure subgroup is precobalanced is equivalent to the group being in $\mathfrak{R}$ for finite rank groups.

The next lemma is useful later.

Lemma: For an exact sequence of torsion-free groups $E: 0 \to A \xrightarrow{\alpha} B \longrightarrow C \to 0$ with $A \in \mathfrak{R}$, E is precobalanced if and only if 1_A has a C-semi-extension with respect to α.

Proof:($\Rightarrow$) Clear from the theorem.

($\Leftarrow$)   Let (α', H, f') be a C-semi-extension of 1_A with respect to α. Then Corollary 1 implies that the image of A in H is precobalanced. Thus for any $f:A \to X$ with $X \in \mathfrak{R}$, there is a C-semi-extension (β, G, g) of f with respect to α'. Then $(\beta\alpha', G, gf')$ is readily shown to be a C-semi-extension of f with respect to α. □

An interesting phenomenon is illustrated in the following.

Example 1: For every $n > 1$, there is a group of rank n such that every pure subgroup with rank greater than one is precobalanced, while every pure rank 1 subgroup is not precobalanced.

Proof: Let p be a prime and B be a pure subgroup of the p-adic integers of rank n. Then if A is any pure subgroup of B with rank greater than one, every rank one image of A is isomorphic to $\mathbb{Q}$. Thus A is cobalanced and hence precobalanced in B. If A is a pure rank one subgroup of B, then A is isomorphic to $\mathbb{Z}_p$, the integers localized at p. Let α be the inclusion map of A into B. Let $(\alpha', \bigoplus_{i=1}^{n} H_i, f)$ be such that α' is a monomorphism of A into $\bigoplus_{i=1}^{n} H_i$, each H_i rank one, and $f:B \to \bigoplus_{i=1}^{n} H_i$ such that $\alpha f = \alpha'$. Without loss of generality, we can assume that for each i the composition of f and the canonical projection of the sum onto H_i is not zero. Since these are rank one images of B, they are isomorphic to $\mathbb{Q}$. Thus $\bigoplus_{i=1}^{n} H_i$ is divisible and so α' is not a pure monomorphism. Thus 1_A does not have a C-semi-extension with respect to α, and so A is not precobalanced in B. □

§2: The class $\mathfrak{R}^*$. In light of the Example 1 and Corollary 1 , we let $\mathfrak{R}^*$ be the collection of torsion-free groups such that every pure finite rank subgroup is precobalanced.

Theorem 2: For a torsion-free group A , the following are equivalent:

(1) $A \in \mathfrak{R}^*$;

(2)   Every pure rank one subgroup of A is precobalanced;

(3)   Every pure finite rank subgroup of A is a Butler group and is precobalanced.

Proof: (1) implies (2) and (3) implies (1) are evident.

For (2) implies (3), we proceed by induction on the rank of the pure subgroups. Rank one is taken care of in the hypothesis. Let $n > 1$ and A be a group such that all pure subgroups of rank less than n are precobalanced and are Butler groups, and let B be a pure subgroup of A of rank n . Let K be a corank one subgroup of B . Then K is pure in B (and so in A) of rank less than n . Thus K is precobalanced in A which implies it is precobalanced in B . That is, the sequence $0 \to K \longrightarrow B \longrightarrow B/K \to 0$ is precobalanced with K and B/K in $\mathfrak{R}$. Theorem 5 in [GR] implies that $B \in \mathfrak{R}$.

By Theorem 1.5 of [AV] , there are pure rank one subgroups $B_1 , \cdots, B_n$ of B and a balanced epimorphism $\theta : \bigoplus_{i=1}^{n} B_i \to B$. Let $\beta_i : B_i \to B$, $\alpha : B \to A$, and $\theta_i : B_i \to \bigoplus_{j=1}^{n} B_j$ be the inclusion maps. (Thus $\beta_i = \theta\theta_i$.) Because each B_i is also a pure subgroup of A and so precobalanced in A , there is for each i a C-semi-extension $(\gamma_i , H_i , \beta_i')$ of β_i with respect to $\alpha\beta_i$. That is, the diagrams

$$
\begin{array}{ccc}
B_i & \xrightarrow{\ \alpha\beta_i\ } & A \\[2pt]
\beta_i \downarrow & & \downarrow \beta_i' \\[2pt]
B & \xrightarrow{\ \gamma_i\ } & H_i
\end{array}
$$

commute with the γ_i's pure monomorphisms. The γ_i's and β_i' 's induce maps γ and β' from B and A , respectively, into $\bigoplus_{j=1}^{n} H_j = H$ satisfying $\pi_i\gamma = \gamma_i$ and $\pi_i\beta' = \beta_i'$ where π_i is the projection map of H onto H_i for each i . It is straightforward to show that γ is a pure monomorphism. For each i , we have that

$$\pi_i\gamma\theta\theta_i = \gamma_i\beta_i = \beta_i'\alpha\beta_i = \pi_i\beta' \alpha\theta\theta_i$$

The universal mapping property of coproducts implies that $\gamma\theta = \beta'\alpha\theta$. Since θ is an epimorphism, $\gamma = \beta'\alpha$. Therefore, (γ , H , β') is a C-semi-extension of 1_B with respect to α. The lemma implies that B is precobalanced in A. $\square$

The following is immediate.

Corollary 2: Every group in $\mathfrak{R}^*$ is a finitely Butler group. In particular, in the finite rank case, $\mathfrak{R}^*$ and $\mathfrak{R}$ coincide.

It is evident that if a subgroup is precobalanced in a summand of a group, it is precobalanced in the group. The group A is said to be (*Baer*) *separable* if every pure rank one subgroup is a subgroup of a finite rank completely decomposable summand of A. (This is actually equivalent to the definition. c.f., [F].) Thus separable groups are in $\mathfrak{R}^*$.

If $0 \to K \to F \to \mathbb{Q} \to 0$ is a free resolution of the rationals, then K, F, and $\mathbb{Q}$ are all in $\mathfrak{R}^*$. We will show later in that K is not precobalanced in F. (In fact, it is also not prebalanced in F.) Thus finite rank is necessary in the theorem.

We conclude this section with some closure properties of the class $\mathfrak{R}^*$.

Theorem 3: The class $\mathfrak{R}^*$ is closed with respect to pure subgroups and direct sums.

Proof: Let $A \in \mathfrak{R}^*$ and B be a pure subgroup of A. If X is a pure rank one subgroup of B, then it is also pure in A. Clearly any C-semi-extension of 1_X with respect to the inclusion of X into A can be factored through B to give a C-semi-extension of 1_X with respect to the inclusion of X into B.

Let $\{A_i : i \in I\} \subset \mathfrak{R}^*$ and X be a pure rank one subgroup of $\bigoplus \{A_i : i \in I\} = A$. Let π_i be the projection of A onto A_i and $J = \{i \in I : \pi_i(X) \neq 0\}$. Then J is a finite set and X is pure in $\bigoplus \{A_i : i \in J\} = A_J$. For each $i \in J$, let X_i be the pure rank one subgroup of A_i generated by $\pi_i(X)$. By Lemma 1, there are C-semi-extensions $(\alpha_i, H_i, \gamma_i)$ of 1_{X_i} with respect to the inclusion into A_i. These will induce a C-semi-extension $(\alpha, \bigoplus \{H_i : i \in J\}, \gamma)$ of the identity of $\bigoplus \{X_i : i \in J\} = X_J$ with respect to the inclusion of X_J into A_J.

The inclusion, θ, of X into A_J factors through the obvious inclusion, β, of X into X_J. Thus X is pure in X_J. This implies that $(\alpha\beta, \bigoplus \{H_i : i \in J\}, \gamma)$ is a C-semi-extension of 1_X with the respect to θ. Hence X is precobalanced in A_J. Since A_J is a summand of A, X is precobalanced in A. $\square$

The comments preceding the theorem imply:

Corollary 3: $\mathfrak{R}^*$ contains the class of pure subgroups of separable groups and hence the class of pure subgroups of completely decomposable groups.

§3: Countable groups in $\mathfrak{R}^*$. We start this section with an example due to Arnold of a countable finitely Butler group that is not a pure subgroup of a separable group.

Example 2: Let $\mathcal{W}$ be the collection of finite words on the alphabet $\{0,1\}$ (we let -1 represent the empty word) and $\mathcal{W}_n$ be the subset of words of length n. We let $X_{-1} = \mathbb{Z}$ and for each $w \in \mathcal{W}$ choose $X_w \subseteq \mathbb{Q}$ satisfying $X_w = X_{w0} \cap X_{w1}$ and if τ is a type such that $\text{type}(X_w) \le \tau$ for all $w \in \mathcal{W}$, then $\tau = \text{type}(\mathbb{Q})$. (The method in which this can be accomplished is detailed in [A].) For each n, let $G_n = \bigoplus \{X_w : w \in \mathcal{W}_n\}$ and $\beta_n : G_n \to G_{n+1}$ be the map induced by the diagonal maps from X_w to $X_{w0} \oplus X_{w1}$ for $w \in \mathcal{W}_n$. Let G be the direct limit of $\{G_n, \beta_n : n = 0, 1, \cdots\}$ and ψ_n the maps of G_n into G. Then G is a countable group and each ψ_n is a pure monomorphism. It can be shown that any pure finite rank subgroup is contained in $\text{Im}(\psi_n)$ for some n and so is in $\mathfrak{R}$. Thus G is a finitely Butler group.

Let B be a pure finite rank subgroup of G and let n_0 be the smallest index such that $B \subseteq \text{Im}(\psi_{n_0})$. Suppose $\alpha : B \to \bigoplus_{i=1}^n H_i$ and $\gamma : G \to \bigoplus_{i=1}^n H_i$ are such that γ restricted to B is α, $\pi_i \alpha \ne 0$ for any i (where π_i is the projection map onto H_i), and each H_i is a rank one group. Then for each i and $n \ge n_0$, the map $\pi_i \gamma \psi_n$ is not zero. This implies $\text{type}(X_w) \le \text{type}(H_i)$ for all $w \in \mathcal{W}$. Thus each H_i is divisible and so α cannot be a pure monomorphism. Hence B is not precobalanced in G. Therefore, $G \notin \mathfrak{R}^*$. $\square$

It is interesting to note that G in the example satisfies $(*)$, while no pure finite rank subgroup is precobalanced in G. Thus even in the countable case the class $\mathfrak{R}^*$ is properly contained in the class of finitely Butler groups. Since there are uncountable separable groups that are not pure subgroups of completely decomposable groups, the class $\mathfrak{R}^*$ will properly contain the class of pure subgroups of completely decomposable groups. The countable case is more satisfying.

Theorem 4: For any countable torsion-free group A, $A \in \mathfrak{R}^*$ if and only if A is a pure subgroup of a completely decomposable group.

Proof:($\Leftarrow$) This is a special case of Corollary 3.

($\Rightarrow$) Let A be a countable group in $\mathfrak{R}^*$. Because A is countable there is a chain of pure finite rank subgroups $A_1 \subset A_2 \subset \cdots \subset A_n \subset \cdots$ such that $\bigcup_{n=1}^{\infty} A_n = A$. By Theorem 1, each A_n is in $\mathfrak{R}$ and is precobalanced in A as well as in A_{n+1}. Let $\alpha_n : A_n \to A$ and $\beta_n : A_n \to A_{n+1}$ be the inclusion maps. Thus $\alpha_{n+1}\beta_n = \alpha_n$.

Since A_1 is precobalanced in A, there is an $H_1 \in C$ and maps $\rho_1 : A_1 \to H_1$ and $\theta_1 : A \to H_1$ such that ρ_1 is a pure monomorphism and $\theta_1 \alpha_1 = \rho_1$.

Consider the pushout diagram

$$
\begin{array}{ccc}
A_1 & \xrightarrow{\beta_1} & A_2 \\
{\scriptstyle \rho_1}\downarrow & {\scriptstyle \beta_1'} & \downarrow{\scriptstyle \rho_1'} \\
H_1 & \xrightarrow{\beta_1'} & H_1' \, .
\end{array}
$$

Then β_1' and ρ_1' are pure monomorphisms and $H_1' \in \mathfrak{R}$. Also, $\theta_1 \alpha_2 : A_2 \to H_1$ satisfies $\theta_1 \alpha_2 \beta_1 = \theta_1 \alpha_1 = \rho_1$. Thus there is a map $s_1' : H_1' \to H_1$ such that $s_1' \beta_1' = 1_{H_1}$ and $s_1' \rho_1' = \theta_1 \alpha_2$. This implies that $H_1' = H_1 \oplus K_1$ where $K_1 \cong H_1'/H_1 \in \mathfrak{R}$. Let s_1'' be the projection of H_1' onto K_1. By Theorem 1, there is a $G_1 \in C$, a pure monomorphism $\gamma_1 : K_1 \to G_1$, and a map $\lambda_1 : A \to G_1$ such that $\gamma_1 s_1'' \rho_1' = \alpha_2 \lambda_1$.

Let $H_2 = H_1 \oplus G_1$. Then the map $\delta_1 = (1_{H_1}, \gamma_1)$ is a pure imbedding of H_1' into H_2. This implies that $\rho_2 = \delta_1 \rho_1'$ is a pure imbedding of A_2 into H_2. Then $\rho_2 \beta_1 = \delta_1 \rho_1' \beta_1 = \delta_1 \beta_1' \rho_1$. Letting $s_1 : H_2 \to H_1$ and $\pi_1 : H_2 \to G_1$ be the canonical projections, there is an unique $\theta_2 : A \to H_2$ such that $\pi_1 \theta_2 = \lambda_1$ and $s_1 \theta_2 = \theta_1$. Thus $\pi_1(\theta_2 \alpha_2) = \lambda_1 \alpha_2 = \gamma_1 s_1'' \rho_1' = \pi_1 \delta_1 \rho_1' = \pi_1 \rho_2$ and $s_1(\theta_2 \alpha_2) = \theta_1 \alpha_2 = s_1' \rho_1' = s_1 \delta_1 \rho_1' = s_1 \rho_2$. Hence $\theta_2 \alpha_2 = \rho_2$. Also $\theta_2 \alpha_1 = \theta_2 \alpha_2 \beta_1 = \rho_2 \beta_1 = \delta_1 \beta_1' \rho_1$.

We then repeat this procedure so that for each $n \geq 2$ there is a $G_{n-1} \in C$ and maps $\rho_n : A_n \to H_n = H_1 \oplus G_1 \oplus \cdots \oplus G_{n-1}$, $\theta_n : A \to H_n$, and $w_n : H_{n-1} \to H_n$ where ρ_n is a pure monomorphism, w_n is the canonical inclusion, $\rho_n \alpha_n = \theta_n$, and $\rho_{n+1}\beta_n = w_n \rho_n$. Let $H = H_1 \oplus \bigoplus_{n=1}^{\infty} G_n$ and μ_n be the canonical imbedding of H_n into H. Then $\mu_n \rho_n$ is a pure imbedding of A_n into H that satisfies $\mu_{n+1}\rho_{n+1}\beta_n = \mu_n \rho_n$. Thus there is a map $\rho : A \to H$ such that $\rho \alpha_n = \mu_n \rho_n$. This map is easily shown to be a pure monomorphism. $\square$

It may be the case that $\mathfrak{R}^*$ is precisely the class of pure subgroups of separable groups. A major stumbling block to a proof of this is illustrated in the next example.

Example 3: There is a countable group $F \in \mathfrak{R}^*$ with a corank one subgroup K not precobalanced in F. Thus F has a family of precobalanced finite rank subgroups

$K_1 \subset K_2 \subset \cdots \subset K_n \subset \cdots$, such that their union is not precobalanced in F .

Proof: Let $E: 0 \to K \xrightarrow{\alpha} F \longrightarrow Q \to 0$ be a free resolution of the rationals with F countable. Let B be a rank two group homogeneous of type($\mathbb{Z}$) such that all of its rank one images are isomorphic to Q . (See [VW] for details on the construction of such groups.) Then there is an exact sequence $E': 0 \to \mathbb{Z} \xrightarrow{f} B \longrightarrow Q \to 0$. As in Example 1, the sequence is easily shown not to be precobalanced.

Applying $\mathrm{Hom}(\ ,\mathbb{Z})$ to E , we have that the induced sequence $\mathrm{Hom}(K,\mathbb{Z}) \longrightarrow \mathrm{Ext}(Q,\mathbb{Z}) \longrightarrow \mathrm{Ext}(F,\mathbb{Z}) = 0$ is exact. This implies that there is an $g \in \mathrm{Hom}(K,\mathbb{Z})$ such that B can be viewed as the pushout of g and α with f as the map from $\mathbb{Z}$ to B in the pushout diagram. Thus for (α', H, g') such that $H \in C$ and the diagram

$$
\begin{array}{ccc}
K & \xrightarrow{\alpha} & F \\
{\scriptstyle g}\big\downarrow & & \big\downarrow{\scriptstyle g'} \\
\mathbb{Z} & \xrightarrow{\alpha'} & H
\end{array}
$$

commutes, there is an $h: B \longrightarrow H$ such that, in particular, $hf = \alpha'$. The map α' can not be a pure monomorphism since this would imply that E' is precobalanced. Therefore, g has no C-semi-extension and thus K is not precobalanced in F .

We can write K as the union of a smooth chain of pure finite rank completely decomposable groups $K_1 \subset K_2 \subset \cdots \subset K_n \subset \cdots$. Each K_n is also pure in F and so precobalanced in F . $\square$

The author would like to thank Professor Rangaswamy, whose helpful suggestions lead to the proof of Theorem 4 .

References

[A] Arnold, D., *Notes on Butler groups and balanced extensions*, Boll. Unione Mat. Ital., A 5 (1986), pp. 175 − 184 .

[AV] _____ , and C., Vinsonhaler, *Pure subgroups of finite rank completely decomposable groups II* , **Abelian Group Theory Proceedings**, Honolulu 1982/83, Lecture Notes 1006, Springer−Verlag, New York, pp. 97 − 143 .

[B] Bican, L., *Splitting in Abelian groups*, **Czech. Math. J.**, 28 (1978), pp. 356 − 364 .

[BS] _____ , and L., Salce, *Butler Groups of infinite rank*, Abelian Group Theory Proc. , Honolulu 1982/83, Lecture Notes 1006, Springer−Verlag, New York, pp. 171 − 189 .

[Bu] Butler, M. C. R., *A class of torsion—free abelian groups*, Proc. London Math. Soc., 15 (1965), pp. 680—698 .

[F]   Fuchs, L., Infinite Abelian Groups, vol. 2, Academic Press, San Francisco, 1970 .

[FM] _____ , and C. Metelli, *Countable Butler groups*, preprint.

[FV] Fuchs, L. and G. Viljoen, *Note on the extension of Butler groups*, Bull. Aust. Math. Soc., 41 (1990), pp. 117 — 122 .

[GR] Giovannitti, A. and K. M. Rangaswamy, *Precobalanced subgroups of Abelian Groups*, Comm. in Alg., 19 (1) (1991), pp. 249 — 269 .

[VW] Vinsonhaler, V. and W. Wickless, *G—Injective groups*, Comm. in Alg., 16(4) (1988), pp. 743—753 .

Quasi-Summands of a Certain Class of
Butler Groups

H. PAT GOETERS AND WILLIAM ULLERY

Department of Mathematics
Auburn University, Alabama, U.S.A.

1. Preliminaries. Recall that a Butler group is a torsion-free homomorphic image of a finite rank completely decomposable group. If C is a finite rank completely decomposable group with A a rank-1 pure subgroup, we follow the lead of the L. Fuchs and C. Metelli [FM] and call the Butler group C/A a $\mathcal{B}^{(1)}$-*group*. In this paper we answer a question raised in [FM]; namely, we show that any quasi-summand of a $\mathcal{B}^{(1)}$-group is quasi-isomorphic to a $\mathcal{B}^{(1)}$-group.

Let n be a positive integer and set $\bar{n} = \{1, \ldots, n\}$. For each $i \in \bar{n}$, let A_i be a nonzero subgroup of the additive group of rationals Q. If $G = Coker\{\Delta : \bigcap_{i \in \bar{n}} A_i \to \oplus_{i \in \bar{n}} A_i\}$, where Δ is the diagonal map, we write $G = G[A_1, \ldots, A_n]$ and call G an *AV-group*. A finite rank torsion-free group is a *qAV-group* if it is quasi-isomorphic to an *AV*-group. It is not difficult to see that if B is a $\mathcal{B}^{(1)}$-group, then $B \sim G \oplus C$ where G is a *qAV*-group, C is completely decomposable and $\sim$ indicates quasi-isomorphism. For this reason, we concentrate most of our efforts on a study of quasi-decompositions of *qAV*-groups. In fact, we will show that any quasi-summand of *qAV*-group is again a *qAV*-group (Theorem 2).

At this point, it is convenient to introduce some notation and terminology which will remain in force throughout. First and foremost, all groups considered are torsion-free abelian groups of finite rank. For each $i \in \bar{n}$, A_i is a nonzero subgroup of Q of type τ_i. If I is a nonempty subset of $\bar{n}$, say $I = \{i_1, \ldots, i_k\}$, we write $\tau_I = \tau_{i_1} \wedge \cdots \wedge \tau_{i_k}$ and $G[A_I]$ for the *AV*-group $G[A_{i_1}, \ldots, A_{i_k}]$. If $B_i \cong A_i$ for all i, it is easily seen that $G[A_1, \ldots, A_n] \sim G[B_1, \ldots, B_n]$. With this in mind, we sometimes write $[\tau_1, \ldots, \tau_n]$ for $G[A_1, \ldots, A_n]$. Since we shall only be concerned with "quasi" results, there is no ambiguity in the abbreviated notation.

It is well known that each strongly indecomposable quasi-summand of a qAV-group is again a qAV-group. The difficulty in considering more general quasi-summands is that the class of qAV-groups is not closed under the formation of direct sums, as observed in [FM]. We shall discuss this phenomenon in more detail in the next section. Nevertheless, it will soon be apparent that some way of combining direct sums of qAV-groups is needed. We will find that our first result is well-suited for this purpose.

Lemma 1. *Suppose $n \geq 2$, $1 \leq s < r \leq n$ and $\tau_s \vee \tau_r = \tau_I \vee \tau_J$, where $I = \{1, \ldots, s\}$ and $J = \{r, \ldots, n\}$. Then, $G[A_I] \oplus G[A_J] \sim [\tau_1, \ldots, \tau_{s-1}, \tau_s \wedge \tau_r, \tau_{r+1}, \ldots, \tau_n]$.*

Proof: If $a_i \in A_i$ for each $i \in \bar{n}$, write $\langle a_1, \ldots, a_n \rangle$ for the natural image of $(a_1, \ldots, a_n) \in A_1 \oplus \cdots \oplus A_n$ in $G[A_1, \ldots, A_n]$. Define a mapping $f : G[A_1, \ldots, A_{s-1}, A_s \cap A_r, A_{r+1}, \ldots, A_r] \to G[A_I] \oplus G[A_J]$ by $f(\langle a_1, \ldots, a_{s-1}, a, a_{r+1}, \ldots, a_n \rangle) = (\langle a_1, \ldots, a_{s-1}, a \rangle, \langle a, a_{r+1}, \ldots, a_n \rangle)$. It is easily verified that f is a well-defined monomorphism. Moreover, the condition $\tau_s \vee \tau_r = \tau_I \vee \tau_J$ implies the existence of a positive integer m with $m(A_s + A_r) \subseteq (\bigcap_{i \in I} A_i) + (\bigcap_{j \in J} A_j)$. Note that $m(G[A_I] \oplus G[A_J]) \subseteq Imf$, thereby showing that f is a quasi-isomorphism. $\square$

A number of authors have stated conditions under which a qAV-group is not strongly indecomposable. Among them are D. Arnold and C. Vinsonhaler [AV1, AV2], L. Fuchs and C. Metelli [FM], P. Hill and C. Megibben [HM], and W. Y. Lee [L]. The version stated below is proved by the present authors in [GU]. Therefore, we omit the proof.

Lemma 2. *Suppose $n \geq 3$ and $G \sim [\tau_1, \ldots, \tau_n]$. Then G is not strongly indecomposable if and only if there exist $r \in \bar{n}$ and subsets I and J of $\bar{n}$ such that $I \bigcup J = \bar{n}$, $I \bigcap J = \{r\}$, $\min\{|I|, |J|\} \geq 2$ and $\tau_r = \tau_I \vee \tau_J$. In this case, $G \sim G[A_I] \oplus G[A_J]$.* $\square$

We remark that duals of Lemmas 1 and 2 are true in the corank 1 case. These dual results, for example, have been utilized in studying the quasi-representing graphs of Arnold and Vinsonhaler.

2. Strongly Indecomposable Quasi-Summands.
In this section we present several results concerning strongly indecomposable quasi-summands of qAV-groups.

Proposition 1. *Suppose G is a qAV-group and X_1 and X_2 are distinct strongly indecomposable quasi-summands of G. Then, $X_1 \oplus X_2$ is a qAV-group.*

Proof: Suppose $G \sim [\tau_1, \ldots, \tau_n]$. Then, $n \geq 3$ and of course G is not strongly indecomposable. We induct on n. If $n = 3$, then $G \sim X_1 \oplus X_2$ and the result is verified. So assume $n \geq 4$. By Lemma 2, there exist $r \in \overline{n}$ and subsets I and J of $\overline{n}$ satisfying the conditions there set forth, such that $G \sim G[A_I] \oplus G[A_J]$. If $X_1 \oplus X_2$ is a quasi-summand of a single one of $G[A_I]$ or $G[A_J]$, the result follows by induction. Therefore, we may assume X_1 is a quasi-summand of $G[A_I]$ and X_2 is a quasi-summand of $G[A_J]$. If both $X_1 \sim G[A_I]$ and $X_2 \sim G[A_J]$, the result follows. Thus, we may assume $G[A_I]$ is not strongly indecomposable. By Lemma 2, there exists $s \in I$ and subsets I_1 and J_1 of I such that $I_1 \cup J_1 = I$, $I_1 \cap J_1 = \{s\}$, $\min\{|I_1|, |J_1|\} \geq 2$ and $\tau_s = \tau_{I_1} \vee \tau_{J_1}$. If $s = r$, it is a routine application of Lemma 2 to see that $G[A_{I_1}] \oplus G[A_J]$ and $G[A_{J_1}] \oplus G[A_J]$ are qAV-groups. Since $X_1 \oplus X_2$ is a quasi-summand of one of these groups, the result follows by induction.

So, we may assume that $s \neq r$. After reindexing if necessary, we may assume $1 < s < r$, $I_1 = \{1, \ldots, s\}$, $J_1 = \{s, \ldots, r\}$ and $J = \{r, \ldots, n\}$ with $\tau_s = \tau_{I_1} \vee \tau_{J_1}$ and $\tau_r = \tau_{I_1 \cup J_1} \vee \tau_J$. Moreover, $G \sim G[A_{I_1}] \oplus G[A_{J_1}] \oplus G[A_J]$. If X_1 is a quasi-summand of $G[A_{J_1}]$, then $\tau_r = \tau_{J_1} \vee \tau_J$ and Lemma 2 imply that $X_1 \oplus X_2$ is a quasi-summand of the qAV-group $G[A_{J_1}] \oplus G[A_J]$. In this case, the result follows by induction. If X_1 is a quasi-summand of $G[A_{I_1}]$, then $\tau_s \vee \tau_r = \tau_{I_1} \vee \tau_{J_1} \vee \tau_I \vee \tau_J = \tau_{I_1} \vee (\tau_s \wedge \cdots \wedge \tau_r) \vee \tau_J \tau_{I_1} \vee (\tau_s \wedge \cdots \wedge \tau_n) \vee (\tau_1 \wedge \ldots \tau_r) \vee \tau_J = \tau_{I_1} \vee \tau_J$. Therefore, by Lemma 1, $G[A_{I_1}] \oplus G[A_J]$ is a qAV-group containing $X_1 \oplus X_2$ as a quasi-summand. A final application of the induction hypothesis yields the result. $\square$

We now turn our attention to the example of Fuchs and Metelli mentioned in the first section.

Example. [FM]. *Let $p_1, \ldots, p_6$ be distinct prime numbers in the additive group of integers Z, and set $A_i = Z[1/p_i]$. Then, $G[A_1, A_2, A_3] \oplus G[A_4, A_5, A_6]$ is not a qAV-group.*

Note in the above that each summand is strongly indecomposable and has inner type equal to type (Z). Thus, our next result generalizes the example.

Proposition 2. *Suppose X_1 and X_2 are strongly indecomposable groups with $IT(X_1) \leq IT(X_2)$. If $X_1 \oplus X_2$ is a qAV-group, then rank $X_2 = 1$.*

Proof: Suppose $X_1 \oplus X_2 \sim [\tau_1 \ldots, \tau_n]$. Without loss we may assume that the τ_i's are cotrimmed (i.e., $\tau_i \geq \bigwedge_{j \neq i} \tau_j$ for all $i \in \overline{n}$). By Lemma 2, there exist $r \in \overline{n}$ and subsets I and J of $\overline{n}$ such that $I \cup J = \overline{n}$, $I \cap J = \{r\}$, $\min\{|I|, |J|\} \geq 2$ and $\tau_r = \tau_I \vee \tau_J$. Thus, $[\tau_1, \ldots, \tau_n] \sim G[A_I] \oplus G[A_J]$ and we may assume $X_1 \sim G[A_I]$ and $X_2 \sim G[A_J]$. Then,

$IT(X_1) = \tau_{I-\{r\}}$ and $IT(X_2) = \tau_{J-\{r\}}$. Therefore, $\tau_r \leq \tau_{I-\{r\}} \vee \tau_{J-\{r\}} = \tau_{J-\{r\}}$ and we conclude that X_2 is almost completely decomposable.   □

Applying Propositions 1 and 2, we obtain

Theorem 1. *Suppose G is a qAV-group and $G \sim X_1 \oplus \cdots \oplus X_k$ with X_i strongly indecomposable for each $i \in \bar{k}$. If $IT(X_i) \leq IT(X_j)$ for distinct $i, j \in \bar{k}$, then rank $X_j = 1$. In particular, if G has no rank-1 quasi-summands, the inner types $IT(X_1), \ldots, IT(X_k)$ are pairwise incomparable.*   □

Even though Theorem 1 may be of some interest in its own right, we shall only make use of the following consequence.

Corollary 1. *Suppose X_1 and X_2 are distinct strongly indecomposable quasi-summands of a qAV-group. If rank $X_1 \geq 2$, then X_1 and X_2 are not quasi-isomorphic.*   □

Our final result of this section appears in a somewhat different form in [AV3].

Lemma 3. [AV3]. *Suppose $n \geq 3$ and $G \sim [\tau_1, \ldots, \tau_n]$ where $\tau_i \geq \bigwedge_{j \neq i} \tau_j$ for all $i \in \bar{n}$. If X is a rank-1 quasi-summand of G, then there exist types $\sigma_1, \ldots, \sigma_{n-1}$ with the following properties.*

(a) $G \sim X \oplus [\sigma_1, \ldots, \sigma_{n-1}]$.

(b) $\sigma_1 \wedge \cdots \wedge \sigma_{n-1} \geq \tau_1 \wedge \cdots \wedge \tau_n$.

(c) *For each $i \in \bar{n}$, there exists $j \in \overline{n-1}$ such that $\sigma_j \leq \tau_i$.*

Proof: Observe that $n \geq 3$ and induct on n. If $n = 3$, an application of Lemma 2 yields the result. So assume $n \geq 4$. After reindexing if necessary, we may apply Lemma 2 and write $G \sim [\tau_1, \ldots, \tau_r] \oplus [\tau_r, \ldots, \tau_n]$, where $1 < r < n$, $\tau_r = (\tau_1 \wedge \cdots \wedge \tau_r) \vee (\tau_r \wedge \cdots \wedge \tau_n)$, and X is a quasi-summand of $[\tau_1, \ldots, \tau_r]$. We consider two cases.

Case 1. $X \sim [\tau_1, \ldots, \tau_r]$. Then $r = 2$ and $G \sim [\tau_1, \tau_2] \oplus [\tau_2, \ldots, \tau_n]$. Set $\sigma_i = \tau_{i+1}$, $1 \leq i < n - 1$. Clearly conditions (a) and (b) hold for these choices of the σ_i's. Moreover, $\sigma_1 = \tau_2 = (\tau_1 \wedge \tau_2) \vee (\tau_2 \wedge \cdots \wedge \tau_n) \leq \tau_1 \vee (\tau_2 \wedge \cdots \wedge \tau_n) = \tau_1$. It now follows that condition (c) also holds.

Case 2. rank $[\tau_1, \ldots, \tau_r] > 1$. In this case, set $\tau_r' = \tau_r \vee (\wedge\{\tau_i : 1 \leq i \leq r - 1\})$. Then, $[\tau_1, \ldots, \tau_r] \sim [\tau_1, \ldots, \tau_{r-1}, \tau_r']$, because for $1 \leq i \leq r - 1$, $\tau_i \geq (\wedge\{\tau_j : j \in \bar{r} - \{i, r\}\}) \wedge \tau_r$. Thus, the induction hypothesis applies and there exist types $\gamma_1, \ldots, \gamma_{r-1}$ such that (a') $[\tau_1, \ldots, \tau_{r-1}, \tau_r'] \sim X \oplus [\gamma_1, \ldots, \gamma_{r-1}]$

(b') $\gamma_1 \wedge \cdots \wedge \gamma_{r-1} \geq \tau_1 \wedge \cdots \wedge \tau_{r-1} \wedge \tau_r'$

(c') $\gamma_{r-1} \leq \tau_r'$, and for each $i \in \overline{r-1}$, there exists $j \in \overline{r-1}$ such that $\gamma_j \leq \tau_i$.

Then, $\gamma_{r-1} \vee \tau_r \leq \gamma_{r-1} \vee \tau_r' = \tau_r' = \tau_r \vee (\tau_1 \wedge \cdots \wedge \tau_{r-1}) = (\tau_1 \wedge \cdots \wedge \tau_r) \vee (\tau_r \wedge \cdots \wedge \tau_n) \vee (\tau_1 \wedge \cdots \wedge \tau_{r-1}) = (\tau_1 \wedge \cdots \wedge \tau_{r-1}) \vee (\tau_r \wedge \cdots \wedge \tau_n)$. Therefore, $\gamma_{r-1} \vee \tau_r \leq (\tau_1 \wedge \cdots \wedge \tau_{r-1} \wedge \tau_r') \vee (\tau_r \wedge \cdots \wedge \tau_n) \leq (\gamma_1 \wedge \cdots \wedge \gamma_{r-1}) \wedge (\tau_r \wedge \cdots \wedge \tau_n)$ and Lemma 1 implies that $G \sim X \oplus [\gamma_1, \ldots, \gamma_{r-2}, \gamma_{r-1} \wedge \tau_r, \tau_{r+1}, \ldots, \tau_n]$. Set

$$
\sigma_i = \begin{cases} \gamma_i & \text{if } 1 \leq i \leq r-2. \\ \gamma_{r-1} \wedge \tau_r & \text{if } i = r-1. \\ \tau_{i+1} & \text{if } r \leq i \leq n-1. \end{cases}
$$

Using these choices and conditions (b') and (c'), it is now easily checked that (a) - (c) hold.

$\square$

3. The Quasi-Summand Theorem. Suppose $G \sim [\tau_1, \ldots, \tau_n]$ is not strongly indecomposable and X is a strongly indecomposable quasi-summand of G. Call (r, I, J) a *triple for* X *in* G if each of the following four conditions are satisfied.

(1) $r \in \overline{n}$.

(2) I and J are subsets of $\overline{n}$ with $I \bigcup J = \overline{n}$, $I \bigcap J = \{r\}$, and $\min\{|I|, |J|\} \geq 2$.

(3) $\tau_r = \tau_I \vee \tau_J$ (and hence $G \sim G[A_I] \oplus G[A_J]$).

(4) X is a quasi-summand of $G[A_I]$.

In view of Lemma 2, every strongly indecomposable quasi-summand of G has a triple in G. Therefore, among all triples (r, I, J) for X in G, there exists one with $|I|$ as small as possible. Call such a triple with $|I|$ minimal a *minimal triple for* X *in* G.

Set $I_0 = \overline{n}$. A sequence of triples $\{(r_i, I_i, J_i) : 1 \leq i \leq k\}$ is called an *exhibiting sequence for* X *in* G if (r_i, I_i, J_i) is a minimal triple for X in $G[A_{I_{i-1}}]$ for all $i \in \overline{k}$, and $X \sim G[A_{I_k}]$. In this case, the positive integer k is the *length* of the exhibiting sequence. Since $|I_{i-1}| > |I_i|$ for all $i \in \overline{k}$, it is clear that each strongly indecomposable quasi-summand of G has an exhibiting sequence.

The next two Lemmas point out properties of exhibiting sequences which will prove to be the key ingredients needed in the proof of the desired quasi-summand result.

Lemma 4. *Suppose* $G \sim [\tau_1, \ldots, \tau_n]$ *is not strongly indecomposable and let* X *be a strongly indecomposable quasi-summand of* G *with rank* $X \geq 2$. *If* $\{(r_i, I_i, J_i) : 1 \leq i \leq k\}$ *is an* *exhibiting sequence for X in G, then for each $i \in \overline{k}$, $r_1, \ldots, r_i \in I_i$. Thus, for every i and j in $\overline{k}$, $r_j \in I_i$.*

Proof: We induct on the length k of the exhibiting sequence. If $k = 1$, the result is clear, so we may assume $k \geq 2$. Observe that $\{(r_i, I_i, J_i) : 2 \leq i \leq k\}$ is an exhibiting sequence for X in $G[A_{I_1}]$ of length $k-1$. Therefore, by the induction hypothesis, $r_2, \ldots, r_i \in I_i$ whenever $2 \leq i \leq k$. Observe also that $r_1, \ldots, r_k \in I_1$.

Suppose to the contrary that there exists $i \geq 2$ with $r_1 \notin I_i$. Select such an $i \in \overline{k}$ minimal with respect to the property $r_1 \notin I_i$. Since $r_1 \in I_{i-1}$ and $I_{i-1} = I_i \bigcup J_i$, we have $r_1 \in J_i \backslash I_i$. We now break the remainder of the proof into several steps.

Step 1. We claim that $i \geq 3$. Indeed, if $i = 2$, then $\tau_{r_2} = \tau_{I_2} \vee \tau_{J_2} = \tau_{I_2} \vee (\tau_{J_2} \wedge \tau_{r_1}) = \tau_{I_2} \vee (\tau_{J_2} \wedge (\tau_{I_1} \vee \tau_{J_1})) = \tau_{I_2} \vee \tau_{J_1 \cup J_2}$. Set $I = (I_2 - (I_2 \bigcap (J_1 \bigcup J_2))) \bigcup \{r_2\}$. Note that $\tau_{r_2} = \tau_I \vee \tau_{J_1 \cup J_2}$, $I \bigcup (J_1 \bigcup J_2) = \overline{n}$ and $I \bigcap (J_1 \bigcup J_2) = \{r_2\}$. Also, $|I| \geq 2$, since otherwise $I = \{r_2\}$ and $\overline{n} = I \bigcup (J_1 \bigcup J_2) = J_1 \bigcup J_2$. But this is clearly impossible since there exists $r \in I_1 \backslash (J_1 \bigcup J_2)$, $I_1 \bigcup J_1 = \overline{n}$ and $J_2 \subsetneq I_1$. Therefore $|I| \geq 2$. Note also that X is a quasi-summand of $G[A_I]$ (since otherwise, X is a quasi-summand of $G[A_{J_1 \cup J_2}]$ and $\tau_{r_1} = \tau_{I_1} \vee \tau_{J_1} \leq \tau_{J_2} \vee \tau_{J_1} \leq \tau_{r_1}$. Therefore $G[A_{J_1 \cup J_2}] \sim G[A_{J_1}] \oplus G[A_{J_2}]$ and X is a quasi-summand of $G[A_{J_1}]$ or $G[A_{J_2}]$. But, in view of Corollary 1, this is impossible since X is a quasi-summand of both $G[A_{I_1}]$ and $G[A_{I_2}]$ and rank $X \geq 2$). Therefore $(r_2, I, J_1 \bigcup J_2)$ is a triple for X in G with $|I| \leq |I_2| < |I_1|$. But this contradicts the fact that (r_1, I_1, J_1) is a minimal triple for X in G. Therefore, $i \geq 3$ as claimed.

Step 2. We claim that $\tau_{r_i} = \tau_{I_i \cup J_2 \cup \cdots \cup J_{i-1}} \vee \tau_{J_i \cup J_1}$. Indeed, $\tau_{r_i} = \tau_{I_i} \vee \tau_{J_i} = (\tau_{I_i} \wedge \tau_{r_2} \wedge \cdots \wedge \tau_{r_{i-1}}) \vee (\tau_{J_i} \wedge \tau_{r_1}) = ((\tau_{I_2} \vee \tau_{I_i \cup J_2}) \wedge \cdots \wedge (\tau_{I_{i-1}} \vee \tau_{I_i \cup J_{i-1}})) \vee \tau_{J_i \cup I_1} \vee \tau_{J_i \cup J_1}$ Since $I_j \subseteq I_1$ whenever $j \in \overline{k}$, $\tau_{J_i \cup I_1} \leq \tau_{I_j} \vee \tau_{I_i \cup J_j}$ for $2 \leq j \leq i-1$. Moreover $\tau_{I_{i-1}} \geq \tau_{I_j}$ for $2 \leq j \leq i-1$ (since $i \geq 3$). Thus, $\tau_{r_i} \leq ((\tau_{I_{i-1}} \vee \tau_{I_i \cup J_2}) \wedge (\tau_{I_{i-1}} \vee \tau_{I_i \cup J_3}) \wedge \cdots \wedge (\tau_{I_{i-1}} \vee \tau_{I_i \cup J_{i-1}})) \vee \tau_{J_i \cup J_1} = \tau_{I_{i-1}} \vee (\tau_{I_i \cup J_2 \cup \cdots \cup J_{i-1}}) \vee \tau_{J_i \cup J_1} = (\tau_{I_{i-1}} \wedge \tau_{r_1}) \vee (\tau_{I_i \cup J_2 \cup \cdots \cup J_{i-1}}) \vee \tau_{J_i \cup J_1} = \tau_{I_1} \vee \tau_{I_{i-1} \cup J_1} \vee \tau_{I_i \cup J_2 \cup \cdots \cup J_{i-1}} \vee \tau_{J_i \cup J_1}$. Note $i \geq 3$ implies that $I_i \bigcup J_2 \bigcup \cdots \bigcup J_{i-1} \subseteq I_1$. Also, $J_i \bigcup J_1 \subseteq I_{i-1} \bigcup J_1$. Therefore, $\tau_{r_i} \leq \tau_{I_i \cup J_2 \cup \cdots \cup J_{i-1}} \vee \tau_{J_i \cup J_1} \leq \tau_{r_i}$, and the claim is established.

Step 3. Set $I = ((I_i \bigcup J_2 \bigcup \cdots \bigcup J_{i-1}) - ((I_i \bigcup J_2 \bigcup \cdots \bigcup J_{i-1}) \bigcap (J_i \bigcup J_1))) \bigcup \{r_i\}$. We claim that $(r_i, I, J_i \bigcup J_1)$ is a triple for X in G. First, $I \bigcup (J_i \bigcup J_1) = J_1 \bigcup \cdots \bigcup J_{i-1} \bigcup J_i \bigcup I_i = J_1 \bigcup I_1 = \overline{n}$. Moreover $I \bigcap (J_i \bigcup J_1) = \{r_i\}$ and, from Step 2, $\tau_{r_i} = \tau_I \vee \tau_{J_i \cup J_1}$. As in step 1, we see that $|I| \geq 2$ and X is a quasi-summand of $G[A_I]$. Thus $(r_i, I, J_i \cup J_1)$ is a triple for X in G.

Step 4. Since $(r_i, I, J_i \cup J_1)$ is a triple for X in G and (r_1, I_1, J_1) is a minimal triple for X in G, we see that $|J_i \cup J_1| \leq |J_1|$. Thus, $J_i \subseteq I_1 \cap J_1 = \{r_1\}$, contradicting $|J_i| \geq 2$. □

Lemma 5. *Suppose $G \sim [\tau_1, \ldots, \tau_n]$ is not strongly indecomposable and let X be a strongly indecomposable quasi-summand of G with rank $X \geq 2$. If $\{(r_i, I_i, J_i) : 1 \leq i \leq k\}$ is an exhibiting sequence for X in G with $k \geq 2$, then for $2 \leq i \leq k$,*

$$\tau_{r_i} \vee (\tau_{r_1} \wedge \cdots \wedge \tau_{r_{i-1}}) = \tau_{J_i} \vee (\tau_{J_1} \wedge \cdots \wedge \tau_{J_{i-1}}).$$

Proof: By Lemma 4, $r_1, \ldots, r_{i-1} \in I_i$. Thus, $\tau_{I_i} \leq \tau_{r_1} \wedge \cdots \wedge \tau_{r_{i-1}}$ and $\tau_{r_i} \vee (\tau_{r_1} \wedge \cdots \wedge \tau_{r_{i-1}}) = \tau_{I_i} \vee \tau_{J_i} \vee (\tau_{r_1} \wedge \cdots \wedge \tau_{r_{i-1}}) = \tau_{J_i} \vee (\tau_{r_1} \wedge \cdots \wedge \tau_{r_{i-1}}) = \tau_{J_i} \vee ((\tau_{I_1} \vee \tau_{J_1}) \wedge \cdots \wedge (\tau_{I_{i-1}} \vee \tau_{J_{i-1}})) = (\tau_{J_i} \vee \tau_{I_1} \vee \tau_{J_1}) \wedge \cdots \wedge (\tau_{J_i} \vee \tau_{I_{i-1}} \vee \tau_{J_{i-1}}) = (\tau_{J_i} \vee \tau_{J_1}) \wedge \cdots \wedge (\tau_{J_i} \vee \tau_{J_{i-1}}) = \tau_{J_i} \vee (\tau_{J_1} \wedge \cdots \wedge \tau_{J_{i-1}})$.
□

For convenience in the proof of the next result, we introduce some further notation. If $L = \{\ell_1, \ldots, \ell_j\}$ is a nonempty subset of $\overline{n}$ and if σ is a type, we write $[\tau(L), \sigma]$ for $[\tau_{\ell_1}, \ldots, \tau_{\ell_j}, \sigma]$.

Proposition 3. *Suppose G is a qAV-group which is not strongly indecomposable. If X is a strongly indecomposable quasi-summand of G, there exists a qAV-group H such that $G \sim X \oplus H$.*

Proof: By Lemma 3, we may assume rank $X \geq 2$. Set $G \sim [\tau_1, \ldots, \tau_n]$ and let $\{(r_i, I_i, J_i) : 1 \leq i \leq k\}$ be an exhibiting sequence for X in G. In view of Lemma 2, we may assume $k \geq 2$.

We show by induction on i that $G \sim G[A_{I_i}] \oplus [\tau((J_1 - \{r_1\}) \cup \cdots \cup (J_i - \{r_i\})), \tau_{r_1} \wedge \cdots \wedge \tau_{r_i}]$. If $i = 1$, then $G \sim G[A_{I_1}] \oplus G[A_{J_1}] \sim G[A_{I_1}] \oplus [\tau(J_1 - \{r_1\}), \tau_{r_1}]$, thereby establishing the result. Thus, we may assume $i \geq 2$ and that $G \sim G[A_{I_{i-1}}] \oplus [\tau((J_1 - \{r_1\}) \cup \cdots \cup (J_{i-1} - \{r_{i-1}\})), \tau_{r_1} \wedge \cdots \wedge \tau_{r_{i-1}}]$. Then, $G \sim G[A_{I_i}] \oplus G[A_{J_i}] \oplus [\tau((J_1 - \{r_1\}) \cup \cdots \cup (J_{i-1} - \{r_{i-1}\})), \tau_{r_1} \wedge \cdots \wedge \tau_{r_{i-1}}]$. By Lemma 5, $\tau_{r_i} \vee (\tau_{r_1} \wedge \cdots \wedge \tau_{r_{i-1}}) = \tau_{J_i} \vee (\tau_{J_1} \wedge \cdots \wedge \tau_{J_{i-1}}) = \tau_{J_i} \vee (\tau_{J_1 - \{r_1\}} \wedge \cdots \wedge \tau_{J_{i-1} - \{r_{i-1}\}} \wedge (\tau_{r_1} \wedge \cdots \wedge \tau_{r_{i-1}}))$. Therefore, by Lemma 1, $G \sim G[A_{I_i}] \oplus [\tau((J_1 - \{r_1\}) \cup \cdots \cup (J_i - \{r_i\})), \tau_{r_1} \wedge \cdots \wedge \tau_{r_i}]$ and the induction is complete. Taking $i = k$ completes the proof. □

We now have the necessary ingredients for the proof of our main result.

Theorem 2. *If G is a qAV-group, then any quasi-summand of G is a qAV-group.*

Proof: Let K be a quasi-summand of G. We show that K is a qAV-group by induction on rank G. If rank $G = 1$, the result is clear. Therefore we may assume rank $G \geq 2$. If G is strongly indecomposable or if rank $K = $ rank G, then $K \sim G$ and we are done. Consequently, we may assume that G is not strongly indecomposable and rank $K < $ rank G. In this case there exists a strongly indecomposable quasi-summand X of G such that $G \sim X \oplus H$ and K is a quasi-summand of H. Since H is a qAV-group by Proposition 3 and rank $H < $ rank G, an application of the induction hypothesis completes the proof. $\square$

Corollary 2. *If G is a $\mathcal{B}^{(1)}$-group, then any quasi-summand of G is quasi-isomorphic to a $\mathcal{B}^{(1)}$-group.*

Proof: As remarked in section one, $G \sim H \oplus C$ where H is a qAV-group and C is completely decomposable. The result is now clear in view of Theorem 2 and the observation that the direct sum of a qAV-group and a completely decomposable group is quasi-isomorphic to a $\mathcal{B}^{(1)}$-group. $\square$

References

[AV1] D. Arnold and C. Vinsonhaler, Invariants for a class of torsion-free abelian groups, Proc. Amer. Math. Soc. **105** (1989), 293-300.

[AV2] D. Arnold and C. Vinsonhaler, Duality and invariants for Butler groups, Pacific J. Math. **148** (1991), 1-10.

[AV3] D. Arnold and C. Vinsonhaler, Invariants for two classes of torsion-free groups, preprint.

[FM] L. Fuchs and C. Metelli, On a class of Butler groups, Manuscripte Math. **71** (1991), 1-28.

[GU] H. P. Goeters and W. Ullery, Butler groups and lattices of types, Comment. Math. Univ. Carolinae **31** (1990), 613-619.

[HM] P. Hill and C. Megibben, The classification of certain Butler groups, J. Algebra, to appear.

[L] W. Y. Lee, Codiagonal Butler groups, Chinese J. Math. **17** (1989), 259-271.

Abelian Groups Whose Semi-Endomorphisms Form a Ring

JUTTA HAUSEN

University of Houston
Houston, Texas, 77204-3476, U.S.A

1. INTRODUCTION. Throughout, the word *ring* is used to mean ring with identity and all modules will be unital left modules.

Given a module M over a ring R, a *semi-endomorphism* (or R-homogeneous map) of M is a function $f : M \to M$ such that, for all $r \in R$ and for all $a \in M$, $f(ra) = rf(a)$. The set of all semi-endomorphisms of M is denoted by $sE_R(M)$, and $sE_R(M)$ is a zero-symmetric abelian near-ring with identity under the usual operations of point-wise addition and composition of functions [P]. This near-ring, sometimes called the centralizer near-ring determined by R and M [MS1-MS3, MvdW1, MvdW2] obviously contains the ring $E_R(M)$ of all R-endomorphisms of M:

$$(1.1) \qquad E_R(M) \subseteq sE_R(M),$$

and satisfies all the ring axioms with the possible exception of the left distributive law, i.e.

$$(1.2) \qquad sE_R(M) \text{ is a ring} \iff \forall f, g, h \in sE_R(M), \ f(g+h) = fg + fh.$$

In a recent paper [FMP], P. Fuchs, C. Maxson and G. Pilz consider the class $\mathcal{C}$ of all rings R such that, for every R-module M, $sE_R(M)$ is, in fact, a ring. They go on to prove the surprising fact that a ring R belongs to $\mathcal{C}$ (if and) only if $sE_R(M) = E_R(M)$ for all R-modules M. The authors determine some necessary conditions for a ring R to belong the class $\mathcal{C}$. It is shown that no commutative ring is in $\mathcal{C}$ and that every ring in $\mathcal{C}$ must have zero divisors. Thus, the ring $\mathbf{Z}$ of integers certainly (as is clear after a moment's reflection) is not a member of $\mathcal{C}$.

The investigations in [FMP] give rise to related questions. It will be convenient to call the R-module M *semi-endomorphal* if $sE_R(M)$ is a ring. If $sE_R(M) = E_R(M)$, i.e. if every semi-endomorphism of M is an endomorphism, then M is said to be *endomorphal*.

175

Problem 1.3. *Given a ring R, find necessary and/or sufficient conditions for an R-module M to be semi-endomorphal.*

Problem 1.4. *Given a ring R, find necessary and/or sufficient conditions for an R-module M to be endomorphal.*

Problem 1.5. *Describe the rings R over which there exist semi-endomorphal modules which are not endomorphal.*

This note considers these questions, first for modules over arbitrary rings R, then for the special case that $R = \mathbf{Z}$. An abelian group G will turn out to be semi-endomorphal if and only if either (i) G is locally cyclic, or (ii) G is torsion-free and has the property that no two independent elements have comparable type. Abelian groups satisfying (ii) are said to be *absolutely anisotropic*. Those abelian groups that are semi-endomorphal but not endomorphal are shown to be precisely the absolutely anisotropic torsion-free groups of rank at least two. The existence of such groups was established by R. A. Beaumont and R. S. Pierce [BP]. Thus, with regard to 1.5, there exist semi-endomorphal $\mathbf{Z}$-modules which are not endomorphal. In contrast, if R is a principal ideal domain with only finitely many prime ideals then every semi-endomorphal R-module is endomorphal. For M a module over an arbitray ring, sufficient conditions for M to be semi-endomorphal will be given. Every semi-endomorphal R-module known to the author satisfies these conditions. Whether, in turn, these conditions are necessary, is posed as

Problem 1.6. *Given a ring R and a semi-endomorphal R-module M, does there exist a family $\{X_i\}_{i \in I}$ of submodules X_i of M such that (i) $M = \bigcup_{i \in I} X_i$ and (ii) for each $f \in sE_R(M)$ and for each $i \in I$, the restriction $f|X_i \in Hom_R(X_i, X_i)$?*

2. NECESSARY CONDITIONS. The problem of classifying the semi-endomorphal modules is greatly aided by the existence of non-trivial direct decompositions (cf. [FMP, MvdW1]). This is due to two facts that are easily established.

Lemma 2.1. *Direct summands of (semi-)endomorphal modules are (semi-)endomorphal.*

Lemma 2.2. *Suppose that M is an R-module and A and B are submodules of M such that $M = A \oplus B$. If M is semi-endomorphal and $f \in sE_R(M)$, then $f(a+b) = f(a) + f(b)$ for all $a \in A$ and all $b \in B$.*

Proof. Let $\pi : M \to A$ denote the projection onto A along B. By (1.2), $f = f \cdot 1_M = f\pi + f(1_M - \pi)$ which implies the stated equation.

A general method for constructing semi-endomorphisms was given by Maxson and van der Walt in [MvdW1]. We modify and extend their construction. Let $X \subseteq M$ be a subset of the R-module M. Then X is said to be *R-closed* if, for all $r \in R$ and all $x \in X$, $rx \in X$. If X is R-closed, an *R-homogeneous map* from X to M is defined to be a function $\varphi : X \to M$ such that $\varphi(rx) = r\varphi(x)$ for all $r \in R$ and all $x \in X$. Finally, X is said to be *strongly R-pure* if $a \in M$, $r \in R$ and $0 \neq ra \in X$ imply $a \in X$.

Remark. If $X \leq G$ is a subgroup of the abelian group G such that the quotient group G/X is torsion-free then X is a $\mathbf{Z}$-closed and strongly $\mathbf{Z}$-pure subset of G.

Lemma 2.3. *Let X be a subset of the R-module M which is R-closed and strongly R-pure, and let $\varphi : X \to M$ be an R-homogeneous map. Define a function $f : M \to M$*

by

$$f(a) = \begin{cases} \varphi(a) & \text{if } a \in X \\ 0 & \text{otherwise.} \end{cases}$$

Then f is a semi-endomorphism of M.

Proof. Note that $f(0) = \varphi(0) = 0$. Let $r \in R$ and $a \in M$. If $a \in X$, then $f(ra) = \varphi(ra) = r\varphi(a) = rf(a)$. Suppose $a \notin X$. Then $f(a) = 0$ and either $ra \notin X$ or $ra = 0$. In either case, $f(ra) = 0 = rf(a)$.

Proposition 2.4. *A semi-endomorphal abelian group is either torsion or torsion-free.*

Proof. Assume, by way of contradiction, that G is a semi-endomorphal abelian group which is mixed. Then G has a decomposition $G = A \oplus B$ into subgroups A and B such that A contains an element $a \neq 0$ of finite order and B contains an element b of infinite order. Let $X = tG$ denote the maximal torsion subgroup of G. By the Remark, the previous lemma is applicable: let $\varphi = 1_G|X$ and consider the semi-endomorphism f constructed in 2.3. Since the element $a+b$ has infinite order, the definition of f together with 2.2 imply

$$0 = f(a+b) = f(a) + f(b) = f(a) = \varphi(a) = a$$

which is a contradiction.

The set of all positive rational primes shall be denoted by Π.

Proposition 2.5. *Let G be an abelian torsion group. If G is semi-endomorphal then $G \leq \bigoplus_{p \in \Pi} Z(p^\infty)$.*

Proof. Assume not. Then there exists an abelian torsion group which is semi-endomorphal and has p-rank at least two for some prime p. By 2.1, there is a semi-endomorphal abelian p-group G such that $G = A \oplus B$ with $A \neq 0 \neq B$. Let $X = \{a+b \mid a \in A, b \in B, o(a) = o(b)\}$. One verifies that X is $\mathbf{Z}$-closed and strongly $\mathbf{Z}$-pure in G. Apply 2.3 with $\varphi = 1_G|X$. That results in $f \in sE_{\mathbf{Z}}(G)$. Choose elements $x \in A$ and $y \in B$ such that both have order p. Then, using 2.2 and the definition of f,

$$0 \neq x + y = f(x+y) = f(x) + f(y) = 0$$

which is the desired contradiction.

Beaumont and Pierce call a torsion-free abelian group G *completely anisotropic* if no two independent elements of G have the same type [BP, p. 28]. This term was modified above: G is said to be *absolutely anisotropic* if no two independent elements have *comparable* type. The existence of absolutely anisotropic torsion-free groups of rank two was established in [BP]. Indeed, if $T = \{\tau_0, \tau_1, \tau_2, \dots\}$ is a countably infinite set of types such that the types with positive subscripts are pairwise incomparable and $\tau_i \cap \tau_j = \tau_0$ whenever $i \neq j$, then there exists a completely anisotropic rank-two torsion-free group G whose type set $T(G)$ is contained in $T \setminus \{\tau_0\}$ [ibid, 7.10, and p. 30]. Such a group must be absolutely anisotropic.

Proposition 2.6. *Let G be a torsion-free abelian group. If G is semi-endomorphal then G is absolutely anisotropic.*

Proof. Assume the contrary. Then there exists a semi-endomorphal torsion-free abelian group G which has two (distinct) pure rank-one subgroups X and Y such that the type of X is less than or equal to the type of Y. Choose $0 \neq \varphi \in Hom(X,Y)$, let $0 \neq x \in X$ and put $\varphi(x) = y$. By the Remark, 2.3 is applicable and yields a map $f \in sE_{\mathbf{Z}}(G)$. Using 1.2 and the independence of x and y it follows that

$$0 = f(y + x) = f(f(x) + 1_G(x)) = f(f + 1_G)(x) = ff(x) + f(x) = f(y) + f(x) = y.$$

This contradiction completes the proof.

The only torsion-free endomorphal abelian groups are those of rank one:

Proposition 2.7. *Let G be a torsion-free abelian group. If G is endomorphal then $G \leq \mathbf{Q}$.*

Proof. Assume G is torsion-free but not a subgroup of the rationals. Then there exist two pure rank-one subgroups A and B of G such that $A \cap B = 0$. Let $X = A \cup B$. Then X is a $\mathbf{Z}$-closed and strongly $\mathbf{Z}$-pure subset of G. Apply 2.3 with $\varphi = 1_G|X$ and call the resulting semi-endomorphism f. Pick $0 \neq a \in A$ and $0 \neq b \in B$. Then

$$f(a + b) = 0 \neq a + b = f(a) + f(b).$$

Thus, G is not endomorphal.

3. THE CHARACTERIZATION FOR $R = \mathbf{Z}$. The R-module M is said to be *locally cyclic* if, given any two elements x and y in M, there exist $a \in M$ and ring elements r and s in R such that $x = ra$ and $y = sa$. One easily proves

Lemma 3.1. *Let $X \leq M$ be a submodule of M and let $f \in sE_R(M)$. If X is locally cyclic then the restriction $f|X \in Hom_R(X,M)$.*

This gives rise to a class of endomorphal modules:

Corollary 3.2. *Every locally cyclic module is endomorphal.*

An abelian group G is locally cyclic if and only if either $G \leq \bigoplus_{p \in \Pi} Z(p^\infty)$, or G is torsion-free of rank one [FI, p. 84, Exercise 5]. Thus

Corollary 3.3. *If G is an abelian group such that $G \leq \bigoplus_{p \in \Pi} Z(p^\infty)$ or $G \leq \mathbf{Q}$, then G is endomorphal.*

Combining 3.3 with 2.5 results in

Theorem 3.4. *The following properties of the abelian torsion group G are equivalent.*

(1) *G is semi-endomorphal.*
(2) *G is endomorphal.*
(3) *$G \leq \bigoplus_{p \in \Pi} Z(p^\infty)$.*

Maxson and Smith give an example of a module V which is semi-endomorphal but not endomorphal [MS3, Example 2.1]. For this example, K may be any field, V is a three-dimensional vector space over K, and R is the subring of the matrix ring $M_3(K)$ consisting of all matrices of the form

$$\begin{pmatrix} a & b & c \\ 0 & a & 0 \\ 0 & 0 & a \end{pmatrix}$$

where $a, b, c \in K$. Then V, considered as a left R-module in the usual way, has the desired properties. One verifies that, for each $f \in sE_R(V)$ and for each $v \in V$, $f(v) \in Rv$. Thus, the set $\{Rv\}_{v \in V}$ of cyclic submodules of V is a covering of V such that, by 3.1, $f|Rv \in Hom_R(Rv, Rv)$. This method provides a general source for modules of this kind as stated in

Lemma 3.5. *If the R-module M has a family $\{X_i\}_{i \in I}$ of submodules X_i such that (i) $M = \bigcup_{i \in I} X_i$, and (ii) for each $f \in sE_R(M)$ and for each $i \in I$, $f|X_i \in Hom_R(X_i, X_i)$, then M is semi-endomorphal.*

Proof. In order to verify (1.2), let $f, g, h \in sE_R(M)$ and let $a \in M$. Then $a \in X_i$ for some $i \in I$. By hypothesis, $g(a), h(a) \in X_i$ and $f|X_i$ is a homomorphism. Thus $f(g + h)(a) = f(g(a) + h(a)) = fg(a) + fh(a) = (fg + fh)(a)$ as desired.

Using 3.5, one proves

Proposition 3.6. *Every torsion-free absolutely anisotropic abelian group is semi-endomorphal.*

Proof. Let G be an absolutely anisotropic torsion-free group and let $\{X_i\}_{i \in I}$ be the set of all pure rank-one subgroups of G. Since $Hom(X_i, X_j) = 0$ whenever $i \neq j$, 3.1 implies $f|X_i \in Hom(X_i, X_i)$ for all $f \in sE_{\mathbf{Z}}(G)$ and all $i \in I$. Apply 3.5.

Thus, all semi-endomorphal abelian groups are determined:

Theorem 3.7. *An abelian group G is semi-endomorphal if and only if either (i) $G \leq \bigoplus_{p \in \Pi} Z(p^\infty)$, or (ii) G is torsion-free absolutely anisotropic.*

Proof. Combine 2.4, 2.5 and 2.6 with 3.3 and 3.6.

As remarked earlier, there do exist torsion-free abelian groups of rank two which are absolutely anisotropic [BP]. Thus, over the ring $R = \mathbf{Z}$, semi-endomorphal modules need not be endomorphal.

Theorem 3.8. *An abelian group G is semi-endomorphal but not endomorphal if and only if G is an absolutely anisotropic torsion-free group of rank at least two.*

Proof. The *if*-part follows from 3.6 and 2.7; the proof is completed using 3.7 and 3.3.

4. MODULES OVER PRINCIPAL IDEAL DOMAINS. The theory of modules over principal ideal domains is almost identical to that of $\mathbf{Z}$-modules [K, p. 36]. One difference may be caused by the abundance of prime elements or the sparsity thereof. For example, if R is a principal ideal domain which has only finitely many prime ideals, then its quotient field $Q(R)$ has, up to isomorphism, only finitely many submodules, i.e. the type set $T(M)$ of any torsion-free R-module M is necessarily finite. This affects the existence of absolutely anisotropic modules where, analogous to the group case, an R-module M is said to be *absolutely anisotropic* if M is torsion-free and no two independent elements of M have comparable type.

Proposition 4.1. *Let R be a principal ideal domain and let $M \neq 0$ be a torsion-free absolutely anisotropic R-module. If R has only finitely many prime ideals then M has rank one.*

Proof. Assume, by way of contradiction, that M satisfies the hypothesis but has rank at least two. Choose two independent elements x and y in M. If R were a field then x

and y would have equal type. Thus, R is not a field. In particular, R is infinite. It follows that the set $S = \{x + ry \mid r \in R\}$ contains infinitely many elements. One easily verifies that any two elements in S are independent. By hypothesis, the type set $T(M)$ of M must be infinite which implies R has infinitely many prime ideals. This contradiction completes the proof.

COROLLARY 4.2. *If R is a principal ideal domain with the property that there exist semi-endomorphal R-modules which are not endomorphal then R has infinitely many (pairwise nonassociate) primes.*

Let Π be a maximal set of pairwise nonassociate prime elements in R. For each $p \in \Pi$, let $R(p^\infty)$ denote the p-primary component of the quotient module $Q(R)/R$. The submodules of $Q(R)$ and of $Q(R)/R = \bigoplus_{p \in \Pi} R(p^\infty)$ are locally cyclic; every R-module is either torsion-free or has a nonzero submodule of $R(p^\infty)$ as a direct summand; thus, the analogues of 2.4 and 2.5 hold when G is an R-module. So do 3.4, 3.7 and 3.8 with the caveat that there exist no absolutely anisotropic R-modules other than the submodules of $Q(R)$ when the spectrum of R is finite. In summary,

THEOREM 4.3. *A module M over a principal ideal domain R is semi-endomorphal if and only if either (i) $M \le \bigoplus_{p \in \Pi} R(p^\infty)$ (in which case M is endomorphal), or (ii) M is torsion-free absolutely anisotropic.*

Acknowledgement. The author would like to express her appreciation to C.J. Maxson and A.P.J. van der Walt for graciously sharing preprints of their work.

REFERENCES

[BP]   R.A. Beaumont and R.S. Pierce, "Torsion Free Groups of Rank Two", Memoirs of the Amer. Math. Soc., Providence, 1961.

[FI]   L. Fuchs, "Infinite Abelian Groups", Vol. I, Academic Press, New York, 1970.

[FII] L. Fuchs, "Infinite Abelian Groups", Vol. II, Academic Press, New York, 1973.

[K]   I. Kaplansky, "Infinite Abelian Groups", Revised ed., The University of Michigan Press, Ann Arbor, 1969.

[FMP] P. Fuchs, C.J. Maxson and G. Pilz, *On rings for which homogeneous maps are linear*, Proc. Amer. Math. Soc. 112 (1991), 1-7.

[MS1] C.J. Maxson and K.C. Smith, *Simple near-ring centralizers of finite rings*, Proc. Amer. Math. Soc. 75 (1979), 8-12.

[MS2] C.J. Maxson and K.C. Smith, *Near-ring centralizers*, Proc. Ninth Annual USL Math. Conf. Research Series 48, Univ. Southwestern Louisiana, April 1979, pp. 49-58.

[MS3] C.J. Maxson and K.C. Smith, *Centralizer near-rings that are endomorphism rings*, Proc. Amer. Math. Soc. 80 (1980), 189-195.

[MvdW1] C.J. Maxson and A.P.J. van der Walt, *Centralizer near-rings over free ring modules*, J. Austral. Math. Soc. (to appear).

[MvdW2] C.J. Maxson and A.P.J. van der Walt, *Homogeneous maps as piecewise endomorphisms* (pre-print).

[P]   G. Pilz, "Near-rings", 2nd ed., North Holland, Amsterdam, 1983.

Equivalence Theorems for
Torsion-free Groups

Paul Hill and Charles Megibben*

Department of Mathematics   Department of Mathematics
Auburn University, USA   Vanderbilt University, USA

1. Introduction. The role of equivalence theorems in elucidating the structure of infinite abelian groups has become increasingly important in recent years. In particular, we cite the classification of *A-groups* in [H], general results on the structure of isotype subgroups of totally projective p-groups in [HM1], and a uniqueness theorem for a class of local mixed groups in [HLM]. Two subgroups H and H' of the group G are said to be *equivalent* provided there is an automorphism ψ of G such that $\psi(H) = H'$. By an *equivalence theorem*, we mean any theorem that asserts under appropriate hypotheses that two subgroups of a given group are equivalent. Usually such theorems appear in a slightly more general guise, namely, H and H' are subgroups of groups G and G', respectively, and an isomorphism $\psi : G \longrightarrow G'$ is shown to exist such that $\psi(H) = H'$. Under these circumstances, $G/H \cong G'/H'$ and in many instances ψ is constructed so as to *induce* a given isomorphism $\phi : G/H \longrightarrow G'/H'$ in the sense that $\phi(x + H) = \psi(x) + H'$ for all $x \in G$.

One possible approach to proving two groups isomorphic is to imbed them as subgroups of a common group of known structure in such a manner that the subgroups are equivalent. This is a surprisingly effective strategy and, in addition to the papers cited above, we mention the proof of the uniqueness theorem for strongly indecomposable corank 1 Butler groups as given in [HM7]. One should also recall that this was the method utilized by Warfield in his well-known classification of *S-groups* [W1]. Equivalence theorems date back at least to the 1950's when J. Erdős [E] proved that two pure subgroups H and H' of a free abelian group G are equivalent if and only if $G/H \cong G/H'$ and rank $H =$ rank H'. (That his theorem can be generalized to pure subgroups of a completely decomposable homogeneous group is crucial to the main result of §3 below.) Erdős's theorem is, of course, but a very special case of the following definitive result.

---
* The authors were supported, respectively, by NSF grants DMS-90-96243 and DMS-89-00235.

1980 Mathematics Classification (1985 Revision), Primary 20K15, 20K20; Secondary 20K30.

THEOREM 1.1. [HM5] Let H and H' be subgroups of the free abelian groups G and G', respectively, with $G/H \cong G'/H'$. Then there is an isomorphism $\psi : G \longrightarrow G'$ such that $\psi(H) = H'$ if and only if, for each prime p,

$$\dim\left(\frac{H + pG}{pG}\right) = \dim\left(\frac{H' + pG'}{pG'}\right).$$

Furthermore, when H and H' have infinite rank, ψ can be chosen so as to induce any given isomorphism $\phi : G/H \longrightarrow G'/H'$.

The foregoing theorem resolved a long-standing problem in [F1] and was used in [HM5] to give, among other things, a new proof of the stacked bases theorem [CG]. The restriction on ranks in the final assertion of Theorem 1.1 is essential since when G is a finite rank free group and H is not pure in G, then G/H will have more automorphisms than G. We also note that Theorem 2.2 below suggests the relevance of generalizing Theorem 1.1 to the context where H is a *regular subgroup* of a completely decomposable homogeneous group G. Except for situations covered by the preceding theorem, other equivalence theorems known to us involve the hypothesis that H and H' are isotype subgroups of the abelian groups G and G', respectively. Moreover, in such theorems, the containing groups G and G' are almost invariably required to be *simply presented* [W2]. We know of but a single exception to this rule; namely, our paper [HM8] in which an equivalence theorem is established, under the assumption of Martin's Axiom, for a special class of primary groups that are neither separable nor simply presented. In the setting of local groups, definitive theorems on the equivalence of isotype subgroups of simply presented groups are available (see [HM1] and [HM6]); but for global groups, the theory is not as well-developed. We mention, however, our forthcoming paper [HM9] in which a specialized equivalence theorem for global mixed groups is exploited to advance the classification theory of countable mixed abelian groups. But in the present paper, we shall limit ourselves to torsion-free groups where, as is well known, the simply presented groups are precisely the completely decomposable ones. Thus we shall henceforth deal exclusively with the following question: *If H and H' are pure subgroups of the completely decomposable torsion-free groups G and G', respectively, then under what conditions will there exist an isomorphism $\psi : G \longrightarrow G'$ such that $\psi(H) = H'$?* Although we do not as yet have a complete answer to this question, we give in §2 (see Theorem 2.2) *a necessary and sufficient condition for a given isomorphism $\phi : G/H \longrightarrow G'/H'$ to be induced by an isomorphism $\psi : G \longrightarrow G'$.* Then in §3, we use this result to solve the general problem in the special case where H and H' are *weakly $*$-pure* subgroups (see Definition 3.1).

From this point on, G and G' will denote torsion-free abelian groups. With but exceptions noted below, we shall follow the standard notation and terminology of [F2]. By a *height* (or *characteristic*), we mean a sequence $s = (s_p)_{p \in \mathbf{P}}$ where $\mathbf{P}$ denotes the set of rational primes and each s_p is either a nonnegative integer or the symbol ∞. When x is an element of G, then we indicate its *height* by $|x| = (|x|_p)_{p \in \mathbf{P}}$ where $|x|_p$ is the p-height of x as computed in G. Heights are, of course, ordered pointwise, and with each height s we associate the fully-invariant subgroup $G(s) = \{x \in G : |x| \geq s\}$. As is customary, two heights s and t are said to be *equivalent* provided (1) $s_p = t_p$ for all but finitely many primes p and (2) $s_p = \infty$ if and only if $t_p = \infty$. With each height s, we also associate the fully-invariant subgroup

$$G(s^*) = \langle x \in G(s) : |x| \text{ is not equivalent to } s \rangle.$$

We allow positive integers to operate on heights in the obvious manner, and hence $nG(s) = G(ns)$ for all positive integers n and all heights s. Equivalence classes of heights are called *types* and will be denoted by Greek letters. If σ is a type, let $G(\sigma) = \bigcup_{s \in \sigma} G(s)$ and $G(\sigma^*) = \bigcup_{s \in \sigma} G(s^*)$. The fully-invariant subgroup $G(\sigma)$ is a pure subgroup of G, but $G(\sigma^*)$ is not necessarily pure in G.

We first prove a useful computational lemma that will be needed in both §2 and §3 below.

LEMMA 1.2. Let H be a pure subgroup of the torsion-free group G and suppose g is an element of G such that $|g| \geq t$ where $t \in \tau$. If $\langle g \rangle_* \subseteq H(\tau) + G(\tau^*)$, then $g = h + z$ where $h \in H(t)$ and $z \in G(t) \cap G(\tau^*)$.

PROOF. Write $g = h' + z'$ where $h' \in H(\tau)$ and $z' \in G(\tau^*)$. Let $s = |z'|$ and observe that $Q = \{p \in P : t_p > s_p \neq \infty\}$ is necessarily finite. Take $n = \prod_{p \in Q} p^{(t_p - s_p)}$ and $\ell = \prod_{p \in Q} p^{s_p}$. Clearly n is the smallest positive integer such that $ns \geq t$. Since $n\ell = \prod_{p \in Q} p^{t_p}$, there is a $g_1 \in G$ such that $n\ell g_1 = g$. By hypothesis, $g_1 = h_1 + z_1$ where $h_1 \in H(\tau)$ and $z_1 \in G(\tau^*)$. Let $s' = |nz_1|$ and choose m to be the smallest positive integer such that $ms' \geq t'$ where t' is the height defined by

$$t'_p = \begin{cases} t_p - s_p & \text{for } p \in Q, \\ t_p & \text{for } p \notin Q. \end{cases}$$

The existence of such an m is guaranteed by the fact that $t' \in \tau$ and $nz_1 \in G(\tau^*)$. Next notice that $\gcd(m,n) = 1$. Indeed if $p|n$, then $s'_p \geq t'_p$ because $p^{t_p - s_p}$ divides n. Therefore we can choose integers λ and μ such that $1 = \lambda m + \mu n$. Then

$$g = \lambda m(n\ell g_1) + \mu n g = \lambda \ell m n(h_1 + z_1) + \mu n(h' + z') = h + z$$

where $h \in H(\tau)$ and $z = \lambda\ell(mnz_1) + \mu n z' \in G(\tau^*)$. But $|\ell m n z_1| = \ell m s' \geq \ell t' = t$ and $|n z'| = ns \geq t$; that is, $z \in G(t) \cap G(\tau^*)$ and consequently $h = g - z \in H \cap G(t) = H(t)$.

We shall require one further preliminary result before proceeding to §2. For this purpose, we shall need to recall certain concepts from [HM3]. If s is a height and p is a prime, then we let $G(s^*, p) = G(s^*) + G(ps)$. An element $x \in G$ is said to be *primitive* provided $x \notin G(s^*, p)$ for each prime p and each height s equivalent to $|x|$ for which $s_p = |x|_p$ and $|x|_p \neq \infty$. When G is completely decomposable, an element $x \in G$ is primitive if and only if $\langle x \rangle_*$ is a direct summand of G [HM3, Corollary 2.9]. A direct sum $A = \bigoplus_{i \in I} A_i$ of an independent family of subgroups of G is called a *-valued coproduct* if $A \cap F = \bigoplus_{i \in I} A_i \cap F$ for all fully-invariant subgroups F of the form $G(s), G(s^*)$ or $G(s^*, p)$. A subgroup M of G is a *knice subgroup* if for each finite subset S of G there exist primitive elements $y_1, y_2, \cdots, y_m$ such that $N = M \oplus \langle y_1 \rangle \oplus \langle y_2 \rangle \oplus \cdots \oplus \langle y_m \rangle$ is a *-valued coproduct with all elements of S having finite order modulo N. If M is both pure and knice, the y_i's can be chosen so that $S \subseteq N$. The prototypical example of a pure knice subgroup is a direct summand of a completely decomposable group; and in fact, the completely decomposable groups are precisely the torsion-free groups satisfying *the third axiom of countability* with respect to pure knice subgroup [HM3, Theorem 5.2]. Our next result is the torsion-free analogue of Proposition 2.8 in [HM2].

PROPOSITION 1.3. Let M be a pure knice subgroups of the torsion-free group and let x be an element of $G(s)$. Then the following conditions are equivalent:

(1) x is a primitive element with $|x| = s$ and $M \oplus \langle x \rangle$ is a $*$-valued coproduct.

(2) $x \notin M + G(s^*, p)$ whenever p is a prime for which $s_p \neq \infty$.

PROOF. That (1) implies (2) is a trivial consequence of the definition of the concepts involved. Conversely, let us assume that condition (2) is satisfied. In particular, $x \notin G(ps)$ when $s_p \neq \infty$ and therefore $|x| = s$. Since M is a pure knice subgroup of G, condition (2) and lemmas 2.7 and 2.3 in [HM3] imply that we can write $x = m + y$ where $m \in M$, y is a primitive element with $|y| = s$ and $M \oplus \langle y \rangle$ is a $*$-valued coproduct. Finally, an application of Lemma 2.5 in [HM3] allows us to conclude that x is primitive and that $M \oplus \langle x \rangle$ is a $*$-valued coproduct.

2. Main theorem. Throughout this section, H and H' are pure subgroups of the completely decomposable torsion-free groups G and G', respectively, and $\phi : G/H \longrightarrow G'/H'$ is a given isomorphism. We shall determine necessary and sufficient conditions for ϕ to be induced by an isomorphism $\psi : G \longrightarrow G'$. An obvious necessary condition for ϕ to be induced by such a ψ is that ϕ *respects heights* in the sense that

$$\phi((G(s) + H)/H) = (G'(s) + H')/H'$$

for all heights s.

LEMMA 2.1. Let $\phi : G/H \longrightarrow G'/H'$ be an isomorphism that respects heights. Then, for all heights s and all types σ,

    (i) $\phi((G(s^*) + H)/H) = (G'(s^*) + H')/H'$,

    (ii) $\phi((G(\sigma) + H)/H) = (G'(\sigma) + H')/H'$,

    (iii) $\phi((G(\sigma^*) + H)/H) = (G'(\sigma^*) + H')/H'$.

Moreover, for each type σ, there is an induced isomorphism

$$\phi_\sigma : \frac{G(\sigma)/G(\sigma^*)}{(H(\sigma) + G(\sigma^*))/G(\sigma^*)} \longrightarrow \frac{G'(\sigma)/G'(\sigma^*)}{(H'(\sigma) + G'(\sigma^*))/G'(\sigma^*)}.$$

PROOF. (i), (ii) and (iii) are straightforward consequences of ϕ respecting heights. The final assertion of the lemma follows form (ii), (iii) and the natural isomorphisms

$$\frac{G(\sigma)}{H(\sigma) + G(\sigma^*)} \cong \frac{G(\sigma) + H}{G(\sigma^*) + H} \quad \text{and} \quad \frac{G'(\sigma)}{H'(\sigma) + G'(\sigma^*)} \cong \frac{G'(\sigma) + H'}{G'(\sigma^*) + H'}.$$

If the isomorphism $\phi : G/H \longrightarrow G'/H'$ respects heights and if $\psi : G \longrightarrow G'$ is an isomorphism that induces ϕ, then for any $x \in G(\sigma)$

$$\phi_\sigma \left[x + G(\sigma^*) + \left(\frac{H(\sigma) + G(\sigma^*)}{G(\sigma^*)} \right) \right] = \psi(x) + G'(\sigma^*) + \left(\frac{H'(\sigma) + G'(\sigma^*)}{G'(\sigma^*)} \right)$$

In other words, if ψ induces ϕ, then each isomorphism ϕ_σ is in turn induced by the isomorphism $\psi_\sigma : G(\sigma)/G(\sigma^*) \longrightarrow G'(\sigma)/G'(\sigma^*)$ given by $\psi_\sigma(x + G(\sigma^*)) = \psi(x) + G'(\sigma^*)$. Thus we have an immediate necessary condition for a height respecting isomorphism $\phi : G/H \longrightarrow G'/H'$ to be induced by an isomorphism between G and G', namely, that for each type σ the map ϕ_σ is induced by an isomorphism from $G(\sigma)/G(\sigma^*)$ to $G'(\sigma)/G'(\sigma^*)$. As we establish in our next theorem, this condition is also sufficient.

THEOREM 2.2. Let H and H' be pure subgroups of the completely decomposable torsion-free abelian groups G and G', respectively, and suppose that $\phi : G/H \longrightarrow G'/H'$ is an isomorphism that respects heights. Then a necessary and sufficient condition for ϕ to be induced by an isomorphism from G to G' is that, for each type σ, there exists an isomorphism $\psi_\sigma : G(\sigma)/G(\sigma^*) \longrightarrow G'(\sigma)/G'(\sigma^*)$ that induces the isomorphism ϕ_σ.

PROOF. We have already proved that the condition is necessary, so we need only establish its sufficiency. Assume that the desired ψ_σ exists for each σ. We begin our proof with three simple observations.

  (a) If $\psi_\sigma(x + G(\sigma^*)) = v + G'(\sigma^*)$ where $x \in G(\sigma)$ and if $\phi(x + H) = w + H'$ where $w \in G'(\sigma)$, then $w - v \in H'(\sigma) + G'(\sigma^*)$.
  (b) Each isomorphism $\psi_\sigma : G(\sigma)/G(\sigma^*) \longrightarrow G'(\sigma)/G'(\sigma^*)$ respects heights.
  (c) If $s \in \sigma$, then $G(s) \cap G(\sigma^*) = G(s^*)$.

Observation (a) follows immediately from the definition of ϕ_σ and the hypothesis that ψ_σ induces ϕ_σ; (b) is a trivial consequence of the fact that $G(\sigma^*)$ and $G'(\sigma^*)$ are direct summands of $G(\sigma)$ and $G'(\sigma)$, respectively; while (c) is easily proved once one notes that $G(\sigma^*)$ is both a direct summand of G and a direct sum of rank 1 groups each of type $> \sigma$.

Consider the family of all isomorphisms $\pi : M \longrightarrow M'$ where M and M' are pure knice subgroups of G and G', respectively, and the following two conditions are satisfied:

  (1) $\pi(m) + H' = \phi(m + H)$ for all $m \in M$.
  (2) For each type σ, $\pi(m) + G'(\sigma^*) = \psi_\sigma(m + G(\sigma^*))$ for all $m \in M \cap G(\sigma)$.

Observe that these two conditions are inductive, that is, both are preserved under ascending unions. Also recall that if S is any finite subset of G, then there is a *-valued coproduct $N = M \oplus \langle x_1 \rangle \oplus \cdots \oplus \langle x_n \rangle$ where the x_i's are primitive in G and $S \subseteq N$. Moreover, by [HM3, Proposition 4.2], $N_* = M \oplus \langle x_1 \rangle_* \oplus \cdots \oplus \langle x_n \rangle_*$ is a pure knice subgroup of G. Since both G and G' satisfy the third axiom of countability with respect to pure knice subgroups, the familiar countable combinatorics associated with generalizations of Ulm's theorem lead us to the conclusion that it suffices to prove the following:

  (3) If $N = M \oplus \langle x \rangle$ is a *-valued coproduct with x a primitive element in G such that $|x| = t \in \tau$, then there exists a primitive element $x' \in G'$ such that $|x'| = t$, $\phi(x + H) = x' + H'$, $\psi_\tau(x + G(\tau^*)) = x' + G'(\tau^*)$ and $N' = M' \oplus \langle x' \rangle$ is a *-valued coproduct in G'.

Indeed if such an x' can be found, then we obtain an isomorphism $\Theta : N_* \longrightarrow N'_*$ extending $\pi : M \longrightarrow M'$ by taking $\Theta(x) = x'$ and the conditions (1) and (2) above are satisfied by the triple (Θ, N_*, N'_*)

Let us assume then that the hypotheses of (3) are satisfied. Since ϕ respects heights, there will exist a $w \in G'(t)$ such that $\phi(x + H) = w + H'$. Let $\psi_\tau(x + G(\tau^*)) = v + G'(\tau^*)$, and note that $w - v \in H'(\tau) + G'(\tau^*)$ by (a) and we can assume that $v \in G'(t)$ by (b). Thus $w - v = h' + z' \in G'(t)$ where $h' \in H'(\tau)$ and $z' \in G'(\tau^*)$. Although Lemma 1.2 is not directly applicable here, its proof is easily adapted to the present context. Indeed if $s = |z'|$ and if n and ℓ are defined exactly as in the proof of that lemma, then choose $x_1 \in G$ so that $n\ell x_1 = x$. Next select w_1 and v_1 in G' so that $n\ell w_1 = w$ and $n\ell v_1 = v$. Since H' and $G'(\tau^*)$ are pure subgroups, $\phi(x_1 + H) = w_1 + H'$ and $\psi_\tau(x_1 + G(\tau^*)) = v_1 + G'(\tau^*)$.

Then $n\ell(w_1 - v_1) = w - v$ with $w_1 - v_1 \in H'(\tau) + G'(\tau^*)$ by (a). So without loss of generality, we may assume that $h' \in H'(t)$ and $z' \in G'(t) \cap G'(\tau^*) = G'(t^*)$, by observation (c). We claim now that $x' = w - h' = v + z'$ is the desired element. At least it is clear that $|x'| \geq t$, $x' + H' = w + H' = \phi(x + H)$ and $x' + G'(\tau^*) = v + G'(\tau^*) = \psi_\tau(x + G(\tau^*))$. By Proposition 1.3, it suffices to show that $x' \notin M' + G'(t^*, p)$ whenever p is a prime for which $t_p \neq \infty$. Suppose, however, by way of contradiction, that $x' \in M' + G'(t^*, p) = \pi(M) + G'(t^*) + G'(pt)$ where $t_p \neq \infty$. Then there is an $m \in M$, a $z'_1 \in G'(t^*)$ and a $g' \in G'(t)$ such that $x' = \pi(m) + z'_1 + pg'$. Since π preserves heights, $m \in M \cap G(t)$ and hence by (2) $\psi_\tau(x - m + G(\tau^*)) = x' - \pi(m) + G'(\tau^*) = pg' + G'(\tau^*)$. But ψ_τ^{-1} respects heights and therefore there is a $g \in G(t)$ and a $z \in G(\tau^*)$ such that $x - m = pg + z$, that is, $x = m + z + pg \in M + G(t^*, p)$ since $z \in G(t) \cap G(\tau^*) = G(t^*)$. By Proposition 1.3 again, this contradicts the fact that $M \oplus \langle x \rangle$ is a *-valuated coproduct. The proof of the theorem is complete.

Unfortunately, even when there is an isomorphism $\psi : G \longrightarrow G'$ such that $\psi(H) = H'$, it can still be the case that a particular height respecting isomorphism $\phi : G/H \longrightarrow G'/H'$ fails to be induced by any isomorphism between G and G'. In fact, as we shall see in Example 2.3 below, this can even happen when $H = H'$ and $G = G'$. There is no analogue for this sort of behavior among local groups; see, for example, the statement of Theorem 2.1 in [HM6]. Ultimately the explanation for the phenomenon occurring for torsion-free global groups is the fact that the homogeneous group $(H(\sigma) + G(\sigma^*))/G(\sigma^*)$ may have finite rank and the failure of ϕ_σ to be induced by an isomorphism between $G(\sigma)/G(\sigma^*)$ and $G'(\sigma)/G'(\sigma^*)$ is already inherent in the corresponding limitation we noted in Theorem 1.1.

EXAMPLE 2.3. Suppose $G = \mathbf{Z} \oplus \mathbf{Z}_5$ where $\mathbf{Z}_5$ denotes the ring of integers localized at the prime ideal generated by 5, and let H be the cyclic subgroup generated by the ordered pair $(5,1)$. Then H is a pure subgroup of G because G/H is torsion-free. Indeed if $n(a,b) = k(5,1)$ for some positive integers n, then $na = 5k = 5nb$ and $(a,b) = (5b,b) \in H$. Since G/H is a rank 1 group in which $(0,1) + H$ is divisible by all integers prime to 5 and $(1,0) + H$ is not divisible by 5, $G/H \cong \mathbf{Z}_5$. Therefore multiplication by 2 yields an automorphism ϕ of G/H. Moreover, ϕ respects heights. The inclusion $\phi((G(s) + H)/H) \subseteq (G(s) + H)/H$ for all heights s is obvious. On the other hand, if $|(a,b)| \geq s$ and if we let $(a',b') = (-2a, \frac{1}{2}(b-a))$, then $(a,b) + H = (5a,a) + (-4a, b-a) + H = \phi((a',b') + H)$ and it suffices to show that $|(a',b')| \geq |(a,b)|$. If $p \neq 5$, then $|(a',b')|_p = |a'|_p = |2a|_p \geq |a|_p = |(a,b)|_p$; while if $p = 5$, then $|(a',b')|_p = \min\{|2a|_p, |\frac{1}{2}(b-a)|_p\} = \min\{|a|_p, |b-a|_p\} = \min\{|a|_p, |b|_p\} = |(a,b)|_p$. Now let σ be the type determined by the height $t = (t_p)_{p \in \mathbf{P}}$ where $t_p = 0$ for all p. Then $G(\sigma) = G$, whereas $G(\sigma^*) = \{(0,b) : b \in \mathbf{Z}_5\}$ and $H(\sigma) = H$. Hence $\overline{G} = G(\sigma)/G(\sigma^*) \cong \mathbf{Z}$ and $\overline{H} = (H(\sigma) + G(\sigma^*))/G(\sigma^*) = 5\overline{G}$. Obviously the only automorphisms of $\overline{G}$ are multiplication by 1 and -1, and it is quickly verified that ϕ_σ is the automorphism of $\overline{G}/\overline{H}$ given by multiplication by 2. Not surprisingly, ϕ_σ fails to be induced by either automorphism of $\overline{G}$. In greater detail, if $x = (1,0)$ and $\overline{x} = (1,0) + G(\sigma^*)$, then $\phi_\sigma(\overline{x} + \overline{H}) = 2\overline{x} + \overline{H}$ can equal neither $\overline{x} + \overline{H}$ nor $-\overline{x} + \overline{H}$ because 5 divides neither 1 nor 3.

With slightly more restrictive constructions, we can exhibit automorphisms $\phi : G/H \longrightarrow G/H$ that satisfy properties more restrictive than respecting heights and which still fail to be induced by any automorphism of G. For example, one can require that for each $x \in G$ there exist a y such that $\phi(x + H) = y + H$ and $|x| = |y|$. In all such examples, the operative condition appears to be the fact that $G(\sigma)/(H(\sigma) + G(\sigma^*))$ can be finite when $(H(\sigma) + G(\sigma^*))/G(\sigma^*)$ has finite rank. In the next section, we show, however, that if a stronger form of purity is imposed on the subgroups H and H', then further appropriate conditions on how H and H' are imbedded in G and G' will insure that each height respecting isomorphism $\phi : G/H \longrightarrow G'/H'$ is induced by an isomorphism between G and G'. Indeed for this more restrictive version of purity, we obtain an exact analogue to the equivalence theorem for mixed local groups [HM6].

3. The equivalence theorem for weakly $*$-pure subgroups. As in the preceding section, H and H' will denote pure subgroups of completely decomposable torsion-free groups G and G', respectively. We shall now introduce a stronger form of purity, the impact of which will be to make $(H(\sigma) + G(\sigma^*))/G(\sigma^*)$ and $(H'(\sigma) + G'(\sigma^*))/G'(\sigma^*)$ pure subgroups of $G(\sigma)/G(\sigma^*)$ and $G'(\sigma)/G'(\sigma^*)$, respectively, for each type σ. Recall that in [HM3] we called a pure subgroup H of an arbitrary torsion-free G a $*$-*pure subgroup* provided $H \cap G(s^*) = H(s^*)$ and $H \cap G(s^*, p) = H(s^*, p)$ for all heights s and all primes p. In many applications, this is too stringent a form of purity; for example, in [DR] it is shown that a finite rank $*$-pure subgroup of a completely decomposable group is necessarily a direct summand. (For infinite rank subgroups, however, $*$-purity is not so restrictive, as is illustrated by an example in [HM4].) On the other hand, the following weakening of $*$-purity is relevant to the study of finite rank torsion-free groups (see Proposition 1.7 in [HM7]).

DEFINITION 3.1. A pure subgroup H of a torsion-free abelian group G is said to be *weakly $*$-pure* provided $H \cap G(s^*, p) = H \cap G(s^*) + H(ps)$ for all heights s and all primes p.

Notice that the equation in 3.1 expresses a distributivity condition since it can be rewritten as
$$H \cap (G(s^*) + G(ps)) = H \cap G(s^*) + H \cap G(ps).$$

Furthermore, a weakly $*$-pure subgroup H is $*$-pure in G if and only if $H \cap G(s^*) = H(s^*)$ for all heights s.

LEMMA 3.2. Let H be a pure subgroup of the completely decomposable torsion-free group G. Then H is a weakly $*$-pure subgroup of G if and only if, for all types σ, the quotient group $G(\sigma)/(H(\sigma) + G(\sigma^*))$ is torsion-free.

PROOF. First assume that H is weakly $*$-pure in G. To prove that $G(\sigma)/(H(\sigma) + G(\sigma^*))$ is torsion-free, it is enough to show that it contains no element of any prime order p. Suppose then that $g \in G(\sigma)$ and that p is a prime such that pg is in $H(\sigma) + G(\sigma^*)$. Then for some appropriate $s \in \sigma$, we can write $pg = h - z$ where $h \in H(s)$ and $z \in G(s^*)$, that is, $h \in H \cap (G(s^*) + G(ps)) = H \cap G(s^*) + H(ps)$. Thus $h = z_1 + ph_1$ where $z_1 \in G(s^*)$ and $h_1 \in H(s)$. Consequently, $g - h_1$ is an element of $G(\sigma)$ such that $p(g - h_1) = z_1 - z \in G(\sigma^*)$. But since G is completely decomposable, $G(\sigma)/G(\sigma^*)$ is torsion-free and therefore $g - h_1 \in G(\sigma^*)$; in other words, $g \in H(\sigma) + G(\sigma^*)$ and we conclude that $G(\sigma)/(H(\sigma) + G(\sigma^*))$ contains no element of order p.

Conversely, assume that each quotient group $G(\sigma)/(H(\sigma) + G(\sigma^*))$ is torsion-free; or equivalently, that each $H(\sigma) + G(\sigma^*)$ is a pure subgroup of G. Now suppose that $h \in H \cap G(s^*, p)$ and write $h = z + pg$ where $z \in G(s^*)$ and $g \in G(s)$. Thus $\langle g \rangle \cap (H(\sigma) + G(\sigma^*)) \neq 0$ where $s \in \sigma$. By purity, $\langle g \rangle_* \subseteq H(\sigma) + G(\sigma^*)$ and Lemma 1.2 implies that $g = h_1 + z_1$

where $h_1 \in H(s)$ and $z_1 \in G(s) \cap G(\sigma^*)$. But then $z_1 \in G(s^*)$, by observation (c) in the proof of Theorem 2.2, and finally $h = (z + pz_1) + ph_1 \in H \cap G(s^*) + H(ps)$, as desired.

THEOREM 3.3. Let H and H' be weakly $*$-pure subgroups of the completely decomposable torsion-free groups G and G', respectively. Then there is an isomorphism $\psi : G \longrightarrow G'$ such that $\psi(H) = H'$ if and only if
(1) there is an isomorphism $\phi : G/H \longrightarrow G'/H'$ that respects heights,
(2) for all types σ,
$$\operatorname{rank}\left(\frac{H \cap G(\sigma)}{H \cap G(\sigma^*)}\right) = \operatorname{rank}\left(\frac{H' \cap G'(\sigma)}{H' \cap G'(\sigma^*)}\right).$$

We have, of course, already noted the necessity of condition (1) in the theorem above; while obviously any isomorphism $\psi : G \longrightarrow G'$ with $\psi(H) = H'$ will induce, for each type σ, an isomorphism
$$\frac{H \cap G(\sigma)}{H \cap G(\sigma^*)} \cong \frac{H' \cap G'(\sigma)}{H' \cap G'(\sigma^*)}.$$

So in addition to the general hypothesis of Theorem 3.3, we assume also that condition (2) is satisfied and that $\phi : G/H \longrightarrow G'/H'$ is a fixed but arbitrary isomorphism that respects heights. We shall exploit Theorem 2.2 to show that ϕ is induced by an isomorphism between G and G', in other words, we shall prove under these circumstances that each ϕ_σ is induced by an appropriate isomorphism $\psi_\sigma : G(\sigma)/G(\sigma^*) \longrightarrow G'(\sigma)/G'(\sigma^*)$. First observe that each ϕ_σ respects heights; indeed this is a routine consequence of the definition of ϕ_σ and the facts that $G(\sigma^*)$ and $G'(\sigma^*)$ are direct summands of $G(\sigma)$ and $G'(\sigma)$, respectively. The added advantage of weak $*$-purity is that $(H(\sigma) + G(\sigma^*))/G(\sigma^*)$ and $(H'(\sigma) + G'(\sigma^*))/G'(\sigma^*)$ are pure subgroups of the completely decomposable homogeneous groups $G(\sigma)/G(\sigma^*)$ and $G'(\sigma)/G'(\sigma^*)$, respectively; while condition (2) and the obvious isomorphisms yield
$$\operatorname{rank}\left(\frac{H(\sigma) + G(\sigma^*)}{G(\sigma^*)}\right) = \operatorname{rank}\left(\frac{H'(\sigma) + G'(\sigma^*)}{G'(\sigma^*)}\right).$$

Consequently, the existence of the desired ψ_σ's is an immediate corollary of the following generalization of Erdős's equivalence theorem.

THEOREM 3.4. Let H and H' be pure subgroups of the completely decomposable homogeneous groups G and G', respectively, and suppose that rank H = rank H'. If $\phi : G/H \longrightarrow G'/H'$ is an isomorphism that respects heights, then there is an isomorphism $\psi : G \longrightarrow G'$ that induces ϕ.

PROOF. Suppose G and G' are homogeneous of type σ. Then, by [F2, Theorem 86.6], H and H' are also completely decomposable and homogeneous of type σ, and hence rank H = rank H' implies $H \cong H'$. But, of course, we are required to prove more than the fact that H and H' are isomorphic, namely, we need to show that these subgroups are isomorphic under an isomorphism $\psi : G \longrightarrow G'$ that induces the given ϕ. Our proof requires the following well-known and essentially trivial fact: *If B/A is a rank 1 torsion-free group and if $x \in B \backslash A$, then a necessary and sufficient condition for $B = A \oplus \langle x \rangle_*$ is that $|x| = |x + A|$ where these heights are computed in B and B/A, respectively.* We first consider the special case where the subgroups H and H' have finite rank. Then, as a consequence of [F2, Lemma 86.8], H and H' are direct summands of G and G', respectively. Therefore we can write $G = H \oplus B$ and $G' = H' \oplus B'$. Moreover, B is also a completely decomposable homogeneous group of type σ [F2, Theorem 86.7] and hence we can write $B = \oplus_{i \in I} \langle x_i \rangle_*$ where $|x_i| = s$ for all $i \in I$ and s is any fixed height in σ. Since we may assume $H' \neq 0$, we can select a fixed $h' \in H'$ with $|h'| = s$. Next, for each $i \in I$, choose $w_i \in G'$ such that $\phi(x_i + H) = w_i + H'$ and $|w_i| \geq s$. Notice that we have invoked the fact that ϕ respects heights in making such a choice of w_i. Furthermore, since replacing w_i by its component in B' neither decreases heights nor changes the coset modulo H', there is no loss of generality in assuming that each w_i lies in B'. Now, for each $i \in I$, let $x_i' = h' + w_i$ and observe that $|x_i'| = |h'| \wedge |w_i| = s = |x_i| = |x_i + H| = |\phi(x_i + H)| = |x_i' + H|$. It follows then, from [F2, Lemma 9.4] and the observation above, that $G' = H' \oplus \oplus_{i \in I} \langle x_i' \rangle_*$. Finally, since $H \cong H'$, there is an isomorphism $\psi : G \longrightarrow G'$ such that $\psi(H) = H'$ and $\psi(x_i) = x_i'$ for all $i \in I$. Clearly ψ induces ϕ.

We now deal with the case where the subgroups H and H' have infinite rank, say, rank $H = \text{rank } H' = r \geq \aleph_0$. Then H and H' are contained in direct summands K and K' of G and G', respectively, with rank $K = r = \text{rank } K'$. By a simple back-and-forth argument, we may furthermore assume that K and K' are chosen so that $\phi(K/H) = K'/H'$. Next we observe that it suffices to prove that there is an isomorphism $\psi : K \longrightarrow K'$ that induces the restriction of ϕ to K/H. The point is that H' will still contain an element h' with $|h'| = s \in \sigma$ and any complement B of K can be handled exactly as in the previous case. Therefore, without loss of generality, we henceforth assume that rank $G = \text{rank } H$ and rank $G' = \text{rank } H'$. The key to completing the proof in the special case presently under consideration is the following observation: *Suppose that $\pi : M \longrightarrow M'$ is an isomorphism where (1) M and M' are direct summands of G and G', respectively, with rank $M < r$, and (2) $\phi(z + H) = \pi(z) + H'$ for all $z \in M$. If $N = M \oplus \langle x \rangle_*$ is also a direct summand of G, then there is an isomorphism $\theta : N \longrightarrow N'$ extending π such that conditions (1) and (2) remain intact when θ, N and N' replace π, M and M', respectively.* To establish this, let $s = |x|$ and choose $w \in G'(s)$ such that $\phi(x + H) = w + H'$. Then there is a direct decomposition $G' = M' \oplus U \oplus V$ where U has finite rank and $w \in M' \oplus U$. Notice, since rank $(M' \oplus U) < r$ that H' has nontrivial intersection with V and hence the pure subgroup $H' \cap V$ will contain an element h' with $|h'| = s$. Let $x' = w + h'$ and observe that $|x'| = |w| \wedge |h'| = s$ and $\phi(x + H) = x' + H'$. The existence of the desired θ will be evident once it is shown that $N' = M' \oplus \langle x' \rangle_*$ is a direct summand of G'. But this follows from [HM3, Lemma 2.5] and the fact that $M' \oplus U \oplus \langle h' \rangle_*$ is a direct summand.

The remainder of the proof consists of routine infinite combinatorics. For example, when $r = \aleph_0$, the fundamental observation in the preceding paragraph and the standard back-and-forth argument associated with the proof of Ulm's theorem suffice to construct, via successive finite rank extensions, an isomorphism $\psi : G \longrightarrow G'$ that induces ϕ. If $r > \aleph_0$, then we fix direct decompositions $G = \oplus_{\alpha < r} A_\alpha$ and $G' = \oplus_{\alpha < r} A_\alpha'$ where each A_α and A_α' is a rank 1 group of type σ. For $I \subseteq r$, we let $G(I) = \oplus_{\alpha \in I} A_\alpha$ and $G'(I) = \oplus_{\alpha \in I} A_\alpha'$. In this case, one constructs inductively a family $\{I_\alpha\}_{\alpha < r}$ of subsets of r and associated isomorphisms $\pi_\alpha : G(I_\alpha) \longrightarrow G'(I_\alpha)$ such that the following conditions are satisfied:

    (a) $\phi(x + H) = \pi_\alpha(x) + H'$ for all $x \in G(I_\alpha)$.
    (b) $A_\alpha \subseteq G(I_{\alpha+1})$ and $A_\alpha' \subseteq G'(I_{\alpha+1})$.
    (c) π_β is an extension of π_α when $\beta > \alpha$.

The desired $\psi : G \longrightarrow G'$ is, of course, the supremum of all the π_α's. As for the mechanics involved in the construction of the π_α's and I_α's, notice that the choices are automatic when α is a limit; while in the transition from π_α and I_α to $\pi_{\alpha+1}$ and $I_{\alpha+1}$, the same back-and-forth methods required in the countable rank case are adequate to obtain the desired $I_{\alpha+1}$ as a countable enlargement of I_α.

Some consequences of Theorem 3.3 in the finite rank setting are considered in [HM7]. We close this paper with a couple of corollaries that are mainly of interest in the context of infinite rank groups.

COROLLARY 3.5. Suppose that H and H' are pure knice subgroups of the completely decomposable torsion-free groups G and G', respectively. Then there is an isomorphism $\psi : G \longrightarrow G'$ such that $\psi(H) = H'$ if and only if $G/H \cong G'/H'$ and, for all types σ, rank $(H(\sigma)/H(\sigma^*)) =$ rank $(H'(\sigma)/H'(\sigma^*))$.

PROOF. Since pure knice subgroups are $*$-pure [HM3, Proposition 4.7], it follows that $H \cap G(\sigma^*) = H(\sigma^*)$ and $H' \cap G'(\sigma^*) = H'(\sigma^*)$. Furthermore, pure knice subgroups are balanced [HM3, Theorem 4.3] and therefore any isomorphism $\phi : G/H \longrightarrow G'/H'$ automatically respects heights. Thus, the hypothesis of Theorem 3.3 are satisfied, and the conclusion follows.

COROLLARY 3.6. Suppose that H and H' are $*$-pure subgroups of the completely decomposable torsion-free groups G and G', respectively. If there exists an isomorphism $\phi : G/H \longrightarrow G'/H'$ that respects heights, then $H \oplus C \cong H' \oplus C$ for an appropriate completely decomposable group C.

PROOF. If C is chosen so that $C(\sigma)/C(\sigma^*)$ has sufficiently large rank for each relevant type σ, then the hypotheses of Theorem 3.3 will be satisfied for the subgroups $H \oplus C$ and $H' \oplus C$ of the completely decomposable groups $G \oplus C$ and $G' \oplus C$, respectively.

REFERENCES

[CG] J. Cohen and H. Gluck, "Stacked bases for modules over principal ideal domains," *J. Algebra,* **14**(1970), 493-505.

[DR] M. Dugas and K. Rangaswamy," On torsion-free abelian k-groups," *Proc. Amer. Math. Soc.* **99**(1987), 403-408.

[E] J. Erdős, "Torsion-free factor groups of free abelian groups and a classification of torsion-free abelian groups," *Publ. Math. Debrecen,* **5**(1957), 172-184.

[F1] L. Fuchs, *Abelian Groups*, Publishing House of the Hungarian Academy of Science, Budapest, 1958.

[F2] L. Fuchs, *Infinite Abelian Groups*, Academic Press, New York, Vol I, 1970 and Vol. II, 1973.

[H] P. Hill, "On the structure of abelian p-groups," *Trans. Amer. Math. Soc.* **288**(1985), 505-525.

[HLM] P. Hill, M. Lane and C. Megibben, "On the structure of p-local groups," *J. of Algebra*, to appear.

[HM1] P. Hill and C. Megibben, "On the theory and classification of abelian p-groups," *Math. Z.* **190**(1985), 17-38.

[HM2] P. Hill and C. Megibben, "Axiom 3 modules," *Trans. Amer. Math. Soc.* **295**(1986), 715-734.

[HM3] P. Hill and C. Megibben, "Torsion free groups," *Trans. Amer. Math. Soc.* **295**(1986), 735-751.

[HM4] P. Hill and C. Megibben, "Pure subgroups of torsion free groups," *Trans. Amer. Math. Soc.* **303**(1987), 765-778.

[HM5] P. Hill and C. Megibben, "Generalizations of the stacked bases theorem," *Trans. Amer. Math. Soc.* **312** (1989), 377-402.

[HM6] P. Hill and C. Megibben, "The local equivalence theorem," *Contemporary Math.* **87**(1989), 377-402.

[HM7] P. Hill and C. Megibben, "The classification of certain Butler groups," *J. of Algebra*, to appear.

[HM8] P. Hill and C. Megibben, "Martin's axiom and the structure of abelian p-groups," to appear.

[HM9] P. Hill and C. Megibben, "An equivalence theorem for global mixed groups," to appear.

[W1]   R. Warfield, "A classification theorem for abelian p-groups," *Trans. Amer. Math. Soc.* **210**(1975), 149-168.

[W2]   R. Warfield, "The structure of mixed abelian groups", *Lecture Notes in Math.*, Vol. 616, pp. 1-38, Springer-Verlag, New York, 1977.

Almost Completely Decomposable Groups
with Cyclic Regulator Quotients

Adolf Mader and Otto Mutzbauer

Department of Mathematics
University of Hawaii, USA

Mathematisches Institut
Universität Würzburg, Germany

INTRODUCTION

An **almost completely decomposable group** is a torsion–free abelian group which contains a completely decomposable group as a subgroup of finite index. It will always be assumed that our groups have finite ranks. Such groups have long served as examples for the various complexities and idiosyncrasies inherent in torsion–free abelian group theory, yet in contrast to the completely decomposable groups, this more general class resisted classification and general approaches until recently.

We fix some notation before continuing the story. Let X be an almost completely decomposable group and let $A = \bigoplus_\sigma A_\sigma$ be a completely decomposable subgroup of finite index. Here σ stands for a type, and A_σ is a direct sum of rank–one groups of type σ. If $A_\sigma \neq 0$ then σ is a **critical type** of A (or X) and we let $T_{cr}(A)$ denote the set of all critical types. Lady [La74] pioneered the study of almost completely decomposable groups and introduced the **regulating subgroups**, namely those completely decomposable subgroups which have minimal index in X. The regulating subgroups of the almost completely decomposable group X are exactly the subgroups $A = \sum_{\sigma \in T_{cr}(X)} X_\sigma$ for arbitrary decompositions $X(\sigma) = X_\sigma \oplus A^\#(\sigma)$ where $X(\sigma) = \{x \in X : \text{type}(x) \geq \sigma\}$, $X^*(\sigma) = \sum_{\tau > \sigma} X(\tau)$, and $X^\#(\sigma) = X^*(\sigma)_*$ are the usual type subgroups. Since X is almost completely decomposable, we actually get a direct sum $A = \bigoplus_{\sigma \in T_{cr}(X)} X_\sigma$. Burkhardt [Bt84] introduced the **regulator** $R(X)$ as the intersection of all regulating subgroups and showed that it is a completely decomposable group of finite index in X. In special cases, always under the assumption that $T_{cr}(A)$ is an antichain and $\text{rk } A_\tau = 1$, certain classes of almost completely decomposable groups were classified. Burkhardt [Bt83] dealt with elementary abelian regulator quotients $X/R(X)$, Kozhukhov [Kv83] with the case that $X/R(X)$ is cyclic, and in a very recent paper Dugas and Oxford [DO] extended these results to regulator quotients $X/R(X)$ which are direct sums of cyclic groups of the same order. The classification in the last paper was up to Lady's near–isomorphism rather than isomorphism and this simplified matters. General schemes for classification up to isomorphism are contained in Krapf–Mutzbauer [KM], also Schultz [Sch85], and finally in Mader–Vinsonhaler [MV1]. In the present paper we adopt the approach and philosophy of the last paper

193

and give a classification up to so-called **type–isomorphism** of all almost completely decomposable groups with cyclic regulator quotient 3.8.

In the present study the critical type sets as well as the ranks of the homogeneous components of the regulator are completely general. "Type–isomorphism" is a weakening of isomorphism natural for almost completely decomposable groups, and will be described below. It was introduced in Mader–Vinsonhaler [MV1] and agrees with Lady's near–isomorphism for almost completely decomposable groups with isomorphic regulators and regulator quotients.

Our notation is standard and can be found in Fuchs [F] and Arnold [Ar82LN]. We write maps on the right.

BACKGROUND

Let $A = \bigoplus_{\sigma \in \mathrm{T}_{cr}(A)} A_\sigma$ be a completely decomposable group of finite rank, C a finite abelian group of exponent e, and let $^- : A \to A/eA = \overline{A}$ be the natural epimorphism. We make the identification $\overline{A} = \bigoplus_{\sigma \in \mathrm{T}_{cr}(A)} \overline{A_\sigma}$ and call the epimorphic images $\overline{A(\sigma)}$, $\overline{A^*(\sigma)} = \overline{A^\#(\sigma)}$ the **type subgroups** of $\overline{A}$. It was shown in [MV1], 2.2, 2.3 that every almost completely decomposable group with regulator isomorphic to A and regulator quotient isomorphic to C is isomorphic to a group X_f which is defined by the pull–back diagram

$$
\begin{array}{ccccccccc}
0 & \longrightarrow & A & \xrightarrow{\;\epsilon\;} & X_f & \xrightarrow{\;\phi\;} & C & \longrightarrow & 0 \\
 & & \| & & \downarrow & & f \downarrow & & \\
0 & \longrightarrow & A & \xrightarrow{\;e\;} & A & \xrightarrow{\;\mathrm{nat}\;} & A/eA & \longrightarrow & 0
\end{array}
$$

where $f \in \mathrm{Hom}(C, \overline{A})$ is a **regulated monomorphism** which means that f is a monomorphism with $A\epsilon = \mathrm{R}(X_f)$. The set of regulated monomorphisms is denoted by $\mathrm{ReMon}(C, \overline{A})$. We will only need the following special criterion for belonging to ReMon. It can happen that an almost decomposable group X has a unique regulating subgroup which then is the regulator of X. In this case $\mathrm{R}(X) = \sum_{\tau \in \mathrm{T}_{cr}(A)} X(\tau)$ and we say that X has a **regulating regulator**.

2.1. Lemma. ([MV1], 2.13(c).) *Let* $f : C \to \overline{A}$ *be a monomorphism. Then* $f \in \mathrm{ReMon}(C, \overline{A})$ *and* X_f *has a regulating regulator if and only if* $Cf \cap \overline{A(\tau)} = 0$ *for each* $\tau \in \mathrm{T}_{cr}(A)$. $\square$

There is a natural map $\mathrm{Aut}\, A \to \mathrm{Aut}\, \overline{A} : \alpha \mapsto \overline{\alpha}$ which is induced by $^- : A \to \overline{A}$. The following isomorphism criterion for almost completely decomposable groups was established in [MV1].

2.2. Theorem. (Mader–Vinsonhaler [MV1], 2.15.) *Let* $f, g \in \mathrm{ReMon}(C, \overline{A})$. *Then the following are equivalent.*

(1) $X_f \cong X_g$.
(2) *There is* $\gamma \in \mathrm{Aut}\, C$, $\alpha \in \mathrm{Aut}\, A$ *such that* $g = \gamma f \overline{\alpha}$.
(3) *There is* $\alpha \in \mathrm{Aut}\, A$ *such that* $Cg = Cf\overline{\alpha}$. $\square$

The isomorphism problem is thus entangled with the question of which automorphisms of $\overline{A}$ are induced by automorphisms of A. Krapf and Mutzbauer [KM] proved a criterion covering this situation. It is clear that for every $\alpha \in \mathrm{Aut}\, A$ and every $\tau \in \mathrm{T}_{cr}(A)$, we have $\overline{A(\tau)}\,\overline{\alpha} = \overline{A(\tau)}$. We therefore define the **group of type–automorphisms** of $\overline{A}$ by

$$
\mathrm{TypAut}\, \overline{A} = \{\xi \in \mathrm{Aut}\, \overline{A} \mid \overline{A(\tau)}\xi = \overline{A(\tau)} \text{ for } \tau \in \mathrm{T}_{cr}(A)\}.
$$

If ξ is a map in $\mathrm{TypAut}\, \overline{A}$ then ξ induces an automorphism ξ_τ on the free $Z/e_\tau Z$–modules $\overline{A(\tau)}/A^\#(\tau)$ where $e_\tau = \exp \overline{A(\tau)}/A^\#(\tau)$, a factor of $e = \exp C$. The determinant $\det \xi_\tau \in Z/e_\tau Z$ is available. Let $Z(\tau, e)^\times$ be the subgroup of $(Z/e_\tau Z)^\times$ generated by -1 and all primes p for which $\sigma(p) = \infty$. The Krapf–Mutzbauer criterion can be formulated as follows.

2.3. Theorem. (Krapf–Mutzbauer [KM], 1.3.) *Let* $A = \bigoplus_{\sigma \in T_{cr}(A)} A_\sigma$ *be a completely decomposable group,* e *a positive integer, and* $^- : A \to \overline{A}$. *Then* $\xi \in \operatorname{Aut} A$ *belongs to* $\overline{\operatorname{Aut} A}$, *the subgroup of induced automorphisms, if and only if* $\xi \in \operatorname{TypAut} \overline{A}$ *and* $\det \xi_\tau \in Z(e, \tau)^\times$. $\square$

Note that $Z(e, \tau)^\times$ depends on the infinities of the type τ, and therefore cannot be computed unless the type is computationally controlled. In contrast, the group $\operatorname{TypAut} \overline{A}$ depends only on the type subgroups, i. e. on the poset $T_{cr}(A)$ but not on the types themselves. This motivates the following weakening of isomorphism and explains why it means a considerable simplification.

2.4. Definition. *Let* $f, g \in \operatorname{ReMon}(C, \overline{A})$. *Then* X_f *and* X_g *are called* **type–isomorphic**, *in symbols* $X_f \cong_t X_g$, *if there is* $\xi \in \operatorname{TypAut} \overline{A}$ *and* $\gamma \in \operatorname{Aut} C$ *such that* $g = \gamma f \xi$.

Analogous to 2.2 we have

2.5. Theorem. (Mader–Vinsonhaler [MV1], 4.2.) *Let* $f, g \in \operatorname{ReMon}(C, \overline{A})$. *Then the following are equivalent.*

(1) $X_f \cong_t X_g$.
(2) *There is* $\gamma \in \operatorname{Aut} C$, $\xi \in \operatorname{TypAut} A$ *such that* $g = \gamma f \xi$.
(3) *There is* $\xi \in \operatorname{TypAut} A$ *such that* $Cg = Cf\xi$. $\square$

A further indication of the simplification and appropriateness inherent in the concept of type–isomorphism is the following reduction theorem to the p–primary case.

Let $C = \bigoplus C_p$ and $\overline{A} = \bigoplus \overline{A}_p$ be the primary decompositions. Any $f \in \operatorname{Hom}(C, \overline{A})$ can be interpreted as $f = \oplus f_p$ with $f_p \in \operatorname{Hom}(C_p, \overline{A}_p)$. If $e = p^{e(p)} \cdot e_p$ with $\gcd(p, e_p) = 1$, then $\overline{A}_p = e_p A / e A = \overline{e_p A}$.

We are now prepared for the reduction theorem.

2.6. Theorem. (Mader–Vinsonhaler [MV1], 4.7.)

(1) $f \in \operatorname{ReMon}(C, \overline{A})$ *if and only if for each* p, $f_p \in \operatorname{ReMon}(C_p, \overline{A}_p)$.
(2) *For* $f, g \in \operatorname{ReMon}(C, \overline{A})$, $X_f \cong_t X_g$ *if and only if for each* p, $X_{f_p} \cong_t X_{g_p}$. $\square$

It turns out that type–isomorphism for almost completely decomposable groups with isomorphic regulators and regulator quotients coincides with Lady's near–isomorphism. We write $X \cong_n Y$ if X and Y are nearly isomorphic.

2.7. Theorem. (Mader–Vinsonhaler [MV1], 4.5.) *Let* $f, g \in \operatorname{ReMon}(C, \overline{A})$. *Then* $X_f \cong_t X_g$ *if and only if* $X_f \cong_n X_g$. $\square$

Thus in the following "type–isomorphic" may be read as "nearly isomorphic".

CYCLIC REGULATOR QUOTIENTS

Let

$$C = \langle s \rangle, \quad \exp C = e,$$

and let

$$A = \bigoplus_{\sigma \in T_{cr}(A)} A_\sigma$$

be a completely decomposable group with homogeneous components A_σ. We emphasize that A is completely general so that – in contrast with the rigid case (any two critical types are incomparable) – the homogeneous components of A are not uniquely determined by A.

We begin with a regulator criterion.

3.1. Theorem. *Let C be a cyclic group and $f : C \to \overline{A}$ a monomorphism. Then the following statements are equivalent.*

(1) $f \in \mathrm{ReMon}(C, \overline{A})$.

(2) $A\epsilon$ *is regulating regulator of X_f.*

(3) $A\epsilon = \sum_{\sigma \in \mathrm{T}_{cr}(A)} X_f(\sigma)$.

(4) *For every $\sigma \in \mathrm{T}_{cr}(A)$, $Cf \cap \overline{A(\sigma)} = 0$.*

PROOF. (1) $\Rightarrow$ (2). Since $f \in \mathrm{ReMon}(C, \overline{A})$, $A\epsilon = \mathrm{R}(X_f)$. The regulator $\mathrm{R}(X_f)$ is the intersection of the regulating subgroups which have the same index in X_f. But $X_f / \mathrm{R}(X_f) \cong C$, being cyclic, contains a single subgroup of a given index hence there is a unique regulating subgroup which must equal $\mathrm{R}(X_f)$.

(2) $\Rightarrow$ (1). Since $A\epsilon$ is in particular the regulator of X_f, we have $f \in \mathrm{ReMon}(C, \overline{A})$ by definition.

(2) $\Rightarrow$ (3). It is a general fact that $\mathrm{R}(X) = \sum_{\sigma \in \mathrm{T}_{cr}(X)} X(\sigma)$ if $\mathrm{R}(X)$ is regulating in the almost completely decomposable group X ([Mu91], 1.7).

(3) $\Rightarrow$ (2). $A\epsilon \cong A$ is completely decomposable. Since $\sum_{\sigma \in \mathrm{T}_{cr}(A)} X_f(\sigma)$ contains every regulating subgroup and is completely decomposable, it is itself regulating and the unique regulating subgroup.

(4) $\Leftrightarrow$ (2). 2.1. $\square$

Let

$$\Delta : A = \bigoplus_{\sigma \in \mathrm{T}_{cr}(A)} A_\sigma^\Delta; \quad \overline{A} = \bigoplus_{\sigma \in \mathrm{T}_{cr}(A)} \overline{A_\sigma^\Delta},$$

be some decomposition of A into homogeneous components. Since the decomposition is not unique, the dependence on the selected decomposition becomes a problem. If

$$\mathbf{E} : A = \bigoplus_{\sigma \in \mathrm{T}_{cr}(A)} A_\sigma^{\mathbf{E}}; \quad \overline{A} = \bigoplus_{\sigma \in \mathrm{T}_{cr}(A)} \overline{A_\sigma^{\mathbf{E}}},$$

is another decomposition, then $A(\tau) = A_\tau^\Delta \oplus A^{\#}(\tau) = A_\tau^{\mathbf{E}} \oplus A^{\#}(\tau)$ and $\overline{A(\tau)} = \overline{A_\tau^\Delta} \oplus \overline{A^{\#}(\tau)} = \overline{A_\tau^{\mathbf{E}}} \oplus \overline{A^{\#}(\tau)}$. It is clear that there exists $\xi \in \mathrm{TypAut}\,\overline{A}$ so that $\overline{A_\tau^\Delta}\xi = \overline{A_\tau^{\mathbf{E}}}$. Thus transition to a new decomposition of $\overline{A}$ is achieved by the action of a suitable type–automorphism. We remark that any decomposition $\overline{A} = \bigoplus_{\sigma \in \mathrm{T}_{cr}(A)} B_\sigma$ with $\overline{A(\sigma)} = B_\sigma \oplus \overline{A^{\#}(\sigma)}$, is induced by a decomposition of A ([KM, Lemma 1.1]).

Based on the decomposition Δ, we introduce certain sets of critical types, $T_f^\Delta(d)$, $d \mid e = \exp C$. Recall that $C = \langle s \rangle$. Write

$$sf = \sum_{\sigma \in \mathrm{T}_{cr}(A)} sf_\sigma^\Delta, \quad sf_\sigma^\Delta \in \overline{A_\sigma^\Delta},$$

i. e. sf_σ^Δ is the component of sf in the summand $\overline{A_\sigma^\Delta}$ of the decomposition Δ.

By $|a|$ we denote the order of an element a in a group. We use the symbol $n \parallel m$ to mean that for each prime $p \mid m$, the highest prime power contained in the integer n is strictly smaller than the highest prime power contained in the integer m.

For $1 \neq d \mid e$, let

$$T_f^\Delta(d) = \{\sigma \,:\, |sf_\sigma^\Delta| = d \text{ and for } \tau < \sigma, \ |sf_\tau^\Delta| \parallel d\},$$

and let

$$T_f^\Delta(1) = \{\sigma \,:\, |sf_\sigma^\Delta| = 1 \text{ and } \sigma \text{ is minimal in } \mathrm{T}_{cr}(A)\}.$$

Thus T_f^Δ is a function on the set of divisors of e to the power set of $\mathrm{T}_{cr}(A)$ and

(1) $\qquad \mu \in T_f^\Delta(d)$ if and only if $|sf_\mu^\Delta| = d$ and $(d/p)(sf_\tau^\Delta) = 0$ for $\tau < \mu$, $p \mid d$.

We will show that the sets $T_f^\Delta(d)$ are independent of the choice of the decomposition Δ, and that the sets $T_f(d) = T_f^\Delta(d)$ are type–isomorphism invariants. They constitute, in fact, a complete system of invariants and serve to classify the almost completely decomposable groups with fixed regulator and cyclic regulator quotient up to type–isomorphism 3.8.

In order to motivate the invariants and to preview the proofs, suppose that

$$\overline{A} = \overline{A_{\sigma_1}} \oplus \overline{A_{\sigma_2}} \oplus \overline{A_{\sigma_3}} \oplus \overline{A_{\sigma_4}}, \quad \overline{A_{\sigma_i}} = \langle a_i \rangle, \quad e = |s| = |a_i| = 8.$$

Further assume that

$$\sigma_1 \text{ is incomparable with } \sigma_2, \sigma_3, \sigma_4, \quad \sigma_2 < \sigma_3, \sigma_4, \quad \text{and} \quad \sigma_3, \sigma_4 \text{ are incomparable.}$$

Let

$$f : C \to \overline{A} : sf = a_1 + 2a_2 + 4a_3 + a_4$$
$$g : C \to \overline{A} : sg = a_1 + 2a_2 + a_4.$$

Using 3.1 (4) it is easily checked that $f, g \in \mathrm{ReMon}(C, \overline{A})$. In fact, $4sf = 4a_1 + 4a_4 = 4sg$, so $C[2]f \cap \overline{A(\sigma)} = 0$ for $\sigma \in \{\sigma_1, \sigma_2, \sigma_3, \sigma_4\}$ and therefore $Cf \cap \overline{A(\sigma)} = 0$. The function

$$\xi : \overline{A} \to \overline{A} : a_1\xi = a_1, \ a_2\xi = a_2 + 2a_3, \ a_3\xi = a_3, \ a_4\xi = a_4,$$

is a well–defined map in $\mathrm{TypAut}\,\overline{A}$ and $g\xi = f$. By 2.5 $X_g \cong_t X_f$. Thus passing from X_f to X_g, the support of sf can be reduced from four terms to three without leaving the type-isomorphism class. This was possible since $4a_3$ had order less than or equal to the order of $2a_2$, **and** since $\sigma_2 < \sigma_3$. Although $\sigma_2 < \sigma_4$, the component a_4 of sf cannot be removed in this fashion since requiring $2a_2\xi = 2a_2 + a_4$ would map an element of order 4 to an element of order 8: incompatible orders prevent the elimination of a_4. On the other hand, although $|a_1| = |a_4| \geq |a_2|$, the component a_1 of sf cannot be removed since this would require something like $a_2\xi = a_1$ which violates the type restrictions of elements in $\mathrm{TypAut}\,\overline{A}$.

We now show that the $T_f^\Delta(d)$ are type–isomorphism invariants, and therefore independent of the choice of the decomposition Δ.

3.2. Lemma. *Let*

$$\Delta : A = \sum_{\sigma \in \mathrm{T}_{cr}(A)} A_\sigma^\Delta$$

be a decomposition of A into homogeneous components, and let $\xi \in \mathrm{TypAut}\,\overline{A}$. Then $T_f^\Delta(d) = T_{f\xi}^\Delta(d)$ for each $d \mid e$. The sets of types $T_f^\Delta(d)$ are independent of the choice of the decomposition Δ and we write $T_f(d) = T_f^\Delta(d)$.

PROOF. The transition to a new decomposition of $\overline{A}$ is achieved by the action of a type–isomorphism of $\overline{A}$. Hence the second claim follows from the first.

Let $\tau \in \mathrm{T}_{cr}(A)$. Since $\xi \in \mathrm{TypAut}\,\overline{A}$,

$$(2) \qquad\qquad\qquad |(sf_\tau^\Delta)\xi| = |sf_\tau^\Delta|,$$

and $(sf_\tau^\Delta)\xi \in \overline{A(\tau)}$ since $sf_\tau^\Delta \in \overline{A(\tau)}$. Therefore

$$(3) \qquad\qquad\qquad (sf_\tau^\Delta)\xi = \sum_{\tau \leq \sigma} x_{\tau\sigma}, \quad x_{\tau\sigma} \in \overline{A_\sigma^\Delta}.$$

Furthermore, ξ induces an automorphism ξ_τ of $\frac{A(\tau)}{A^{\#}(\tau)} \cong \overline{A_\tau^\Delta}$ whereby $(sf_\tau^\Delta + \overline{A^{\#}(\tau)})\xi_\tau = x_{\tau\tau} + \overline{A^{\#}(\tau)}$ by (3). Hence $|sf_\tau^\Delta| = |sf_\tau^\Delta + \overline{A^{\#}(\tau)}| = |x_{\tau\tau} + \overline{A^{\#}(\tau)}| = |x_{\tau\tau}|$ so, using (3) and (2),

(4)   for $\tau \leq \sigma$, $|x_{\tau\sigma}|$ divides $|(sf_\tau^\Delta)\xi| = |sf_\tau^\Delta| = |x_{\tau\tau}|$.

We compute the components of $sf\xi$.

$$sf\xi = (\sum_\tau sf_\tau^\Delta)\xi = \sum_\tau (sf_\tau^\Delta \xi) = \sum_\tau \sum_{\tau \leq \sigma} x_{\tau\sigma} = \sum_\sigma \sum_{\tau \leq \sigma} x_{\tau\sigma}$$

hence

(5) $$(sf\xi)_\sigma^\Delta = \sum_{\tau \leq \sigma} x_{\tau\sigma}.$$

We show first that $T_f^\Delta(1) \subset T_{f\xi}^\Delta(1)$. Let $\mu \in T_f^\Delta(1)$, i.e. μ is minimal in $T_{cr}(A)$ and $|sf_\mu^\Delta| = 1$. Then by (5), $(sf\xi)_\mu^\Delta = x_{\mu\mu}$ and by (4), $|x_{\mu\mu}| = |sf_\mu^\Delta| = 1$ so $\mu \in T_{f\xi}^\Delta(1)$.

Next we show that $T_f^\Delta(d) \subset T_{f\xi}^\Delta(d)$ for $1 \neq d \mid e$. Let $\mu \in T_f^\Delta(d)$. For $p \mid d$

$$(d/p)(sf\xi)_\mu^\Delta = (d/p)x_{\mu\mu} + \sum_{\tau < \mu}(d/p)x_{\tau\mu}.$$

Since $\mu \in T_f^\Delta(d)$, we have $(d/p)sf_\mu^\Delta = 0$ for all $\tau < \mu$ by (1), and therefore, by (4), $(d/p)x_{\tau\mu} = 0$. On the other hand, again by (4), $|x_{\mu\mu}| = |sf_\mu^\Delta| = d$, so $(d/p)(sf\xi)_\mu^\Delta = (d/p)x_{\mu\mu} \neq 0$. This implies that $|(sf\xi)_\mu^\Delta| = d$.

Suppose that $\nu < \mu (\in T_f^\Delta(d))$. Then $|sf_\nu^\Delta| \parallel d$ and, if $\tau \leq \nu (<\mu)$, then $|sf_\tau^\Delta| \parallel d$, so $|x_{\tau\nu}| \parallel d$ by (4). Hence, for every $p \mid d$,

$$(d/p)(sf\xi)_\nu^\Delta = \sum_{\tau \leq \nu}(d/p)x_{\tau\nu} = 0.$$

Thus $\mu \in T_{f\xi}^\Delta(d)$ and $T_f^\Delta(d) \subset T_{f\xi}^\Delta(d)$ is verified.

By symmetry $T_{f\xi}^\Delta(d) = T_f^\Delta(d)$. $\square$

The following "Reduction Lemma" will be used repeatedly to replace a group X_f by an isomorphic (not only type–isomorphic) group X_g for which sg is as simple as possible.

3.3. Lemma. *Assume that $e = p^m$, i. e. C is primary cyclic. Let $f \in \mathrm{ReMon}(C, \overline{A})$. With respect to the fixed decomposition Δ we write*

$$sf = \sum_{\sigma \in T_{cr}(A)} sf_\sigma, \quad sf_\sigma \in \overline{A_\sigma^\Delta}.$$

Fix $\tau \in T_{cr}(A)$ and let

$$S_\tau = \{\sigma \in T_{cr}(A) : \sigma > \tau \text{ and } |sf_\sigma| \leq |sf_\tau|\}.$$

Then there is $g \in \mathrm{ReMon}(C, \overline{A})$ and $\overline{\alpha} \in \mathrm{Aut} A$ such that $g = f\overline{\alpha}$ and $sg_\sigma = 0$ for all $\sigma \in S_\tau$ and $sg_\sigma = sf_\sigma$ for $\sigma \notin S_\tau$. In particular, $X_g \cong X_f$. Furthermore, $\overline{\alpha}$ induces the identity on each $\overline{A(\rho)/A^{\#}(\rho)}$.

PROOF. Embed $\langle sf_\tau \rangle$ into a direct summand $\langle x_\tau \rangle$ of $\overline{A_\tau}$ such that $sf_\tau = p^i x_\tau$ for some i. Write $\overline{A_\tau} = \langle x_\tau \rangle \oplus B_\tau$. For every $\sigma \in S_\tau$ choose $y_\sigma \in \overline{A_\sigma}$ such that $p^i y_\sigma = sf_\sigma$. Since $|sf_\sigma| \leq |sf_\tau|$, such elements y_σ exist. We have $\overline{A} = \langle x_\tau \rangle \oplus B_\tau \oplus (\bigoplus_{\rho \neq \tau} \overline{A_\rho})$. Define $\xi : \overline{A} \to \overline{A}$ by stipulating that

$$x_\tau \xi = x_\tau + \sum_{\sigma \in S_\tau} y_\sigma, \quad \xi = 1 \text{ on } B_\tau \oplus \bigoplus_{\rho \neq \tau} \overline{A_\sigma}.$$

It is obvious that ξ is a well–defined automorphism of $\overline{A}$ and that $\xi \in \text{TypAut}\,\overline{A}$. It is also clear that ξ_ρ, the induced map on $\overline{A(\rho)/A^{\#}(\rho)}$, is the identity, $\det \xi_\rho = 1$ and therefore $\xi = \overline{\alpha}$ for some $\alpha \in \text{Aut}\,A$ by 2.3.

Define
$$g : C \to \overline{A} : \begin{cases} sg_\sigma = 0 & \sigma \in S_\tau \\ sg_\sigma = sf_\sigma & \sigma \notin S_\tau \end{cases}$$

Then $g\xi = f$, as is easily checked, and g is a monomorphism. It remains to show that $g \in \text{ReMon}(C, \overline{A})$. We utilize the criterion 3.1, and note first that 3.1(4) (applied to g) is equivalent to $s_0 g \notin \overline{A(\sigma)}$ where $s_0 = p^{m-1}s$. If $|sg_\tau| < p^m$ then $s_0 g = s_0 f$ and $g \in \text{ReMon}(C, \overline{A})$ just as $f \in \text{ReMon}(C, \overline{A})$. So suppose $|sg_\tau| = p^m$. We show that $s_0 g \notin \overline{A(\sigma)}$ for $\sigma \in \text{T}_{\text{cr}}(A)$ by distinguishing four cases.

Case I. $\sigma = \tau$. Since $s_0 f \notin \overline{A(\tau)}$ there is $\rho \not\geq \tau$ such that $0 \neq s_0 f_\rho = s_0 g_\rho$, so $s_0 g \notin \overline{A(\tau)}$.

Case II. $\sigma \geq \tau$. Then $\overline{A(\sigma)} \subset \overline{A(\tau)}$ and $s_0 g \notin \overline{A(\tau)}$ implies $s_0 g \notin \overline{A(\sigma)}$.

Case III. $\sigma \not\geq \tau$ and $\sigma \not\leq \tau$, i. e. τ, σ are incomparable. Here $s_0 g \notin \overline{A(\sigma)}$ since $s_0 g_\tau \neq 0$.

Case IV. $\sigma \not\geq \tau$ and $\sigma \leq \tau$, i. e. $\sigma < \tau$. Since $s_0 f \notin \overline{A(\sigma)}$, there is $\rho \not\geq \sigma$ such that $s_0 f_\rho \neq 0$. Since $\rho \not\geq \tau > \sigma$, $s_0 g_\rho = s_0 f_\rho \neq 0$ so $s_0 g \notin \overline{A(\sigma)}$. $\square$

We can use the Reduction Lemma 3.3 repeatedly in order to achieve that $sf_\sigma^\Delta = 0$ for $\sigma \notin \bigcup_{d|e} T_f(d)$ by replacing f within its isomorphism class. The final stage is described in the following proposition.

3.4. Proposition. Let $f \in \text{ReMon}(C, \overline{A})$. Then the isomorphism class of X_f contains a group X_g, $g \in \text{ReMon}(C, \overline{A})$, such that $sg_\sigma = 0$ for $\sigma \notin \bigcup_{d|e} T_g(d)$.

PROOF. We use the primary decompositions of C and $\overline{A}$ in order to write $f = \oplus_{p|e} f_p$ with $f_p \in \text{ReMon}(C_p, \overline{A}_p)$ (see 2.6). Assume that the Reduction Lemma 3.3 has been applied to f_p for each $\tau \in \text{T}_{\text{cr}}(A)$. The result is a function $g_p \in \text{ReMon}(C_p, \overline{A}_p)$ such that for every $\tau \in \text{T}_{\text{cr}}(A)$, if $\sigma \in \text{T}_{\text{cr}}(A), \sigma > \tau$, and $|s_p g_{p\sigma}| \leq |s_p g_{p\tau}|$, then $s_p g_{p\sigma} = 0$. Thus $|s_p g_{p\sigma}| = p^i \neq 1$ and $\sigma > \tau$ imply that $|s_p g_{p\sigma}| > |s_p g_{p\tau}|$, i.e. $|s_p g_{p\tau}| \,\|\, |s_p g_{p\sigma}|$. Hence $\sigma \in T_{g_p}(p^i)$. This means that $s_p g_{p\tau} = 0$ for $\tau \notin \bigcup_{p^i|e} T_{g_p}(p^i)$. Further for each p, we have a map $\xi_p \in \text{TypAut}\,\overline{A}_p$ such that $g_p = f_p \xi_p$ and $(\xi_p)_\tau = 1$ on $\overline{A(\tau)}_p / \overline{A^{\#}(\tau)}_p$.

We now combine these results. We set $g = \oplus_p g_p$ and $\xi = \prod_p \xi_p$. Then $g \in \text{ReMon}(C, \overline{A})$ by the Primary Reduction Theorem 2.6, $\xi \in \text{TypAut}\,\overline{A}$ and ξ induces the identity mapping on $\overline{A(\tau)/A^{\#}(\tau)}$, so $\xi \in \overline{\text{Aut}A}$. Clearly $g = f\xi$ so $X_g \cong X_f$. Since $|sg_\tau| = \prod_p |s_p g_{p\tau}|$, if $|sg_\tau| \neq 1$ then $|s_p g_{p\sigma}| \neq 1$ for some p and it follows from the local results that $\sigma \in \bigcup_{p^i|e} T_{g_p}(p^i)$. Hence $\sigma \in \bigcup_{p,i} T_{g_p}(p^i) = \bigcup_d T_g(d)$. $\square$

It is now easy to state and prove the type–isomorphism theorem.

3.5. Theorem. Let C be a cyclic group of order e, and A a completely decomposable group. For $f, g \in \text{ReMon}(C, \overline{A})$, $X_f \cong_t X_g$ if and only if $T_f = T_g$.

PROOF. By 3.2 the condition $T_f = T_g$ is necessary for $X_f \cong_t X_g$ to hold. Suppose that $T_f(d) = T_g(d)$ for all $d \mid e$. By 3.4 we may and do assume that $sf_\tau = sg_\tau = 0$ for $\tau \notin \bigcup T_f(d) = \bigcup T_g(d)$. This together with $T_f(d) = T_g(d)$ means that $|sf_\sigma| = |sg_\sigma|$ for every $\sigma \in \text{T}_{\text{cr}}(A)$. Choosing $\xi_\sigma \in \text{Aut}\,\overline{A}_\sigma$ such that $sf_\sigma \xi_\sigma = sg_\sigma$, we obtain $\xi = \prod \xi_\sigma \in \text{TypAut}\,\overline{A}$ so that $f\xi = g$. $\square$

Our next goal is to characterize the invariants $T_f(d)$, $d \mid e$, and to obtain complete systems of type–isomorphism invariants for almost completely decomposable groups with cyclic regulator quotients.

3.6. Lemma. Let $f \in \text{ReMon}(C, \overline{A})$ and write $e = \prod_{p|e} p^{e(p)}$. Then

(1) $\bigcup_{p^{e(p)}|d} T_f(d)$ has no lower bound in $\text{T}_{\text{cr}}(A)$ for each $p \mid e$.

(2) If $\sigma \in T_f(d)$, $\tau \in T_f(d')$, and $\sigma < \tau$ then $d \,\|\, d'$.

(3) If $\sigma(p) = \infty$ then $\sigma \notin \bigcup_{p|d} T_f(d)$.

PROOF. Without loss of generality $sf_\sigma = 0$ for $\sigma \notin \bigcup_{d|e} T_f(d)$ by 3.4. We have $\frac{e}{p} sf_\sigma = 0$ for $\sigma \notin \bigcup_{p^{e(p)}|d} T_f(d)$. If μ is a lower bound for $\bigcup_{p^{e(p)}|d} T_f(d)$ in $\mathrm{T_{cr}}(A)$ then $\frac{e}{p} sf \in \overline{A(\mu)}$ which contradicts 3.1. The conditions (2) and (3) follow directly from the definition of $T_f(d)$. □

We call a function T on the set of divisors of e to the power set of $\mathrm{T_{cr}}(A)$ satisfying the conditions of 3.6 a $d|e$-**partition** of $\mathrm{T_{cr}}(A)$. These $d|e$-partitions form a complete system of type–isomorphism invariants for $\mathrm{ReMon}(C, \overline{A})$.

3.7. Theorem. *Let C be cyclic of order e, and A a completely decomposable group. If T is a $d|e$-partition, then there is $f \in \mathrm{ReMon}(C, \overline{A})$ such that $T_f = T$.*

PROOF. As above we write $e = \prod_{p|e} p^{e(p)}$. Each $\overline{A_\sigma}$ is the direct sum of cyclic groups of the same order, and $p^{e(p)}$ divides $\exp \overline{A_\sigma}$ when $\sigma(p) < \infty$. Hence for $\sigma \in T(d)$ we have $d \mid \exp \overline{A_\sigma}$ (property (3) of $d|e$-partitions), and we can pick $x_\sigma \in \overline{A_\sigma}$ such that $|x_\sigma| = d$. Define $f : C \to \overline{A} : sf = \sum_{\sigma \in \mathrm{T_{cr}}(A)} x_\sigma$. Condition (1) of "$d|e$-partition" guarantees that $f \in \mathrm{ReMon}(C, \overline{A})$ and by construction $T_f(d) = T(d)$ for each $d \mid e$. □

Combining 3.5, 3.6, and 3.7, we obtain the main result of this paper. It is striking since it is barely more complicated than the result for the antichain case [MV2].

3.8. Theorem. *Let A be a completely decomposable group with critical typeset $\mathrm{T_{cr}}(A)$, and let C be a cyclic group of exponent e. Then the set of $d|e$-partitions of $\mathrm{T_{cr}}(A)$ is a complete system of type–isomorphism invariants for the almost completely decomposable groups with regulator A and regulator quotient C.* □

Evidently, the more comparable elements there are in $\mathrm{T_{cr}}(A)$, the fewer $d|e$-partitions exist. The extreme with most partitions occurs when $\mathrm{T_{cr}}(A)$ is an antichain.

In the antichain case

$$T_f(d) = \{\sigma \in \mathrm{T_{cr}}(A) \colon |sf_\sigma| = d\},$$

and it is automatic that $sf_\sigma = 0$ for $\sigma \notin \bigcup_{d|e} T_f(d)$.

3.9. Corollary. *([MV2], 2.1.) Let A be a completely decomposable group whose critical typeset $\mathrm{T_{cr}}(A)$ is an antichain, and let C be a cyclic group of exponent e. Let $f, g \in \mathrm{ReMon}(C, \overline{A})$. Then $X_f \cong_t X_g$ if and only if $|Cf_\sigma| = |Cg_\sigma|$ for all $\sigma \in \mathrm{T_{cr}}(A)$.* □

REFERENCES

[Ar82LN]   D. Arnold, *Finite Rank Torsion Free Abelian Groups and Rings*, Lecture Notes in Mathematics **931** (1982), Springer–Verlag.

[Bt83]   R. Burkhardt, *Elementary abelian extensions of finite rigid systems*, Communications in Algebra **11** (1983), 1473–1499.

[Bt84]   R. Burkhardt, *On a special class of almost completely decomposable torsion free abelian groups I*, Abelian groups and modules, C.I.S.M., Udine, Springer Verlag, 1984, pp. 141–150.

[DO]   M. Dugas and E. Oxford, *Near isomorphism invariants for a class of almost completely decomposable groups*, Manuscript (1991).

[F]   L. Fuchs, *Infinite Abelian Groups I,II*, Academic Press, 1970, 1973.

[KM]   K. J. Krapf and O. Mutzbauer, *Classification of almost completely decomposable groups*, Abelian groups and modules C.I.S.M., Udine, Springer Verlag, 1984, pp. 151–161.

[Kv83]   S. F. Kozhukhov, *On a classification of almost completely decomposable torsion-free abelian groups*, Iz. Vuz. Matematika, **27**(10) (1983), 29–36.

[La74]   E. L. Lady, *Almost completely decomposable torsion free abelian groups*, Proc. Amer. Math. Soc. **45** (1974), 41–47.

[MV1]   A. Mader and C. Vinsonhaler, *Classifying almost completely decomposable abelian groups I: General Theory*, Manuscript (1992).

[MV2]   A. Mader and C. Vinsonhaler, *Classifying almost completely decomposable abelian groups II: Applications*, Manuscript (1992).

[Mu91]   O. Mutzbauer, *Regulating subgroups of Butler groups*, Proceedings of the Curaçao Conference 1991 (to appear).

[Sch85]   Ph. Schultz, *Finite extensions of torsion-free groups*, Abelian Group Theory, Proceedings of the 3rd Oberwolfach Conference 1985, Gordon & Breach Science Publ., 1985, pp. 333–350.

Endomorphisms of
Valuated Torsion-free Modules

Warren May*

Department of Mathematics
University of Arizona, Tucson

We want to prove two results, one on endomorphism rings of valuated torsion-free abelian groups, the other on split representations of algebras by mixed abelian groups. Since we also want to allow modules over the p-adic integers, throughout this paper we shall let R denote a principal ideal domain which is not a field, and for which we have selected a representative from every associativity class of primes. When we refer to a prime of R, we mean one of these chosen representatives. A valuated R-module will be an R-module together with a p-valuation for every prime p.

When A.L.S. Corner proved in [2] that every countable reduced torsion-free ring is isomorphic to the endomorphism ring of a reduced torsion-free abelian group, he observed that this failed to be the case for such an uncomplicated p-adic algebra as $\hat{\mathbb{Z}}_p \times \hat{\mathbb{Z}}_p$. The hypothesis of countability was later replaced by Dugas and Göbel [5] with the condition that the algebra be cotorsion-free. We shall show that in the category $\mathcal{V}$ of valuated R-modules, the restriction of cotorsion-free can be removed as long as we are dealing with reduced torsion-free algebras.

Theorem 1. *Let A be a reduced torsion-free R-algebra of rank λ. Then there exists a reduced torsion-free valuated R-module F of rank 2λ such that $\mathrm{End}_{\mathcal{V}}(F) = A$.*

By using a construction of Richman and Walker [10], suitably generalized to the present situation, the valuated structure of F may be regarded as arising from a certain type of embedding into a mixed module. As an application, we obtain

Theorem 2. *Let A be a reduced torsion-free R-algebra of rank λ. Then there exists a reduced R-module M of torsion-free rank 2λ such that the torsion submodule T is a direct sum of cyclic modules, M/T is divisible and $\mathrm{End}_R(M) = A \oplus \mathrm{Hom}(M, T)$.*

As a corollary to Theorem 1, every reduced torsion-free p-adic algebra can be realized by a valuated p-adic module. For λ finite, the rank 2λ in both theorems will be shown to

---
*Supported in part by NSF grant INT 8822604

be minimal in that for every R and $\lambda \geq 2$, there exists an algebra A of rank λ requiring a module of rank 2λ. This will be more or less the example of Corner in [2], but different proofs are necessitated by the different settings. We point out that the novelty in Theorem 2 is achieving the minimal rank for every R. In the case of p-adic modules, previous constructions such as found in [3,4] produce modules of torsion-free rank $\geq 2^{\aleph_0}$ even for finite λ, while the results in [6] yield modules of rank 2λ, but do not realize every algebra A.

If p is a prime, we let $|x|_p$ denote the p-height of a module element. If the prime is clear from context, we shall just write $|x|$. We always let A denote a reduced torsion-free R-algebra, and F an A-module which is reduced and torsion-free as an R-module.

To prove Theorems 1 and 2 we shall need three lemmas. We shall assume the following setting for Lemmas 1 and 2. We suppose that we are given:

    (i) a set Γ and a sequence of integers $0 = n_0 < n_1 < n_2 < \ldots$;
    (ii) decompositions $F = \oplus_{\alpha \in \Gamma} F_{k\alpha} (k \in \omega)$ into A-submodules; and
    (iii) a p-valuation $v_{p\alpha}$ on F for every prime p and $\alpha \in \Gamma$.

Put $I_k = \{m \in \omega : n_k \leq m < n_{k+1}\}$ for $k \in \omega$. We shall further suppose that the valuations satisfy the following two conditions for every prime p, $x_1, x_2 \in F$, $\alpha, \alpha_1, \alpha_2, \in \Gamma$ and $k_1, k_2 \in \omega$:

    (a) if $|x_1|_p \leq |x_2|_p$, then $v_{p\alpha}(x_1) \leq v_{p\alpha}(x_2)$; and
    (b) if $k_1 < k_2$ and $|x_i|_p \in I_{k_i}$, then $v_{p\alpha_1}(x_1) < v_{p\alpha_2}(x_2)$.

We now define, for every prime p, a function v_p on F which will turn out to be a p-valuation. Let $x \in F$. If $|x|_p = \infty$, we define $v_p(x) = \infty$. If $|x|_p \neq \infty$, then there is a unique $k \in \omega$ with $|x|_p \in I_k$, and thus by (ii) a unique expression $x = \Sigma_{\alpha \in \Gamma} x_{k\alpha} (x_{k\alpha} \in F_{k\alpha})$. We now define $v_p(x) = \min_{\alpha \in \Gamma}\{v_{p\alpha}(x_{k\alpha})\}$.

Lemma 1. *Let p be a prime and $x, x_1, x_2 \in F$.*
    (1) If $|x|_p \in I_k$ and $x \in F_{k\alpha}$, then $v_p(x) = v_{p\alpha}(x)$.
    (2) If $k_1 < k_2$ and $|x_i|_p \in I_{k_i}$, then $v_p(x_1) < v_p(x_2)$.
    (3) If $a \in A$, then $v_p(ax) \geq v_p(x)$.

Proof. We drop the subscript p in the proof. We see (1) immediately since $x = x_{k\alpha}$. Note that in general, if $|x| \in I_k$, then $|x| = \min_{\alpha \in \Gamma}\{|x_{k\alpha}|\}$, hence $v(x) = \min\{v_\alpha(x_{k\alpha}) : |x_{k\alpha}| \in I_k\}$. Clearly, (2) follows from (b). To show (3), observe first that $|ax| \geq |x|$. We may suppose $|x| \in I_k$. If $|ax| \notin I_k$, then (2) gives the result; therefore, assume $|ax| \in I_k$. But $ax = \Sigma_\alpha ax_{k\alpha}$, and $|ax_{k\alpha}| \geq |x_{k\alpha}|$ for every α. By (a), we have $v_\alpha(ax_{k\alpha}) \geq v_\alpha(x_{k\alpha})$, from which (3) follows. ∎

Lemma 2. *The function v_p is a p-valuation on F.*

Proof. We will drop the subscript p in the following. Let $x \in F$ and let $u \in R$ be relatively prime to p. We first show $v(ux) = v(x)$ and $v(px) > v(x)$. If $|x| = \infty$, these are clear (using the convention $\infty > \infty$). Thus we may assume $|x| \in I_k$ and $x = \Sigma_\alpha x_{k\alpha}(x_{k\alpha} \in F_{k\alpha})$. Then $ux = \Sigma_\alpha ux_{k\alpha}$, from which $v(ux) = v(x)$ follows easily. Since $|px| > |x|$, Lemma 1(2) shows that $v(px) > v(x)$ unless $|px| \in I_k$. In that case, $px = \Sigma_\alpha px_{k\alpha}$ and $v_\alpha(px_{k\alpha}) > v_\alpha(x_{k\alpha})$, hence $v(px) > v(x)$ again.

Finally, we must show that $v(x + y) \geq \min\{v(x), v(y)\}$. Since it is clear if $|x| = |y| = \infty$, we may assume that $v(x) \leq v(y)$ and $|x| \in I_k$. By Lemma 1(2), $|y| \geq n_k$, hence

$|x+y| \geq n_k$. By the same reference, we are done if $|x+y| \geq n_{k+1}$, thus we may assume that $|x+y| \in I_k$. Express y as $y = \Sigma_\alpha y_{k\alpha}(y_{k\alpha} \in F_{k\alpha})$. Then $v(x+y) = \min_\alpha\{v_\alpha(x_{k\alpha}+y_{k\alpha})\} \geq \min_\alpha\{v_\alpha(x_{k\alpha}), v_\alpha(y_{k\alpha})\}$. If $|y| \in I_k$, then we have $\min\{v(x), v(y)\}$ and we are done. If $|y| \geq n_{k+1}$, then $|y_{k\alpha}| \geq n_{k+1}$ for every α. Thus (b) implies $v_\alpha(y_{k\alpha}) \geq v(x)$, and the proof is finished. ∎

The various v_p make F into a valuated module which we refer to as a segmented direct sum in $\mathcal{V}$. With aid of this idea, we now show that given only F and the decompositions (ii), there exists a valuated module structure on F having certain mapping properties.

Lemma 3. *Let F be an A-module which is reduced and torsion-free as an R-module. Assume we are given decompositions $F = \oplus_{\alpha \in \Gamma} F_{k\alpha}(k \in \omega)$ into A-submodules, where $|\Gamma| \leq 2^{\aleph_0}$. Then there exists a valuated module structure on F such that:*

(1) F is reduced as a valuated module;

(2) the values of the valuations on F are finite or ∞;

(3) the elements of A operate on F as valuated endomorphisms;

(4) if F is embedded in a valuated module M with torsion submodule T such that M/F is torsion, then $\phi(F_{k\alpha}) \subseteq F_{k\alpha}+T$ for every $k \in \omega$, $\alpha \in \Gamma$ and $\phi \in \mathrm{End}_\mathcal{V}(M)$.

Proof. We may assume $|\Gamma| = 2^{\aleph_0}$ by adding trivial summands to the decompositions. Moreover, it is not hard to show that we may reindex the decompositions and may take Γ to be a subset of 2^ω such that if $\alpha, \beta \in \Gamma, \alpha \neq \beta$, then for every $k \in \omega$ there exist infinitely many $\ell \in \beta \setminus \alpha$ such that $F_{\ell\gamma} = F_{k\gamma}$ for every $\gamma \in \Gamma$. To do this, suppose $D_0, D_1, \dots$ are the original decompositions. Choose $\Gamma' \subseteq 2^\omega$ of cardinal $2^{\aleph_0}$ such that $\alpha' \setminus \beta'$ is infinite whenever α' and β' are distinct elements of Γ'. Choose a bijection $\phi = (\phi_1, \phi_2) : \omega \to \omega \times \omega$. Now let the kth decomposition be $D_{\phi_1(k)}$, and put $\Gamma = \{\phi^{-1}(\omega \times \alpha') : \alpha' \in \Gamma'\}$.

Choose the sequence $\{n_i\}$ such that $n_{k+1}-n_k \to \infty$ as $k \to \infty$. By induction on k, one easily constructs strictly monotonic functions g and h from ω to ω such that $g(i) < h(j)$ and $h(n_{k+1} - 1) < g(n_{k+1})$ for all k and $i, j \in I_k$. To meet requirement (iii), we must define a p-valuation $v_{p\alpha}$ on F for every prime p and $\alpha \in \Gamma$. Let $x \in F$. If $|x|_p = \infty$, put $v_{p\alpha}(x) = \infty$. If $|x|_p \in I_k$, put $v_{p\alpha}(x) = g(|x|_p)$ if $k \in \alpha$, or $v_{p\alpha}(x) = h(|x|_p)$ if $k \notin \alpha$. It follows that $v_{p\alpha}(x_1) \leq v_{p\alpha}(x_2)$ if and only if $|x_1|_p \leq |x_2|_p$. Consequently, $v_{p\alpha}$ is a p-valuation on F. Requirements (i)-(iii) and (a) are now met, and it is clear that (b) holds. Defining v_p as before Lemma 1 for every prime p, we have made F into a valuated module. Clearly (2) and (3) are true, and (1) follows since if $x \in F$, $x \neq 0$, then $|x|_p < \infty$ for some p, thus $v_p(x) = v_{p\alpha}(x_{k\alpha}) < \infty$ for some α with $|x_{k\alpha}|_p < \infty$.

Now let M, k, α and ϕ be as in (4). If $z \in F_{k\alpha}$, we must show that $\phi(z) \in F_{k\alpha} + T$. We may choose $r \in R$, $r \neq 0$, with $r\phi(z) \in F$. Put $x = rz$, thus $\phi(x) \in F$. We now let $||_p$ denote p-height in F. We first claim that if p is a prime such that $|\phi(x)|_p < \infty, \beta \neq \alpha$ and $m \geq 0$, then there exist $t \geq 0$ and $\ell \in \beta \setminus \alpha$ such that $F_{\ell\gamma} = F_{k\gamma}(\gamma \in \Gamma), |p^t x|_p, |p^t \phi(x)|_p, m+ t \in I_\ell$ and $v_p(p^t \phi(x)) \geq h(|p^t x|_p)$. Note that $|p^t x|_p$ and $|p^t \phi(x)|_p$ must be finite since $|\phi(x)|_p$ is. By the choice of Γ and $\{n_i\}$, the existence of ℓ and t is clear except for the inequality. Using the facts that ϕ is an endomorphism in $\mathcal{V}$, that $x \in F_{k\alpha} = F_{\ell\alpha}$, and Lemma 1 (1), we obtain $v_p(p^t\phi(x)) \geq v_p(p^t x) = v_{p\alpha}(p^t x) = h(|p^t x|_p)$ as desired.

Next we claim that $\phi(x) \in F_{k\alpha}$. We may write $\phi(x) = \Sigma_\gamma y_{k\gamma}(y_{k\gamma} \in F_{k\gamma})$. For the sake of contradiction, we assume $y_{k\beta} \neq 0$ for some $\beta \neq \alpha$. Since F is reduced, there exists a prime p such that $|y_{k\beta}|_p = m < \infty$. Then $|\phi(x)|_p < \infty$, hence we may take t and ℓ as in the previous claim. We write $p^t\phi(x) = \Sigma_\gamma w_{\ell\gamma}(w_{\ell\gamma} \in F_{\ell\gamma})$. Since

$F_{\ell\gamma} = F_{k\gamma}$ for every γ, we conclude that $w_{\ell\beta} = p^t y_{k\beta}$, hence $|w_{\ell\beta}| = m + t \in I_\ell$. But then $g(m + t) = v_{p\beta}(w_{\ell\beta}) \geq v_p(p^t \phi(x)) \geq h(|p^t x|_p)$, using the previous claim. This contradicts the choice of g and h, proving the claim.

Recall $x = rz$. If we can show that $|\phi(x)|_p \geq |r|_p$ for every prime p (where $|r|_p$ is p-height in R), then $\phi(x) \in r F_{k\alpha}$ and we will have $\phi(z) \in F_{k\alpha} + T$ as desired. We may assume $|\phi(x)|_p < \infty$. Choose any $\beta \neq \alpha$ and $m \geq 0$ and apply the first claim above and Lemma 1(1). We get $h(|p^t \phi(x)|_p) = v_{p\alpha}(p^t \phi(x)) = v_p(p^t \phi(x)) \geq h(|p^t x|_p)$. By the nature of h, we have $|\phi(x)|_p \geq |x|_p = |rz|_p \geq |r|_p$, and we are done. ∎

It will be useful in the proof of Theorem 2 to note that the p-valuation v_p constructed in Lemma 3 has the property that $v_p(p^k x) - k \to \infty$ as $k \to \infty$ for every $x \in F$ with $|x|_p < \infty$. This can easily be seen since $v_p(p^k x) \geq g(|p^k x|_p)$ and $g(n_{k+1}) - g(n_k) \geq 2(n_{k+1} - n_k)$.

Proof of Theorem 1. We shall take F to be a free A-module together with certain decompositions. We use Lemma 3 to make F into a valuated module. By the lemma, $A \subseteq \text{End}_\nu(F)$ and taking $M = F$, we see that every $\phi \in \text{End}_\nu(F)$ will preserve the decompositions. Thus we are reduced to showing that F of the correct rank with suitable decompositions may be chosen which forces $\phi \in A$. These may be obtained by simple well-known means for $\lambda \leq 2^{\aleph_0}$ (see [1]). The upper limit is since we have $2^{\aleph_0}$ summands available in each decomposition. For $\lambda > 2^{\aleph_0}$, we can appeal to the infinite four submodules theorem [8] with $F = \oplus_\lambda A$.

For the convenience of the reader, we give decompositions for $\lambda \leq 2^{\aleph_0}$. First suppose $\lambda = n$ is finite, and let $a_1, \ldots, a_n$ be a torsion-free basis for A. Put $F = Ae \oplus Af$ and $\Delta = \{xe + yf \in F : x + y = 0\}$. We take the $n+2$ decompositions $F = Ae \oplus Af = Ae \oplus \Delta = Ae \oplus A(a_i e + 1f)(1 \leq i \leq n)$. By standard arguments, if ϕ preserves these decompositions, then $\phi \in A$. Now suppose $\aleph_0 \leq \lambda \leq 2^{\aleph_0}$, and that $\{a_i : i < \lambda\}$ is a torsion-free basis for A. Put $F = \oplus_{i<\lambda}(Ae_i \oplus Af_i)$ and $\Delta = \{(x_i e_i + y_i f_i)_i : \Sigma(x_i + y_i) = 0\}$. We take the three decompositions $F = (\oplus_\lambda Ae_i) \oplus (\oplus_\lambda Af_i) = Ae_0 \oplus \Delta = \oplus_\lambda(Ae_i \oplus A(a_i e_i + 1f_i))$. ∎

Proof of Theorem 2. Take F from the proof above. We must modify the construction of Richman and Walker in [10] to embed F as a nice valuated submodule of an A-module $\widetilde{M}$ such that $\widetilde{M}/F$ is a torsion module with simply presented components. In the p-local part of the construction, we must take the free A-module on the given generators rather than the free $\mathbb{Z}_p$-module. The set U will consist of representatives of the cosets of pA in A. The important point is that A is a torsion-free R-module so that a nonzero element of A is either uniquely a power of p times an element of U or else of p-height ∞.

Let M be $\widetilde{M}$ modulo its maximal divisible torsion submodule. Then F is naturally embedded as a valuated submodule of M, and the torsion T of M is reduced. Since F is reduced as a valuated module, M/F is torsion, and T is reduced, it follows that M is reduced. Referring to the construction of $\widetilde{M}$, the p-heights which occur are either finite or ∞ since those are what occur in F. Since the torsion of $\widetilde{M}/F$ has simply presented components and F is nice, it follows that T is a direct sum of cyclic modules. To see that M/T is divisible, it suffices to note that if $x \in F$, p is a prime, and $|x|_p < \infty$, then the remark after Lemma 3 shows that the p-height sequence of x has infinitely many gaps. Since M and F have the same torsion-free rank, we have shown everything except the mapping property of M.

We have $A \oplus \text{Hom}(M, T) \subseteq \text{End}_R(M)$ since M is an A-module and F is a torsion-free faithful A-module. Let $\phi \in \text{End}_R(M)$. Lemma 3 implies that $\phi(F_{k\alpha}) \subseteq F_{k\alpha} + T$ for all k and α. Identifying F with $(F + T)/T$, ϕ induces an endomorphism of F preserving the decompositions. Therefore, there exists $a \in A$ with $(\phi - a)(F) \subseteq T$. Thus $\phi - a \in \mathrm{Hom}(M, T)$ as desired. ∎

Remark. In both theorems one can construct families of modules with maps between distinct modules mapping into torsion (which is 0 in Theorem 1). To do this, in the proof of Lemma 3, the set Γ can be partitioned into $2^{\aleph_0}$ sets, each of which can be used to construct different F's. For $\lambda > 2^{\aleph_0}$, the four submodules theorem provides 2^λ families of decompositions of F which yield different valuated modules.

We now examine limitations to the theorems. In work of Dugas [4] and of Corner and Göbel [3], split realization theorems are given for reduced torsion-free algebras A in which $\mathrm{End}_R(M) = A \oplus Bd(M)$, where $Bd(M)$ is the ideal of bounded endomorphisms of M. We show next that this is impossible to achieve with modules of countable torsion-free rank if there is any unbounded primary torsion. Thus low rank split realization theorems such as Theorem 2 must be of a weaker nature.

Proposition 1. *Let M be a reduced R-module with unbounded p-torsion submodule for some prime p such that $\mathrm{End}_R(M) = A \oplus Bd(M)$. Then M has uncountable torsion-free rank.*

Proof. Suppose for the sake of contradiction that the rank of M is countable. We first claim that there exists a submodule $B \subseteq M$ such that M/B is unbounded reduced p-torsion. If Q denotes the quotient field of R, we have natural maps $M \to \mathrm{Ext}(Q/R, M) \to \mathrm{Ext}(R(p^\infty), M)$. The kernel of the composite is the maximal p-divisible submodule of M. Let $\widehat{R}_p$ be the completion of R in the p-adic topology. Then $\mathrm{Ext}(R(p^\infty), M)$ is an $\widehat{R}_p$-module, thus we may let $\widetilde{M}$ be the $\widehat{R}_p$-submodule generated by the image of M. The p-component of the torsion T of M is injected into $\widetilde{M}$. The torsion-free rank of $\widetilde{M}$ over $\widehat{R}_p$ is countable and $\widetilde{M}$ is reduced. Applying Lemma 11 from [9], there exists a nice torsion-free $\widehat{R}_p$-submodule $\widetilde{B}$ of $\widetilde{M}$ such that $\widetilde{M}/\widetilde{B}$ is p-torsion. Moreover, $\widetilde{M}/\widetilde{B}$ is reduced since $\widetilde{B}$ is nice in $\widetilde{M}$. Let B be the submodule of M which maps into $\widetilde{B}$. Then B satisfies the claim.

Now let $H = \{\alpha \in \mathrm{End}_R(M) : \alpha(B) = 0, \alpha(M) \subseteq T\}$. Then $H \cong \mathrm{Hom}(M/B, T)$, which is complete and has unbounded torsion submodule. The torsion submodule of H is $H \cap Bd(M)$, hence $H/(H \cap Bd(M))$ contains a nontrivial divisible submodule. But then the same would be true of $\mathrm{End}_R(M)/Bd(M)$, and equally well of $\mathrm{End}_R(M)$ and of M, contrary to hypothesis. ∎

We remark that if the torsion submodule of M consists of infinitely many nonzero components, each of which is bounded, then the proposition may fail. Finally, we show that for $\lambda = n \geq 2$ finite, the rank in the theorems cannot be lowered beyond $2n$.

Proposition 2. *Let $n \geq 2$. Then there exists a reduced torsion-free R-algebra A of rank n such that:*
(1) if F is a torsion-free valuated R-module with $\mathrm{End}_V(F) = A$, then rank $M \geq 2n$;
(2) if M is a reduced R-module with totally projective torsion T such that $\mathrm{End}_R(M) = A \oplus \mathrm{Hom}(M, T)$, then torsion-free rank $M \geq 2n$.

Proof. Fix a prime p of R and let R_p be the localization of R at the prime ideal generated by p. For R_p-modules, R-homomorphisms are the same as R_p-homomorphisms, and both notions of rank are the same. Therefore, we may assume that R is a discrete valuation ring with prime p. Let $A = R[q]$, where $q^n = p$ in some algebraic closure of the quotient field of R. Since q has degree n over the quotient field by Eisenstein's criterion, we have that $A = R \oplus Rq \oplus \ldots \oplus Rq^{n-1}$ has rank n. Note for the moment that $Aq = Rp \oplus Rq \oplus \ldots \oplus Rq^{n-1}$, hence $A/Aq \cong R/Rp$ is a simple A-module. We claim that every proper nonzero ideal of A is generated by a positive power of q. Thus A will be a discrete valuation ring with prime q. Let I be such an ideal of A. Since A is integral over R, it follows that $I \cap R$ is a proper nonzero ideal of R (look at the constant term of a minimal integral equation for a nonzero element of I). Thus some power of p, and hence of q, lies in I. Let $q^m \in I$, where we assume $m \geq 1$ is minimal. It will suffice to derive a contradiction from the existence of $\alpha \in I \setminus Aq^m$. We may assume $\alpha q \in Aq^m$, thus $\alpha \in Aq^{m-1} \setminus Aq^m$. But $Aq^{m-1}/Aq^m \cong A/Aq$, which we have noted is simple. Thus $I \supseteq Aq^{m-1}$, contrary to the choice of m, and the claim is shown.

If either F or M has rank $< 2n$, then its rank as an A-module is exactly one since neither can be torsion. For the sake of contradiction, we shall assume that this is the case. For F as in (1), we may then assume that $F = A$. Since the valuation layers of F are A-submodules, they are all of the form Aq^m. For a contradiction, we need only produce $\phi \in \mathrm{End}_R(A)$ such that $\phi(Aq^m) \subseteq Aq^m$ for every m, but $\phi \notin A$. We observe that $Aq^m = \oplus_{0 \leq i < n} Rq^{m+i} = \oplus_{0 \leq i < n} Rp^{k_i}q^i$ for appropriate k_i. We may take ϕ to be the identity map on R and zero on Rq^i for $0 < i < n$ to obtain the desired contradiction.

For M as in (2), we choose $x \in M$ torsion-free. Since M has rank 1 as an A-module, Ax is a nice A-submodule of M with respect to q-height. Observe that if $y \in M$ and $|y|_p = \sigma_0 + k$, where σ_0 is a limit ordinal and $k < \omega$, then $|y|_q = \sigma_0 + nk$. Consequently, Ax is a nice R-submodule of M with respect to p-height. The p-height layers in Ax are of the form $Aq^m x$, therefore the map ϕ constructed above (using $A \cong Ax$) does not decrease p-height. Since M/Ax is totally projective, ϕ can be extended to an endomorphism $\tilde{\phi} \in \mathrm{End}_R(M)$. It is easy to see that $\tilde{\phi} \notin A \oplus \mathrm{Hom}(M, T)$, thus we have a contradiction in this case also. ∎

REFERENCES

1. S. Brenner and M. C. R. Butler, Endomorphism rings of vector spaces and torsion-free abelian groups, *J. London Math. Soc.* **40** (1965), 183-187.

2. A. L. S. Corner, Every countable reduced torsion-free ring is an endomorphism ring, *Proc. London Math. Soc.* (3), **13** (1963), 687-710.

3. A. L. S. Corner and R. Göbel, Prescribing endomorphism algebras, a unified treatment, *Proc. London Math. Soc.* (3), **50** (1985), 447-479.

4. M. Dugas, On the existence of large mixed modules, "Abelian Group Theory," Proc. Honolulu Conf. 1982/83, 412-424, Springer Lecture Notes in Mathematics, Vol. **1006**, Springer, Berlin, 1983.

5. M. Dugas and R. Göbel, Every cotorsion-free algebra is an endomorphism algebra, *Math. Z.* **181** (1982), 451-470.

6. B. Franzen and B. Goldsmith, On endomorphism algebras of mixed modules, *J. London Math. Soc.* **31** (1985), 468-472

7. L. Fuchs, "Infinite Abelian Groups," Vols. I, II, Academic Press, New York, 1970, 1973.

8. R. Göbel and W. May, Four submodules suffice for realizing algebras over commutative rings, *J. Pure Appl. Algebra* **65** (1990), 29-43.

9. W. May, Isomorphism of endomorphism algebras over complete discrete valuation rings, *Math. Z.* **204** (1990), 485-499.

10. F. Richman and E. A. Walker, Valuated groups, *J. Algebra* **56** (1979), 145-167.

11. P. Schultz, The endomorphism ring of a valuated group, "Abelian Group Theory," Proc. Perth Conf. 1987, 75-84, *Contemporary Mathematics*, Vol. **87**, American Mathematical Society, Providence, 1989.

Regulating Subgroups of Butler Groups

OTTO MUTZBAUER

Mathematisches Institut, Universität Würzburg, Germany

1. Introduction

Krapf and Mader showed independently that there are only finitely many regulating subgroups in a Butler group. Both proofs are presented here. An example of nested regulating subgroups is given. This doesn't happen for almost completely decomposable groups.

A Butler group B is said to be *torsionless* if $B^*(t)_* = B^*(t)$ for all types, and *regulating torsionless* if all regulating subgroups are torsionless. For regulating torsionless Butler groups all regulating subgroups are known to have the same index. For such groups the number of regulating subgroups is proved to be $\prod_{t \in T} i^B(t)^{r_t}$, the same as for almost completely decomposable groups, where $T = \{t \mid B^*(t)_* \neq B(t)\}$ is the *critical typeset*, $i^B(t)$ is the index of one, i. e. of all, regulating subgroups of $B(t)$ and r_t is the rank of a type complement, i. e. a direct complement of $A^*(t)_*$ in $A(t)$.

An important step in showing this result is a statement of Mader, which controls the relation between two different type complements of a Butler group. A stronger result, Theorem 3.3, is included which is of interest in itself.

2. Regulating Subgroups

We mainly use the notation of Fuchs [6] and Arnold [1]. A *quasi-summand* of a torsion-free abelian group A is a subgroup B such that there is a second subgroup C with $B \cap C = 0$ and $A/(B \oplus C)$ of finite exponent. If the type subgroup $A^*(t)_*$ of a torsion-free abelian group A is a direct summand of the type subgroup $A(t)$, i. e. $A(t) = A_t \oplus A^*(t)_*$, then the complement A_t is called *type complement* or *t-complement*. It is a homogeneous group of type t. For a torsion-free abelian group A the set $T(A) = \{t \mid A(t) \neq A^*(t)_*\}$ is called *critical typeset*.

If the torsion-free abelian group A has typeset T and if for all $t \in T$ we have $A(t) = A_t \oplus A^*(t)_*$ then subgroups $U = \sum_{t \in T} A_t$ are called *regulating subgroups*. The equation $A(t) = A_t \oplus A^*(t)_*$ is called *Butler equation for* t. If the defining equations for type complements are quasi-equations, i. e. $A(t)/(A_t \oplus A^*(t)_*)$ is of finite exponent, then we use terms like *quasi type complements* and *quasi-regulating subgroups*. Note that quasi type complements A_t are in general neither pure in A nor completely decomposable.

In Butler groups there is a Butler equation for all types t. Thus there are always regulating subgroups and they are of finite index.

Whereas in almost completely decomposable groups all regulating subgroups have the same index, it may happen in Butler groups that regulating subgroups are properly contained in each other.

Following Burkhardt [3] we introduce the intersection of all regulating subgroups and call it the *regulator* of a Butler group. Krapf proved in 1986 [8, Satz 31], and Mader [10] independently in 1990, that a Butler group has only finitely many regulating subgroups. Thus the regulator of a Butler group B is a uniquely determined subgroup $R(B)$ of finite index and the isomorphism type of the quotient is an invariant of the group. Moreover it makes sense to iterate the regulator defining a descending chain of regulators of a Butler group B by $R^{n+1}(B) = R\left(R^n(B)\right)$. It is an open question whether the so called *hyper-regulator* $R^\infty(B) = \bigcap_n R^n(B)$ is always of finite index in B. All the isomorphism types of the (finite) regulator quotients $R^{n+1}(B)/R^n(B)$ are invariants of B. In almost completely decomposable groups the hyperregulator equals the regulator.

For the sake of completeness we include the theorem of Krapf (in a slightly generalized version) and of Mader with the original proofs.

Theorem 2.1. [Krapf] *Let A be a torsion-free abelian group with finite typeset T. Let $A^*(t)$ be a quasi-summand of $A(t)$ and A_t a quasi-t-complement for all $t \in T$. Let $e_t = \exp\left(A(t)/(A_t \oplus A^*(t))\right)$. Let U be a quasi-regulating subgroup. Then the exponent of A/U is finite dividing $\prod_{t \in T} e_t$.*

In particular, a Butler group B has only finitely many regulating subgroups U and the exponent of B/U divides $\prod_{t \in T} |B^*(t)_*/B^*(t)|$.

Proof. Let $T = \{t_1, \ldots, t_k\}$ be an indexing of T such that t_i is minimal in the set $\{t_i, \ldots, t_k\}$ for all $1 \leq i \leq k$. Let $A_l \oplus A^*(t_l)$ be of finite exponent in $A(t_l)$ for all $1 \leq l \leq k$. Let $U = \sum_{l=1}^{k} A_l$ be a quasi-regulating subgroup. Let $A = K_0 \supset K_1 \supset \ldots \supset K_k = U$ be the chain with $K_l = \sum_{m \leq l} A_m + \sum_{m > l} A(t_m)$ for all l. Then since $A^*(t_l) = \sum\{A(t_m) \mid m > l, t_m > t_l\} \subset \sum_{m > l} A(t_m)$, the quotient

$$\frac{K_{l-1}}{K_l} = \frac{\sum_{m < l} A_m + A(t_l) + \sum_{m > l} A(t_m)}{\sum_{m < l} A_m + A_l + A^*(t_l) + \sum_{m > l} A(t_m)}$$

is of finite exponent dividing e_{t_l}. Thus the exponent of U in A divides the product of all e_t. If B is a Butler group, the Butler equations imply that $e_t = \exp\left(B^*(t)_*/B^*(t)\right)$. ∎

Theorem 2.2. [Mader] *Let B be a Butler group with typeset T. Let d be the longest path in the typeset T considered as a poset. Let e be the least common multiple of the exponents of $B^*(t)_*/B^*(t)$ for all $t \in T$. Then the indices of all regulating subgroups divide the order of $B/e^d B$.*

Proof. Let C be a regulating subgroup. Set $e = \text{lcm}\{\exp(B^*(t)_*/B^*(t)) \mid t \in T\}$ and let $d(x)$ denote the depth of $t(x)$ in the type graph (the poset T) of B, i. e. the longest path from a maximal type down to the type of x. We will prove by induction on $d(x)$

that $e^{d(x)}x \in C$. Since T is finite there is a largest depth d and $e^d B \subset C$ so B/C is finite because B has finite rank.

If $d(x) = 0$ then $t(x)$ is maximal in T so $B^*(t(x))_* = 0$ and $x \in B(t(x)) = C_{t(x)} \subset C$. Now suppose $0 \neq x \in B$ and for all $y \in B$ with $t(y) > t(x)$ it is true that $e^{d(y)}y \in C$. Let $t = t(x)$. Then $x \in B(t) = C_t \oplus B^*(t)_*$ so $x = c + y$ with $c \in C_t$ and $y \in B^*(t)_*$. If $t(y) > t$ then $d(y) < d(x)$ and by induction hypothesis $e^{d(x)}x = e^{d(x)}c + e^{d(x)}y \in C$. Otherwise $t(y) = t$ and $d(y) = d(x)$. By choice of e we have $ey \in B^*(t)$ and $ey = \sum_{s>t} y_s$ with $t(y_s) = s$. By induction hypothesis $e^{d(y)}y = \sum_{s>t} e^{d(y)-1}y_s \in C$ so $e^{d(x)}x = e^{d(x)}c + e^{d(x)}y \in C$ as claimed. ∎

The next Proposition gives an example of nested regulating subgroups and answers also the question, whether the regulator can be equal to a regulating subgroup without being isomorphic to each regulating subgroup. The subgroup U in Proposition 2.3 is the regulator of G and a so called *regulator equal* group, i. e. equals its own regulator. Torsionless Butler groups (or B_0-groups), cf. section 4, are regulator equal but the converse is not true, see the example of Arnold-Vinsonhaler [2, Example 4]. By the way, the hyperregulator is always regulator equal and the class of regulator equal Butler groups is especially interesting.

Proposition 2.3. *There is a Butler group containing non-isomorphic regulating subgroups U and V where $U \subset V$ and U is regulator equal.*

Proof. Let A, B, C, D, X, Y be rational groups containing 1 but not p^{-1} for a fixed prime p, such that B, C, D, X and Y form a rigid system. Let $A = B \cap C \not\cong \mathbf{Z} = B \cap D = B \cap X = B \cap Y = C \cap D = C \cap X = C \cap Y = D \cap X = D \cap Y = X \cap Y$. Let the Butler group G be defined by

$$G = Aa + Bpb + Cpc + Dpd + X(a+b+d) + Y(c-d) \subset \mathbf{Q}a \oplus \mathbf{Q}b \oplus \mathbf{Q}c \oplus \mathbf{Q}d.$$

The group G is one of its own regulating subgroups, because up to t_A all critical types are maximal, i. e. $G(t_B) = G_{t_B}$ etc. and $G^*(t_A)_* = \langle Bpb \oplus Cpc, b+c \rangle$ and

$$G(t_A) = Aa \oplus G^*(t_A)_* = A(a+b+c) \oplus G^*(t_A)_*.$$

But also

$$U = A(a+b+c) + Bpb + Cpc + Dpd + X(a+b+d) + Y(c-d)$$

is a regulating subgroup. U is strictly contained in G since $a \notin U$.

Next we assume in addition that none of the rational groups $A, \ldots, Y$ has any q-divisibility, q a prime number. To show that U is not isomorphic to G observe that the incomparability of the relevant types force an assumed isomorphism between G and U to map b to $\pm b$, c to $\pm c$ and d to $\pm d$. Moreover $Y(c-d)$ and $X(a+b+d)$ have to be mapped onto themselves. Hence the signs $\pm$ coincide for c and d. Moreover, there are precisely four possibilities, namely,

$$(a,b,c,d) \rightarrow (\pm a, \pm b, \pm c, \pm d) \quad \text{or} \quad (a,b,c,d) \rightarrow (\pm(a+2b), \mp b, \pm c, \pm d),$$

taking either the upper or the lower signs. In all of these cases U is mapped onto U, contradicting $U \neq G$. This shows the proposition. ∎

A typeset T is said to be *V-free* if for all types $t \in T$ the subsets $\{s \in T \mid s \geq t\}$ are chains. In particular antichains are V-free. The "V"-constellation of the types $\{t(A), t(B), t(C)\}$ in Proposition 2.3 is necessary to obtain more than one regulating subgroup.

Proposition 2.4. *Let B be a Butler group with critical typeset T which is V-free. Then B has a unique regulating subgroup, namely $R = \sum_{t \in T} B(t)$, i. e. the regulator is regulating.*

Proof. Let $X = \sum_{t \in T} B_t$ be regulating in B. By hypothesis $B(t)$ has linearly ordered critical typeset and is completely decomposable by Butler [1, 1.4], thus regulating in itself and $B(t) = \sum_{s \geq t} B_s$. Hence $X = \sum_{t \in T} B(t)$ is the unique regulating subgroup. ∎

Proposition 2.5. *The regulating subgroup W of a Butler group B is the unique maximal regulating subgroup in B if for all critical types $W(t) = B(t)$.*

Proof. Let $V = \sum_{t \in T} B_t$ be any regulating subgroup of B. Then $W(t) = B(t) = B_t \oplus B^*(t)_*$ for all critical types t. Hence $B_t \subset W$ for all critical types t and $V \subset W$. ∎

Proposition 2.5 directly implies the following corollary.

Corollary 2.6. *Let W be a minimal regulating subgroup of the Butler group B with critical typeset T. If $B(t) = W(t)$ for all critical types, then W is the regulator of B and $W = \sum_{t \in T} B(t)$.* ∎

Proposition 2.3 shows that $W(t)$ need not be equal to $B(t)$ for all critical types t if W is the regulator of B.

The next proposition was proved by Mader in an unpublished note. The original proof is included for the sake of completeness.

Proposition 2.7. *[Mader] If the regulator R of the Butler group B is the unique regulating subgroup, then $R = \sum \{ B(t) \mid t \text{ critical} \}$ and the regulator is fully invariant.*

Proof. Let $R = \sum R_t$. Let s be a critical type and $\phi \in \text{Hom}\left(R_s, B^*(s)_* \right)$. Then $R_s(1 + \phi) + \sum_{t \neq s} R_t$ is another regulating subgroup of B and hence equals R. It follows that $\text{Im}\, \phi \subset R \cap B^*(s)_*$. Since R_s has a rank one summand of type s and every element of $B^*(s)_*$ has type $\geq s$ it follows that $B^*(s)_* \subset R$ and therefore $B(s) = R_s \oplus B^*(s)_* \subset R$. ∎

3. Type Complements

There is a very useful result of Mader [9, 2.6] dealing with complements of direct summands which we use in a specialized form, cf. Corollary 3.2. The following corollary could be replaced also by an argument of Metelli [11, 2.2].

Lemma 3.1. *[9, 2.6] Let M be a module, $M = A \oplus B$. Then there is a bijection between the set $\mathcal{B}$ of all direct complements of A in M and the homomorphism group $\text{Hom}(B, A)$ given by $\text{Hom}(B, A) \longrightarrow \mathcal{B}$ where $f \to B(1 + f)$ and $B(1 + f) = \{ b + bf \mid b \in B \}$.* ∎

Corollary 3.2. *[Mader] Let B be a Butler group with critical type t and t-complement B_t.*
(1) *The different t-complements are precisely the groups $B_t(1 + f)$ where*
    $$f \in \text{Hom}\left(B_t, B^*(t)_* \right).$$
(2) *Let $B_t = \bigoplus_i Sx_i$ where $1 \in S \subset \mathbf{Q}$. Then all different t-complements are uniquely represented by $\bigoplus_i S(x_i + y_i)$ where $y_i \in B^*(t)_*$ with $\chi(y_i) \geq \chi(x_i)$ for all i.*

Proof. (1) is obviously a special case of Lemma 3.1.

(2): Let B'_t be another t-complement. Then there is an $f \in \mathrm{Hom}\Big(B_t, B^*(t)_*\Big)$ such that $B'_t = B_t(1 + f) = \bigoplus_i Sx_i(1 + f) = \bigoplus_i S(x_i + x_i f)$. Hence all t-complements are of the claimed form where $y_i = x_i f$.

Conversely, let $y_i \in B^*(t)_*$ with $\chi(y_i) \geq \chi(x_i)$. Then there is an $f : B_t \longrightarrow B^*(t)_*$ given by $x_i f = y_i$ and $B_t(1 + f) = \bigoplus_i S(x_i + y_i)$. Assuming $\bigoplus_i S(x_i + y_i) = \bigoplus_i S(x_i + y'_i)$ we have $B_t(1 + f) = B_t(1 + f')$ with $x_i f = y_i$, $x_i f' = y'_i$ thus $f = f'$, $y_i = y'_i$. ∎

Next we want to develop a method to describe all regulating subgroups in a Butler group in terms of a given regulating subgroup. With this in mind we describe the connection between one t-complement of a Butler group and one of its subgroups of finite index. The following theorem is much stronger than needed here, but it is of interest in itself.

Theorem 3.3. *Let B be a Butler group with critical type t. Let U be a subgroup of finite index. Let X and Y be t-complements of B and U, respectively. Let S be a rational group of type t, containing 1.*

There are elements $x_1, \ldots, x_r \in X$, $y_1, \ldots, y_r \in Y$, $b_1, \ldots, b_r \in B^*(t)_*$ and natural numbers $d_1, \ldots, d_r$ such that
(1) $X = \bigoplus_{i=1}^r Sx_i$, $Y = \bigoplus_{i=1}^r Sy_i$; thus $\chi^B(x_i) = \chi^U(y_i) = \chi^S(1)$;
(2) $d_1 | d_2 | \ldots | d_r$;
(3) $y_i = d_i x_i + b_i$ for all $1 \leq i \leq r$.
Moreover,

$$\frac{B(t)}{U(t) + B^*(t)_*} \cong \bigoplus_{i=1}^r \mathbf{Z}(d_i).$$

Proof. U is a Butler group like B and $X = \bigoplus_{i=1}^r Sz_i \cong Y = \bigoplus_{i=1}^r Sw_i$ are completely decomposable homogeneous of type t.

Let $R = \langle p^{-1} \mid pS = S \rangle_{\mathrm{ring}}$ be the unitary rational ring of type $t : t$. Because of $U(t) = Y \oplus U^*(t)_* \subset B(t) = X \oplus B^*(t)_*$ there is an $r \times r$-matrix (α_{ij}) with entries in S and $c_i \in B^*(t)_*$ such that $w_i = \sum_{j=1}^r \alpha_{ij} z_j + c_i$ for all $1 \leq i \leq r$. In fact, $\alpha_{ij} \in R$ because otherwise the characteristic of w_i could not be equal to the characteristic of z_j.

Suppose different bases $X = \bigoplus_{i=1}^r Sx_i = \bigoplus_{i=1}^r Sz_i$ and $Y = \bigoplus_{i=1}^r Sy_i = \bigoplus_{i=1}^r Sw_i$. Then there are $r \times r$-matrices (λ_{kl}) and (η_{nf}) with entries in S such that $x_k = \sum_{l=1}^r \lambda_{kl} z_l$ and $y_n = \sum_{f=1}^r \eta_{nf} w_f$ for all $1 \leq k, n \leq r$. By the same argument on characteristics we have $\lambda_{kl}, \eta_{nf} \in R$. Moreover, these matrices describe basis transformations hence they are invertible over R. Let $(\hat{\lambda}_{jm}) = (\lambda_{kl})^{-1}$ be the inverse. Then for all $1 \leq n \leq r$

$$y_n = \sum_{f=1}^r \sum_{j=1}^r \sum_{m=1}^r \eta_{nf} \alpha_{fj} \hat{\lambda}_{jm} x_m + \sum_{f=1}^r \eta_{nf} c_f.$$

The ring R is a principal ideal domain, hence by [7, Theorem 3.8] the matrix (α_{fj}) is equivalent to a diagonal matrix $\mathrm{diag}(d_1, \ldots, d_r) = (\eta_{nf})(\alpha_{fj})(\hat{\lambda}_{jm})$, with suitable choice of the relevant bases, where $d_1, \ldots, d_r \in R$ and $d_1 | d_2 | \ldots | d_r$. As $\eta_{nf} \in R$ and $Sc_f = S(w_f - \sum_{j=1}^r \alpha_{fj} z_j) \subset B^*(t)_*$, we have $b_n = \sum_{f=1}^r \eta_{nf} c_f \in B^*(t)_*$. Thus

$$y_i = d_i x_i + b_i \quad \text{for all} \ 1 \leq i \leq r,$$

where $d_i \in R$. The invariant factors d_i cannot be 0, because $y_n \notin B^*(t)_*$. Further, these invariant factors are unique up to multiplication by units of R. The denominators of elements in R are units in R. Hence the invariant factors d_i can be assumed to be natural numbers and (1) to (3) are proved.

We have $b_i \in B^*(t)_*$ and certainly by (3) $\chi^B(b_i) \geq \chi^B(x_i) = \chi^B(x_j)$. Further,

$$\frac{B(t)}{U(t) + B^*(t)_*} = \frac{X \oplus B^*(t)_*}{Y \oplus B^*(t)_*} = \frac{\bigoplus_{i=1}^{r} Sx_i \oplus B^*(t)_*}{\bigoplus_{i=1}^{r} S(d_i x_i + b_i) \oplus B^*(t)_*} \cong \bigoplus_{i=1}^{r} \mathbf{Z}(d_i). \ \blacksquare$$

Corollary 3.4. *Let B be a Butler group with critical type t. Let U be a subgroup of finite index. Suppose that $B^*(t)_* = U^*(t)_*$. Then for each t-complement B_t of B there is a t-complement U_t of U such that $U_t \subset B_t$.*

Proof. Let U'_t be any t-complement of $U^*(t)_*$ in $U(t)$. By 3.3 $B_t = \bigoplus_{i=1}^{r} Sx_i$ and $U'_t = \bigoplus_{i=1}^{r} Sy_i$, where $y_i = d_i x_i + b_i$. Thus $\chi(b_i) \geq \chi(y_i)$ and by Corollary 3.2 $U_t = \bigoplus_{i=1}^{r} S(y_i - b_i)$ is the desired t-complement of U contained in B_t. $\blacksquare$

Without the hypothesis in Corollary 3.4 a type complement of B need not contain a type complement of the subgroup U.

The following corollary is a direct consequence of Corollary 3.2. By the way, the stronger Theorem 3.3 can also be used to prove the corollary, observing that $d_1 = \ldots = d_r = 1$ if regulating subgroups are dealt with.

Corollary 3.5. *Let B be a Butler group with critical typeset T. Let S_t be rational groups of type t, containing 1, for all $t \in T$. Let $B_t = \bigoplus_{i=1}^{r_t} S_t b_{ti}$ be type complements. Let $U = \sum_{t \in T} B_t$ be a fixed regulating subgroup. Let $b'_{ti} \in B^*(t)_*$ with characteristic $\chi(b'_{ti}) \geq \chi(b_{ti})$ for all t and i. Then $U' = \sum_{t \in T} B'_t$ with $B'_t = \bigoplus_{t \in T} S_t(b_{ti} + b'_{ti})$ is a regulating subgroup and all regulating subgroups are obtained this way.* $\blacksquare$

4. Regulating Torsionless Butler Groups

A Butler group B is called *torsionless* or a B_0-group, see [4] and [1], if $B^*(t) = B^*(t)_*$ for all types t. A Butler group B is called *regulating torsionless* if all regulating subgroups are torsionless. A Butler group may contain a torsionless regulating subgroup without being regulating torsionless, see the subgroup U of G in Proposition 2.3. In particular, almost completely decomposable groups are regulating torsionless.

Lemma 4.1. *If B is a torsionless Butler group, then $B(t) = \sum_{s \geq t} B_s$, $B^*(t)_* = B^*(t) = \sum_{s > t} B_s$ for any collection of type complements of B. In particular, all type subgroups $B(t)$, $B^*(t)_*$ are torsionless for all types t.*

Proof. The lemma follows from [1, 2.2]. $\blacksquare$

Proposition 2.3 shows that a Butler group which equals one of its regulating subgroups need not be torsionless.

Lemma 4.2. *In regulating torsionless Butler groups all type subgroups $B(t)$, $B^*(t)_*$ are regulating torsionless for all types t and all regulating subgroups have the same index.*

Proof. The lemma follows from [1, 3.1 (d)] and [1, 4.2]. $\blacksquare$

Corollary 4.3. *A regulating subgroup W of the regulating torsionless Butler group B with critical typeset T is the regulator of B if $W(t) = B(t)$ for all critical types; and then $W = \sum_{t \in T} B(t)$.*

Proof. By Lemma 4.2 and hypothesis all regulating subgroups are minimal and maximal. Thus Corollary 2.6 shows the rest. ∎

Remark 4.4. *Two regulating subgroups U, U' in a Butler group B with critical typeset T are equal if and only if $U(t) = U'(t)$ for all $t \in T$, because for type complements B_t we have $U = \sum_{t \in T} B_t = \sum_{t \in T} U(t)$ and similarly for U'. Hence $U = U'$ if and only if $U(t) = U'(t)$ for all $t \in T$.*

For regulating torsionless Butler groups there is a more detailed criterion for equality of regulating subgroups. Observe that Lemma 4.5 deals with arbitrary regulating subgroups U, U' in view of Corollary 3.5.

Lemma 4.5. *Let B be a regulating torsionless Butler group. Let $U = \sum_{t \in T} B_t$ be a regulating subgroup of B and let $V = \sum B_t(1 + f_t)$, $f_t \in \mathrm{Hom}\,(B_t, B^*(t)_*)$ be another regulating subgroup. Then $U = V$ if and only if $B_t f_t \subset U^*(t)_*$.*

Proof. If $B_t f_t \subset U^*(t)_* \subset U$ then clearly $V \subset U$ and since B is regulating torsionless $V = U$ by Lemma 4.2. Conversely, if $U = V$ then $U(t) = B_t \oplus U^*(t)_* = B_t(1+f_t) \oplus U^*(t)_*$ and hence $B_t f_t \subset U^*(t)_*$. ∎

5. Numbers of Regulating Subgroups in Regulating Torsionless Butler Groups

The number of regulating subgroups in almost completely decomposable groups is dealt with in [5]. The same result holds for regulating torsionless Butler groups, but the methods of proof are different.

Lemma 5.1. *Let U be a torsion-free abelian group. Let $b \in \mathbf{Q}U \setminus U$ and $\mathbf{Q}b \cap U$ be of type $t = t(\mathbf{Q}b \cap U)$. Let $\chi \in t$ with $\chi_p = 0$ for all primes p dividing the (finite) order of $b + U \in \mathbf{Q}U/U$. Then there is an element b' in the coset $b + U \in \mathbf{Q}U/U$ such that $\chi^{\langle U, b \rangle}(b') \geq \chi$.*

Proof. Let k be the order of $b + U$. For all natural numbers m, the characteristic of each $mb \in \langle U, b \rangle = V$ is equivalent to χ. Hence there is a natural number n such that $\chi^V(nmb) \geq \chi$ for all $1 \leq m < k$. Moreover, n may be assumed relatively prime to k, because $\chi_p = 0$ for all primes dividing k. Thus multiplication of the finite group $\langle b + U \rangle$ by n is an automorphism and there is an element $b' = nmb \in b + U$ for suitable m with characteristic $\chi^{\langle U, b \rangle}(b') \geq \chi$. ∎

Let $r_t = \mathrm{rk}\left(B(t)/B^*(t)_*\right)$ be the *critical ranks of B*. For a regulating torsionless Butler group B, for each $t \in T$ let $i^B(t)$ be the index of regulating subgroups in the type subgroup $B^*(t)_*$.

The following theorem may be found without proof in the Diplomarbeit of K.-J. Krapf [8, Satz 44 (a)].

Theorem 5.2. *The number of regulating subgroups of a regulating torsionless Butler group B with critical typeset T is $\prod_{t \in T} i^B(t)^{r_t}$.*

Proof. Let T be the critical typeset of the Butler group B with an indexing $T = \{t_1, \ldots, t_k\}$ such that t_i is maximal in the set $\{t_i, t_{i+1}, \ldots, t_k\}$ for all $1 \leq i \leq k$. Let $U = \sum_{l=1}^{k} B_l$ be a fixed regulating subgroup, where $B(t_l) = B_l \oplus B^*(t_l)_*$. Corollary 3.2 shows that all regulating subgroups of B will be obtained by the following construction. Let $V_1 = B_1 = B(t_1)$ observing that t_1 is a maximal critical type. Let $V_l = B_l' + V_{l-1}$ where $B_l' = B_l(1 + f_l)$ with $f_l \in \text{Hom}(B_l, B^*(t_l)_*)$. All such subgroups $V = V_k$ of B are regulating, and two of them $V = \sum_{l=1}^{k} B_l'$ and $\hat{V} = \sum_{l=1}^{k} B_l''$, where $B_l'' = B_l(1 + f_l')$ for all l, are equal if and only if $f_l - f_l' \in \text{Hom}(B_l, V^*(t_l)_*)$, for all l, using Lemma 4.5 and $B^*(t_l)_* \cap V = V^*(t_l) = V^*(t_l)_*$.

Further, $V^*(t_l) = V_{l-1} \cap B^*(t_l)_*$ because by Lemma 4.1 the type subgroup $V^*(t_l) = V^*(t_l)_* = \sum\{B_m' \mid m < l, t_m > t_l\}$ is pure in V_{l-1} and as a regulating subgroup, see [1, 3.1], of finite index in $B^*(t_l)_*$. Thus we have for a fixed V_{l-1} precisely $|B^*(t_l)_*/V^*(t_l)|^{r_{t_l}}$ different possibilities to construct a group V_l if in each coset $0 \neq \bar{b} = b + V^*(t_l) \in B^*(t_l)_*/V^*(t_l)$ there is an element of characteristic greater or equal than $\chi \in t_l$ if χ has components 0 for all primes dividing the order of $\bar{b}$. This is guaranteed by Lemma 5.1. By Lemma 4.2 $i^B(t_l) = |B^*(t_l)_*/V^*(t_l)|$. Consequently there are $i^B(t_l)^{r_{t_l}}$ possible groups V_l for fixed V_{l-1}.

Now the groups V and $\hat{V}$ are equal if and only if V_l and $\hat{V}_l$ are equal for all l because $V_l = \sum_{m=1}^{l} V(t_m)$ using that V and $\hat{V}$ are torsionless. This proves the theorem. ∎

Corollary 5.3. *The regulating torsionless Butler group B has a regulating regulator if and only if the type subgroups $B(t)$ are torsionless for all critical types.*

Proof. The regulator of B is regulating if and only if there is only one regulating subgroup. By Theorem 5.2 this is equivalent to $i^B(t) = 1$ for all critical types. Hence the type subgroups $B(t)$ are their own regulating subgroups, i. e. torsionless for all critical types. ∎

I thank the referee for several improvements.

References

[1] D. M. Arnold, *Pure subgroups of finite rank completely decomposable groups*, Springer Verlag, Lecture Notes in Mathematics **874** (1984), 1–31.

[2] D. M. Arnold and C. Vinsonhaler, *Endomorphism rings of Butler groups*, J. Austral. Math. Soc., (Series A), **42** (1987), 322 – 329.

[3] R. Burkhardt, *On a special class of almost completely decomposable torsion free abelian groups I*, Springer Verlag, Abelian groups and modules, C.I.S.M., Udine (1984), 141–150.

[4] M. C. R. Butler, *A class of torsion-free abelian groups of finite rank*, Proc. London Math. Soc., **15** (1965), 680–698.

[5] B. Frey and O. Mutzbauer, *Regulierende Untergruppen und der Regulator fast vollständig zerlegbarer Gruppen*, Rend. Sem. Math. Univ. Padova (1992).

[6] L. Fuchs, "Infinite Abelian Groups I+II", Academic Press (1970, 1973).

[7] N. Jacobson, "Basic Algebra I", Freeman and Company (1974).

[8] K.-J. Krapf, *Klassifizierung fast vollständig zerlegbarer abelscher Gruppen und Untersuchung von Butler-Gruppen*, Diplomarbeit (Würzburg, 1986).

[9] A. Mader, *On the automorphism group and the endomorphism ring of abelian groups*, Ann. Univ. Sci. Budapest **8** (1965), 3–12.

[10] A. Mader, *Regulating subgroups of Butler groups*, unpublished notes (1990).

[11] C. Metelli, *On type-related properties of torsionfree abelian groups*, Springer Verlag, Lecture Notes in Mathematics **1006** (1983), 253–267.

Quasi-Realizing Modules

R.S. PIERCE

C. VINSONHALER[1]

University of Arizona, Tucson, AZ 85721

University of Connecticut, Storrs, CT 06269

Introduction. If H is a torsion–free group, then the rational vector space $V = QH \simeq Q \otimes_Z H$ is a faithful module over the quasi–endomorphism algebra $A = QE(H)$. Note that H is a full subgroup of V and A is a subalgebra of End V, the endomorphism algebra of V. This paper is concerned with the following question:

Basic Problem. Let V be a finite dimensional Q–space and suppose that A is a subalgebra of End V. What conditions (on the algebra A or the module $_AV$) imply that there is a full (locally free or quotient divisible) subgroup H of V such that $A = QE(H)$?

We will say that a group H which solves the basic problem **quasi–realizes** the module $_AV$. The more delicate problem of realizing a subring R of End V as End H for a full subgroup H of V will not be considered here.

Section one consists of general conditions for the quasi–realizability of a module. The principal results of the paper are found in Section two. We show that if A has the double centralizer property in End V, then $_AV$ is quasi–realizable by a locally free group (Theorem 2.1). In particular this conclusion is obtained if A is semisimple. Using a different approach, we show that if A is semisimple, then any module $_AV$ is quasi–realizable by a quotient divisible group (Theorem 2.3). Examples demonstrate that there exist modules $_AV$ that are

---

[1]Research supported in part by NSF grant DMS–9022730.

quasi–realizable by a locally free group, but not by a quotient divisible group (Example 2.5); and vice versa, there are modules that are realizable by a quotient divisible group, but not by a locally free group (Example 2.6). Finally, we construct a class of modules that are not quasi–realized by any group (Example 2.7).

Throughout this paper, the notation and hypotheses of the Basic Problem are retained. In particular, V is a Q–space of finite dimension, and A is a subalgebra of the rational algebra $E = \text{End } V$. Additional notation will be introduced as we proceed. In general, our group and ring theoretic notation is consistent with the standards set by Fuchs's books [F]. The set of all rational primes will be denoted by Π.

1 Reductions. A simple observation will play a central role in our discussion.

1.1 Lemma. There is a full, free subgroup G of V and a full, free subring R of A such that $RG = G$.

Proof. Let $\rho_1,...,\rho_n$ be a Q–basis of A such that $\rho_i\rho_j = \sum_{k=1}^n a_{ijk}\rho_k$ with all $a_{ijk} \in Z$. Clearly, a suitable integral multiple of any basis will have this property. Then $R = Z\cdot 1_A + \sum_{k=1}^n Z\rho_k$ is a finitely generated (hence free) full subring of A. Let $u_1,...,u_m$ be a Q–basis of V and define $G = \sum_{i=1}^m Zu_i + \sum_{k=1}^n \sum_{i=1}^m Z\rho_k u_i$. Then G is a full free subgroup of V such that $RG = G$. □

In what follows, R and G are as described in Lemma 1.1. For $p \in \Pi$, and X any of the rings or groups under consideration, denote $\hat{X}_p = \hat{Z}_p \otimes X$, where $\otimes$ abbreviates $\otimes_Z$. The embedding of Z_p in $\hat{Z}_p$ induces an injection of $Z_p \otimes X$ to $\hat{X}_p$. Moreover, the tensor product $Z_p \otimes X$ can be identified with the composite $X_p = Z_pX$. Hence, we can view X_p as a subset of $\hat{X}_p$. Recall that group endomorphisms of full subgroups of V extend uniquely to Q–linear maps of V to itself, and subsequently to $\hat{Q}_p$–linear transformations of $\hat{V}_p$ by $\varphi \to \text{id}_{\hat{Q}_p} \otimes \varphi$. Usually it is harmless to identify φ with its various extensions, thereby simplifying notation.

For convenience, we record several well–known local–global principles that will be used in our discussion. The proofs are elementary and familiar.

1.2 Lemma. (a) If H is a subgroup of V, then $\hat{H}_p \cap V = H_p$. Moreover, $\cap_{p\in\Pi} H_p = H$, so that $\varphi \in \text{End } V$ satisfies $\varphi\hat{H}_p \subseteq \hat{H}_p$ for all $p \in \Pi$ if and only if $\varphi H \subseteq H$.

(b) Let K be a full subgroup of V and for each $p \in \Pi$, assume that $H(p)$ is a

$\hat{Z}_p$–submodule of $\hat{V}_p$ with $K \subseteq H(p)$. If $H = \cap_{p\in\Pi}(H(p) \cap V)$, then $\hat{H}_p = H(p)$ for all p.

1.3 **Lemma.** For $p \in \Pi$, suppose that $H(p)$ is an $\hat{R}_p$–submodule of $\hat{V}_p$ such that $G \subseteq H(p)$. Define $H = \cap_{p\in\Pi}(H(p) \cap V)$. If $\varphi \in$ End V, then $\varphi \notin QE(H)$ if and only if $\varphi H(p) \not\subset H(p)$ for infinitely many $p \in \Pi$ or $\varphi(\text{div } H(p)) \not\subset \text{div } H(p)$ for at least one p.

Proof. By 1.2(b), $\hat{H}_p = H(p)$. Thus, if $\varphi H(p) \not\subset H(p)$ for infinitely many p, then for every nonzero integer n, $n\varphi\hat{H}_p \not\subset \hat{H}_p$ for infinitely many p. Consequently, $n\varphi \notin$ End H for all nonzero n, so that $\varphi \notin QE(H)$. Similarly, if $\varphi \text{div } H(p) \not\subset \text{div } H(p)$ for some p, then $n\varphi\text{div }\hat{H}_p \not\subset \hat{H}_p$ for all nonzero integers n, so that $n\varphi \notin$ End H for all n and $\varphi \notin QE(H)$. For the converse, assume that $\varphi H(p) \subseteq H(p)$ for almost all p and $\varphi\text{div } H(p) \subseteq \text{div } H(p)$ for all p. Since $H(p)$ is a module over the complete discrete valuation domain $\hat{Z}_p$, it follows that $H(p) = M(p) \oplus \text{div } H(p)$, where $M(p)$ is finitely generated. Therefore, the hypotheses imply that there exists a positive integer n such that $n\varphi H(p) \subseteq H(p)$ for all $p \in \Pi$. By 1.2(a), $n\varphi \in$ End H and $\varphi \in QE(H)$. □

We will be interested in the existence of finitely generated $\hat{Z}_p$–submodules $H(p)$ of $\hat{V}_p$ that satisfy $\varphi H(p) \not\subset H(p)$.

1.4 **Lemma.** Let $\varphi \in$ End V and $p \in \Pi$. The following conditions are equivalent.
(a) There exists $w \in G$ such that $\varphi w \notin R_p w$.
(b) There exists $w \in \hat{G}_p$ such that $\varphi w \notin \hat{R}_p w$
(c) There is an $\hat{R}_p$–module $H(p)$ of $\hat{V}_p$ such that $G \subseteq H(p)$, $H(p)$ is finitely generated as a $\hat{Z}_p$–module and $\varphi H(p) \not\subset H(p)$.

Proof. (a) implies (b). Let w be as in (a). If $\varphi w \in \hat{R}_p w$, then since $\varphi \in$ End V and $w \in G$, it follows that $\varphi w \in \hat{R}_p w \cap V = Z_p Rw = R_p w$ by 1.2(a). This contradiction shows that (b) follows from (a).

(b) implies (c). Since R is finitely generated as a group, it follows that $\hat{R}_p w$ is finitely generated as $\hat{Z}_p$–module. Consequently $\hat{R}_p w$ is compact in the p–adic metric. Thus, the p–adic distance from φw to $\hat{R}_p w$ is positive, say $\text{dist}(\varphi w, \hat{R}_p w) > p^{-k}$. In particular, $\varphi w \notin \hat{R}_p w + p^k \hat{G}_p$. Define $H(p) = p^{-k}\hat{R}_p w + \hat{G}_p$. Then $H(p)$ is an $\hat{R}_p$–submodule of $\hat{V}_p$ that is finitely generated as a $\hat{Z}_p$–module, and $H(p)$ and contains G.

Moreover, $p^{-k}w \in H(p)$ but $\varphi(p^{-k}w) \notin H(p)$. Thus, $\varphi H(p) \not\subset H(p)$.

(c) implies (a). Let $H(p)$ be as in (c), and choose $v \in H(p)$ such that $\varphi v \notin H(p)$. Since $H(p)$ is a finitely generated $\hat{Z}_p$–module, it is compact in the p–adic metric and $\text{dist}(\varphi v, H(p)) > 0$. Using the facts that φ is continuous, and (by 1.2(b)) $V \cap H(p)$ is dense in $H(p)$, we can find $u \in V \cap H(p)$ such that $\text{dist}(\varphi u, \varphi v) < \text{dist}(\varphi v, H(p))$. In particular, $\varphi u \notin H(p)$. Therefore, $\varphi u \notin R_p u$, because $u \in H(p)$. Finally, since G is full in V, there exists $n > 0$ such that $w = nu \in G$. Evidently, $\varphi w \notin R_p w$. $\square$

1.5 Proposition. There is a full, locally free subgroup H of V such that $QE(H) = A$ if and only if for each $\varphi \in \text{End } G \setminus A$, there exist infinitely many $p \in \Pi$ such that $\varphi w \notin R_p w$ for some $w \in G$.

Proof. Assume that $QE(H) = A$, where H is full and locally free. Replacing H by $n^{-1}RH$ for a suitable integer n, it can be assumed that $G \subseteq H$ and $R \subseteq \text{End } H$. For $p \in \Pi$, denote $H(p) = \hat{H}_p$. Then $H(p)$ is an $\hat{R}_p$–submodule of $\hat{V}_p$ that is finitely generated as a $\hat{Z}_p$–module; and $G \subseteq H(p)$. By 1.2(a), $H = \cap_{p \in \Pi}(H(p) \cap V)$. It follows from 1.3 that if $\varphi \in \text{End } G \setminus A = \text{End } G \setminus QE(H)$, then $\varphi H(p) \not\subset H(p)$ for infinitely many $p \in \Pi$. By 1.4(a), there are infinitely many $p \in \Pi$ such that $\varphi w \notin R_p w$ for a suitable $w \in G$. Conversely, suppose that for each $\varphi \in \text{End } G \setminus A$, there are infinitely many $p \in \Pi$ such that $\varphi w \notin R_p w$ for some $w \in G$. Let $\{\varphi_n : n < \omega\}$ be an enumeration of $\text{End } G \setminus A$. By the Bernstein–Kuratowski–Sierpinski Theorem [K], there is a family $\{\Sigma_n : n < \omega\}$ of infinite subsets of Π with $\Sigma_n \cap \Sigma_m = \phi$ for $n \neq m$ such that if $p \in \Sigma_n$, then $\varphi_n w \notin R_p w$ for some $w \in G$. By 1.4, for each $p \in \Sigma_n$ there is an $\hat{R}_p$–submodule $H(p)$ of $\hat{V}_p$ with $H(p)$ finitely generated as a $\hat{Z}_p$–module, such that $G \subseteq H(p)$ and $\varphi H(p) \not\subset H(p)$. Note that $\hat{G}_p \subseteq H(p)$ with $H(p)/\hat{G}_p$ finite, so that $G_p \subseteq H(p) \cap V$ and $H(p) \cap V$ is finitely generated as a Z_p–module. If $p \in \Pi \setminus \cup_{n<\omega} \Sigma_n$, let $H(p) = \hat{G}_p$ and denote $H = \cap_{p \in \Pi} H(p) \cap V$. By 1.2(b), $\hat{H}_p = H(p)$ for all $p \in \Pi$. Thus, H is a full, locally free subgroup of V with $R\hat{H}_p \subseteq \hat{H}_p$ for all p. Consequently, $R \subseteq \text{End } H$ by 1.2(a), and $A = QR \subseteq Q\text{End } H = QE(H)$. On the other hand, by 1.3, $\varphi_n \notin QE(H)$ for all $n < \omega$. Since $\text{End } V = Q\text{End } G$, it follows that $QE(H) = A$. $\square$

1.6 Proposition. There is a full, quotient divisible subgroup H of V with $QE(H) = A$ if and only if, for each $p \in \Pi$, there is an $\hat{A}_p$–submodule W_p of $\hat{V}_p$ such that $\cup_{p \in \Pi} \{\varphi \in \text{End } V : \varphi W_p \not\subset W_p\} = \text{End } V \setminus A$.

Proof. Suppose that H exists. Without loss of generality, it can be assumed that $G \subseteq$

H. Then $k(H/G)$ is divisible for some integer $k > 0$ because H is quotient divisible. For $p \in \Pi$, $W_p = \text{div } \hat{H}_p$ is a $\hat{Q}_p$–subspace of $\hat{V}_p$ with $k\hat{H}_p \subseteq W_p + \hat{G}_p \subseteq \hat{H}_p$. Since $RH = H$, it follows that $AW_p = W_p$ and W_p is an $\hat{A}_p$–submodule of $\hat{V}_p$. If $\varphi \in \text{End } V \setminus A = \text{End } V \setminus QE(H)$, then $\ell\varphi H \not\subset H$ for all nonzero integers ℓ. Suppose that $\varphi W_p \subseteq W_p$ for all $p \in \Pi$. Since G is finitely generated and full in V, there exists $\ell > 0$ such that $\ell\varphi G \subseteq G$. Therefore, $\ell k\varphi\hat{H}_p \subseteq \varphi W_p + \hat{Z}_p\ell\varphi G \subseteq W_p + \hat{G}_p \subseteq \hat{H}_p$ for all $p \in \Pi$. Thus, $\ell k\varphi \in \text{End } G$ by 1.2(a), and $\varphi \in QE(H)$, which is contrary to hypothesis. Therefore, $\varphi W_p \not\subset W_p$ for some $p \in \Pi$. This argument shows that $\text{End } V \setminus A = \cup_{p\in\Pi}\{\varphi \in \text{End } V : \varphi W_p \not\subset W_p\}$. For the converse assertion, define $H = \cap_{p\in\Pi}(W_p + \hat{G}_p) \cap V$. By 1.2(b) $\hat{H}_p = W_p + \hat{G}_p$. Moreover, $G \subseteq H$ and H/G is a torsion group whose p–component is isomorphic to $\hat{H}_p/\hat{G}_p \simeq W_p/(W_p \cap \hat{G}_p)$, a divisible group. Hence, H is quotient divisible. If $\varphi \in R$, then $\varphi\hat{H}_p \subseteq \hat{H}_p$ for all p, so that $\varphi \in \text{End } H$ by 1.2(b). Thus, $A = QR \subseteq QE(H)$. If $\varphi \in \text{End } V \setminus A$, then by hypothesis $\varphi W_p \not\subset W_p$ for some $p \in \Pi$. If $\varphi W_p \subseteq W_p + \hat{G}_p$, then $\varphi W_p \subseteq W_p$ since W_p is divisible. It follows that $k\varphi\hat{H}_p \not\subset \hat{H}_p$ for all $k > 0$. Therefore, $\varphi \notin \cap_{p\in\Pi}QE(H_p)$. Because H is quotient divisible, this last intersection is $QE(H)$ by [MV, Lemma 3.3]. $\square$

Let B be an F–space, where F is any field. If $\mathcal{F}$ is a family of subspaces of V, denote $A_V(\mathcal{F}) = \{\varphi \in \text{End}_F V : \varphi W \subseteq W \text{ for all } W \in \mathcal{F}\}$. If $\mathcal{E} \subseteq \text{End}_F V$, denote $S_V(\mathcal{E}) = \{W \leq V : \varphi W \subseteq W \text{ for all } \varphi \in \mathcal{E}\}$. We will also use this notation with V replaced by $\hat{V}_p$; to simplify notation in this case, we write A_p for $A_{\hat{V}_p}$ and S_p for $S_{\hat{V}_p}$.

1.7 Corollary. If there are infinitely many $p \in \Pi$ such that $A_p(S_p(\hat{A}_p)) = \hat{A}_p$, then there is a full, quotient divisible subgroup H of V such that $QE(H) = A$.

Proof. Note that $S_p(\hat{A}_p)$ is the family of all $\hat{A}_p$–submodules of $\hat{V}_p$. Let $\{\varphi_n\}$ be an enumeration of $\text{End } V \setminus A$. By assumption, there is an infinite set $\Sigma \subseteq \Pi$ such that $A_p(S_p(\hat{A}_p)) = \hat{A}_p$ for all $p \in \Sigma$. Consequently, there exist distinct primes $p(n) \in \Sigma$ and $W_n \in S_{p(n)}(\hat{A}_{p(n)})$ such that $\varphi_n(W_n) \not\subset W_n$. For $p \in \Pi\setminus\Sigma$ denote $W_p = \hat{V}_p$. Then $\cup_{p\in\Pi}\{\varphi \in \text{End } V : \varphi W_p \not\subset W_p\} = \text{End } V \setminus A$, so that the corollary follows from 1.6. $\square$

In particular, if $A_V(S_V(A)) = A$, then $A_p(S_p(\hat{A}_p)) = \hat{A}_p$ for all p, so the corollary applies in this case.

1.8 Proposition. Assume that for some $\varphi \in \text{End } V \setminus A$, $\varphi w \in R_p w$ for all $p \in \Pi$ and all $w \in G$. Then $QE(H) \neq A$ for all full subgroups H of V; that is, $_{A}V$ is not quasi–realizable.

Proof. Suppose that $QE(H) = A$. Since R is finitely generated, there exists $n > 0$ such that $nR \subseteq \text{End } H$. Replacing R by $Z \cdot 1 + nR$, if necessary, we can suppose that $R \subseteq \text{End } H$. Thus, $\hat{H}_p$ is an $\hat{R}_p$–module for all $p \in \Pi$. Similarly, it can be assumed that $G \subseteq H$. If $\varphi \in \text{End } V \setminus A$, it follows from 1.2(a) that there exists $p \in \Pi$ such that $\varphi \hat{H}_p \not\subset \hat{H}_p$. In particular, $\varphi u \notin \hat{R}_p u$ for some $u \in \hat{H}_p$. Since $\hat{H}_p/\hat{G}_p$ is a torsion p–group, there exists $k \geq 0$ such that $w = p^k u \in \hat{G}_p$; and $\varphi w \notin \hat{R}_p w$. This conclusion contradicts the hypotheses by 1.4. □

2. **Applications and Examples.** If B is a subalgebra of the Q–algebra E, let $C_E(B)$ denote the **centralizer** of B in E. That is, $C_E(B) = \{\psi \in E : \psi\varphi = \varphi\psi \text{ for all } \varphi \in B\}$. Clearly, $C_E(B)$ is a subalgebra of E, and $B \subseteq C_E(C_E(B))$. The algebra B has the **double centralizer property** in E if $C_E(C_E(B)) = B$. Our interest is of course in the case where $E = \text{End } V$. Note that if V is itself an algebra, then $C_E(V)$ is the set of right multiplications by elements of V; and $C_E(C_E(V)) = V$. In this sense, the first result of this section may be viewed as a generalization of the Q–space version of Butler's theorem on realizing locally free rings [B].

2.1 **Theorem.** Let $E = \text{End } V$. If $C_E(C_E(A)) = A$, then there is a full, locally free subgroup H of V such that $QE(H) = A$.

Proof. If $\varphi \in \text{End } G \setminus A$, then $\varphi \notin C_E(C_E(A))$ so that $\varphi\beta - \beta\varphi \neq 0$ for some $\beta \in C_E(A)$. Replacing β by $n\beta$ for a suitable integer n, we can assume that $\beta \in \text{End } G$. It follows that there exists $x \in G$ such that $y = \varphi\beta x - \beta\varphi x$ is a nonzero element of G. Let $\Sigma = \{p \in \Pi : \text{for some } c_p \in Z, \ y \notin (\beta - c_p)G_p\}$. By the Lemma in [B], Σ is infinite. For $p \in \Sigma$ let $w = (\beta - c_p)x \in G$. Then $\varphi w = y + (\beta - c_p)\varphi x \notin R_p w$. Otherwise, $y = \varphi w - (\beta - c_p)\varphi x \in R_p w + (\beta - c_p)G = R_p(\beta - c_p)x + (\beta - c_p)G = (\beta - c_p)(R_p x + G) \subseteq (\beta - c_p)G_p$, using the fact that $\beta - c_p \in C_E(A) = C_E(R_p)$. This contradiction shows that the hypothesis of 1.5 is satisfied and the theorem follows. □

By a result of Nesbitt and Thrall [NT], if A is a **uniserial algebra**, that is, a product of full matrix algebras over algebras whose lattices of left and right ideals are chains, then A has the double centralizer property in End V. Hence we have the following useful consequence of 2.1.

2.2 **Corollary.** If A is a uniserial subalgebra of End V, then there is a full, locally free subgroup H of V such that $QE(H) = A$.

The corollary applies in case that A is a semisimple Q–algebra. We will show next that semisimple algebras can also be quasi–realized by quotient divisible groups.

If A is a semisimple Q–algebra, then there is an algebraic number field F which is a splitting field for A. That is, $F \otimes_Q A$ is a product of full matrix rings over F. Since every field extension of F is also a splitting field for A, and there are infinitely many primes p such that $\hat{Q}_p$ contains a copy of F, it follows that $\hat{Q}_p$ is a splitting field for A for infinitely many p. (See [P] for an account of these facts.)

2.3 Theorem. If A is semisimple, then there is a full, quotient divisible subgroup H of V such that $QE(H) = A$.

Proof. Let Σ be the set of all $p \in \Pi$ such that $\hat{Q}_p$ is a splitting field for A. By the above remarks, Σ is infinite. If $p \in \Sigma$ then $\hat{A}_p$ is a semisimple $\hat{Q}_p$–algebra such that $\text{End}_{\hat{A}_p} W = \hat{Q}_p$ for every simple $\hat{A}_p$–module W, and $\hat{V}_p$ is a faithful $\hat{A}_p$–module with $\dim_{\hat{Q}_p} \hat{V}_p = \dim V$. The theorem is a consequence of 1.7 and the following technical result.

2.4 Lemma. Let F be a field, B a finite dimensional, semisimple F–algebra, and suppose that W is a B–module. If $\text{End}_B M = F$ for all simple B–submodules M of W, then $A_W(S_W(B)) = B$.

Proof. Denote $E = \text{End}_F W$. It suffices to show that $\varphi\theta = \theta\varphi$ for all $\theta \in C_E(B)$ and $\varphi \in A_W(S_W(B))$. In this case, $\varphi \in C_E(C_E(B)) = B$, since B is semisimple. Therefore, $B \subseteq A_W(S_W(B)) \subseteq B$. Let $v \in W$ be such that Bv is a simple submodule of W. If $B\theta v \cap Bv \neq 0$, then $Bv = B\theta v = \theta Bv$ because Bv is simple. Hence, $\theta \in \text{End}_B(Bv) = F$ by hypothesis, and $\varphi\theta = \theta\varphi$. Suppose that $B\theta v \cap Bv = 0$. Since $\varphi \in A_W(S_W(B))$, there exist $a,b,c \in B$ such that $\varphi v = av$, $\varphi\theta v = b\theta v$ and $\varphi(\theta v + v) = c\theta v + cv$. Hence, $b\theta v + av = c\theta v + cv$, so that $\varphi\theta v = b\theta v = c\theta v = \theta cv = \theta av = \theta\varphi v$, since $B\theta v \cap Bv = 0$. The argument shows that $\varphi\theta v = \theta\varphi v$ for all $v \in W$ such that Bv is simple. However, A is semisimple, so that W is semisimple, and every element of W is a sum $v_1 + \cdots + v_k$ where each Bv_i is simple. Thus, $\varphi\theta w = \theta\varphi w$ for all $w \in W$. That is, $\varphi\theta = \theta\varphi$. $\square$

The following example shows that it isn't possible to generalize 2.3 to include subalgebras of E that have the double centralizer property as in 2.1.

2.5 Example. There is a Q–space V and a subalgebra A of $\text{End } V$ such that $_AV$ is quasi–realizable by a locally free subgroup but not by a quotient divisible subgroup. Let V be an m–dimensional vector space over Q. We may identify $\text{End}_Q V = M_m(Q)$ and regard

V as the space of first columns of $M_m(Q)$, with basis $\eta_i = e_{i1}$, $1 \le i \le m$, where e_{ij} are the usual matrix units. Let ι denote the identity matrix and let $\alpha = e_{12} + e_{23} + \cdots + e_{(m-1)m}$. If $v = c_1\eta_1 + \cdots + c_{k-1}\eta_{(k-1)} + \eta_k$, then $Q[\alpha]v = \oplus_{j \le k} Q\eta_j \equiv V_k$. Indeed, $\alpha^\ell v = c_{\ell+1}\eta_1 + \cdots + c_{k-1}\eta_{k-\ell-1} + \eta_{k-\ell}$ for $\ell < k$, $\alpha^k v = 0$, and the assertion follows by backward induction. It is clear that if $\beta \in M_m(Q)$ is upper triangular, then $\beta V_k = V_k$ for all k. On the other hand, if $\beta = \sum_{i,j} c_{ij} e_{ij}$ and $c_{k\ell} \ne 0$ for some $k > \ell$, then $\beta V_\ell \not\subset V_\ell$. Let $A = Q[\alpha]$. Then $\hat{A}_p = \hat{Q}_p[\alpha]$ and the above computations apply to the $\hat{A}_p$–submodules of $\hat{V}_p$. In particular, $S_p(\hat{A}_p) = \{\hat{V}_{kp} : 0 \le k \le m\}$, and for $\varphi \in$ End V, we have $\varphi \hat{V}_{kp} \subseteq \hat{V}_{kp}$ if and only if φ is upper triangular. Thus, by 1.6, there is no full, quotient divisible subgroup H of V such that QE(H) = A. On the other hand, A is a local algebra, so that the module $_A V$ is quasi–realizable by a locally free group by 2.2. $\square$

If B is the algebra of upper triangular matrices in $M_m(Q)$, the remarks above imply that for all $p \in \Pi$, $A_p(S_p(\hat{B}_p)) = \hat{B}_p$. Thus, by 1.7, there is a full quotient divisible subgroup H of V such that QE(H) = B. This observation shows that 2.3 is not the final word on quasi–realization by quotient divisible groups. Moreover, B does not have the double centralizer property in $E = M_m(Q)$. In fact, $C_E(B) = F$, so that $C_E(C_E(B)) = E$. However, it is not difficult to deduce from 1.5 that B is quasi–realizable by a locally free group. Thus, the result of 2.1 does not characterize the modules that are quasi–realizable by locally free groups.

2.6 Example. There are modules that cannot be quasi–realized by a locally free group, but can be quasi–realized by a quotient divisible group. Let F be an algebraic number field of degree $n \ge 2$ over Q. Define $A = Q \oplus F$ to be the Q–algebra obtained by adding an identity to the nil ring defined over F. Specifically, the multiplication is given by $(r,a)(s,b) = (rs, sa+rb)$. Let $V = Fe_1 \oplus Fe_2$ be a direct sum of two copies of the Q–module F. Then V becomes an A–module under the action $(r,a)(be_1+ce_2) = rbe_1 + (rc+ab)e_2$. Let S be the ring of algebraic integers in F, and set $R = Z \oplus S \subset A$, $G = Se_1 \oplus Se_2$. Plainly, R is a finitely generated subring of A, G is a finitely generated subgroup of V, and RG = G.

(a) $_A V$ is not quasi–realizable by a locally free group. Indeed, let $\varphi \in$ End V be defined by $\varphi(be_1 + ce_2) = be_1 - ce_2$. An easy check shows that $\varphi \in$ End V \ A. If $w = be_1 + ce_2 \in G$, then $\varphi w \in R_p w$ for almost all $p \in \Pi$. In fact, if $b = 0$, then $\varphi w = -w \in R_p w$. If $b \ne 0$, and p is an odd prime such that $pS + bS = S$, then b becomes a unit in S_p and $R_p w = (Z_p \oplus S_p)w$ contains $\varphi w = be_1 - ce_2 = be_1 + (c - 2c)e_2 = (1, -2b^{-1}c)w$. Thus, $\varphi w \in R_p w$ for almost all p, as claimed. By 1.5, there is no full, locally free subgroup H of V such that QE(H) = A.

If $p \in \Pi$, then $\hat{A}_p = \hat{Q}_p \oplus \hat{F}_p$ with nil multiplication on $\hat{F}_p$, and $\hat{V}_p = \hat{F}_p e_1 \oplus \hat{F}_p e_2$ with the induced $\hat{A}_p$–action. The $\hat{A}_p$–submodules of $\hat{V}_p$ are exactly the $\hat{Q}_p$–spaces W of $\hat{V}_p$ such that

(*) $\beta e_1 + \gamma e_2 \in W$ implies $\hat{F}_p \beta e_2 \in W$.

Let $\Sigma \subseteq \Pi$ consist of all primes p such that F splits completely over $\hat{Q}_p$. That is,

$$\hat{F}_p = \hat{Q}_p \eta_1 + \cdots + \hat{Q}_p \eta_n$$

where $\eta_1,...,\eta_n$ is a complete system of primitive orthogonal idempotents in $\hat{F}_p$. In particular, $\hat{Q}_p \eta_i$ is a minimal ideal in $\hat{F}_p$. By well known results of algebraic number theory (see [P]), the set Σ is infinite.

(b) $_A V$ is quasi–realizable by a full quotient divisible subgroup of V.

To prove (b), it suffices by 1.7 to show that $A_p(S_p(\hat{A}_p)) = \hat{A}_p$ for all $p \in \Sigma$. Let $\varphi \in A_p(S_p(\hat{A}_p))$, that is, $\varphi W \subseteq W$ for all $\hat{A}_p$–submodules of $\hat{V}_p$. Using this condition leads to the required conclusion via the following steps:

for $W = \hat{Q}_p \eta_i e_2$, $\varphi W \subseteq W$ implies $\varphi(\eta_i e_2) = \sigma_i \eta_i e_2$ for some $\sigma_i \in \hat{Q}_p$;

for $W = \hat{Q}_p (\Sigma \eta_i e_2)$, $\varphi W \subseteq W$ implies $\sigma_1 = \sigma_2 = \cdots = \sigma_n = \sigma$;

for $W = \hat{Q}_p \eta_i e_1 + \hat{Q}_p \eta_i e_2$, $\varphi W \subseteq W$ implies $\varphi(\eta_i e_1) = \rho_i \eta_i e_1 + \tau_i \eta_i e_2$ for some $\rho_i, \tau_i \in \hat{Q}_p$;

for $W = \hat{Q}_p (\Sigma \eta_i e_1) + \hat{F}_p e_2$, $\varphi W \subseteq W$ implies $\rho_1 = \cdots = \rho_n = \rho$;

for $W = \hat{Q}_p (\eta_i e_1 + \eta_j e_2) + \hat{Q}_p \eta_i e_2$, $i \neq j$, $\varphi W \subseteq W$ implies $\rho = \sigma$.

Summarizing, for $1 \leq i \leq n$, $(\varphi - (\rho,0))(\eta_i e_1) = \tau_i \eta_i e_2$ and $(\varphi - (\rho,0))(\eta_i e_2) = 0$. Let $\alpha = \Sigma_{i=1}^n \tau_i \eta_i \in \hat{F}_p$. If $\beta = \Sigma_{i=1}^n \xi_i \eta_i \in \hat{F}_p$, then $\varphi(\beta e_1) = \Sigma_{i=1}^n \xi_i \tau_i \eta_i e_2 = (\Sigma_{i=1}^n \tau_i \eta_i)(\Sigma_{i=1}^n \xi_i \eta_i) e_2 = \alpha \beta e_2$. That is, φ agrees with the action of (ρ,α) on $\hat{V}_p$. Thus, $\varphi \in A_p(S_p(\hat{A}_p))$, as we asserted. $\square$

Our final example shows that not every module can be quasi–realized.

2.7 **Example.** There is an algebra A and a Q–space V with $A \subseteq \text{End } V$ such that there is no full subgroup H of V such that $QE(H) = A$. The construction is analogous to that used in 2.6. We also use the matrix notation of Example 2.5. Specifically, U is the space of m by 1 matrices over Q, with $m \geq 2$, and $V = Ue_1 \oplus Ue_2$. Let N be the Q–subspace of $M_m(Q)$ spanned by the matrices $\xi_{ij} = e_{ij} + k_{ij} e_{mm}$, where the k_{ij} are integers that will be delimited in the course of our construction. For a start, set $k_{mm} = -1$ so that $\xi_{mm} = 0$.

Impose the nil multiplication on N, and define $A = Q \cdot 1 \oplus N$, where 1 is an identity for the Q–algebra A. The action of A on V is defined as in 2.6 by $(r,v)(u_1 e_1 + u_2 e_2) = r u_1 e_1 + (r u_2 + v u_1) e_2$, where $r \in Q$, $v \in N$, and $u_1, u_2 \in U$. Denote $R = Z \cdot 1 \oplus M$, where $M = \Sigma Z \xi_{ij}$, and $G = \Sigma Z \eta_i e_1 \oplus \Sigma Z \eta_i e_2$. Thus, R is a full, free subring of A, G is a full, free subgroup of V, and $RG = G$. Let $\varphi \in \text{End } V$ be defined by $\varphi(u_1 e_1 + u_2 e_2) = (e_{mm} u_1) e_2$. It follows easily from the definition of A, using $\xi_{mm} = 0$, that $\varphi \notin A$. We will prove that for a suitable choice of the integers k_{ij}, $\varphi w \in Rw$ for all $w \in G$. It will follow from 1.8 that $\varphi \in QE(H)$ for all full subgroups H of V, and therefore $_A V$ is not quasi–realizable. It is sufficient to prove that if $u_1 = a_1 \eta_1 + \cdots + a_{m-1} \eta_{m-1} + a_m \eta_m \in \Sigma Z \eta_i$, then $a_m \eta_m \in M u_1$. Furthermore, it may be assumed without loss of generality that $a_m \neq 0$ and $\gcd(a_1, \ldots, a_m) = 1$. Let $v = \Sigma x_{ij} \xi_{ij}$ be an arbitrary element of M, so that $x_{ij} \in Z$ for all i, j. Since $\xi_{mm} = 0$ we may take $x_{mm} = 0$. By the definition of ξ_{ij},
$$v u_1 = \Sigma_{i < m} (\Sigma_{j \leq m} x_{ij} a_j) \eta_i + (\Sigma_{j < m} x_{mj} a_j + (\Sigma x_{ij} k_{ij}) a_m) \eta_m.$$
The required result is obtained by showing that the system of Diophantine equations,

(a) $\Sigma_{j \leq m} a_j x_{ij} = 0$, $i < m$,

(b) $\Sigma_{j < m} a_j x_{mj} + a_m (\Sigma k_{ij} x_{ij}) = a_m$

has an integral solution x_{ij} with $x_{mm} = 0$. Using (a) in the form $a_m x_{im} = -\Sigma_{j < m} a_j x_{ij}$ and $k_{mm} = -1$, the equation (b) can be rewritten in the form

(c) $\Sigma_{j < m} \Sigma_{i \leq m} (k_{ij} a_m - k_{im} a_j) x_{ij} = a_m$.

Conversely, a solution of (a) and (c) is a solution of (a) and (b). An integral solution of

(d) $\Sigma_{j < m} \Sigma_{i \leq m} (k_{ij} a_m - k_{im} a_j) y_{ij} = 1$

will yield a solution of (a) and (c) by defining $x_{ij} = a_m y_{ij}$ for $i \leq m$, $j < m$, and $x_{im} = -\Sigma_{j < m} a_j y_{ij}$ for $i < m$, and $x_{mm} = 0$. By a well–known elementary property of linear Diophantine equations, (d) has an integral solution if and only if the greatest common divisor of $\{k_{ij} a_m - k_{im} a_j : i \leq m, j < m\}$ is 1. The numbers in this set are the entries of the column

vector $K \begin{bmatrix} a_1 \\ \vdots \\ a_m \end{bmatrix}$, where $K = \begin{bmatrix} -K_m & 0 & \cdots & 0 & K_1 \\ 0 & -K_m & & 0 & K_2 \\ 0 & \cdots & 0 & -K_m & K_{m-1} \end{bmatrix}$ with $K_j = \begin{bmatrix} k_{1j} \\ k_{2j} \\ \vdots \\ k_{mj} \end{bmatrix}$ for $1 \leq j \leq m$.

The $m^2 - m$ by m matrix K has the m by m submatrix

$$K' = \begin{bmatrix} -k_{1m} & 0 & \cdots & 0 & k_{11} \\ 0 & -k_{2m} & & 0 & k_{22} \\ 0 & \cdots & 0 & -k_{m-1\,m} & k_{m-1\,m-1} \\ 0 & & 0 & -k_{mm} & k_{m\,m-1} \end{bmatrix}$$

which is unimodular for special choices of the k_{ij}. For example, if $k_{im} = -1$ for $1 \leq i \leq m$, $k_{m-1\,m-1} = 1$ and $k_{m\,m-1} = 2$, then $\det K' = 1$. In this case, there is an m by $m^2 - m$ matrix L such that LK is the m by m identity matrix. Since $\gcd(a_1, \ldots, a_m) = 1$, there exists a row vector $(b_1,...,b_m) \in Z^m$ such that $(b_1,...,b_m)\begin{bmatrix} a_1 \\ \vdots \\ a_m \end{bmatrix} = \Sigma b_i a_i = 1$. Then $1 =$

$(b_1,...,b_m)LK\begin{bmatrix} a_1 \\ \vdots \\ a_m \end{bmatrix}$.   Therefore, the $m^2 - m$ entries of $(b_1,...,b_m)L =$

$(y_{11},...,y_{m1},...,y_{m\,m-1})$ form a solution of (d). $\square$

References

[B]   M.C.R. Butler, On locally free torsion–free rings of finite rank, J. London Math. Soc. 43(1968), 297–300.

[F]   L. Fuchs, Infinite Abelian Groups, Vol. I and II, Academic Press (1973), New York.

[K]   C. Kuratowski, Sur l'extension de deux theoremes topologiques a la theorie des ensembles, Fundamenta Math. 34(1947), 34–38

[MV]   A. Mader and C. Vinsonhaler, Torsion–free E–modules, J. Algebra 115(1988), 401–411.

[NT]   C.J. Nesbitt and R.M. Thrall, Some ring theorems with applications to unimodular representations, Ann. Math. 47(1946), 551–567.

[P]   R.S. Pierce, Associative Algebras, Springer–Verlag(1982), New York.

[Z]   H. Zassenhaus, Orders as endomorphism rings of modules of the same rank, J. London Math. Soc. 42(1967), 180–182.

Common Extensions of Finitely Additive Measures and a Characterization of Cotorsion Abelian Groups

K. M. RANGASWAMY

Department of Mathematics, University of Colorado
Colorado Springs, CO 80906

and

J. D. REID

Department of Mathematics, Wesleyan University
Middletown, CT 06459-0128

Abstract

We prove that an abelian group G has the property that, for every set X, and every pair α, β of consistent G-valued finitely additive measures on fields A, B of subsets of X, there is a common extension $\gamma : \mathcal{P}(X) \to G$ if and only if G is a cotorsion group. Groups with the common extension property have the so called infinite Chinese remainder property (ICRP). We characterize the groups with this property; and we also characterize the groups G that have a more general Chinese remainder property as those for which G/G^1 is algebraically compact. These results extend or refine various results in the literature.

INTRODUCTION

Let G be an additively written abelian group and let A be a field (i.e. a Boolean Algebra) of subsets of a non-empty set X. A G-*valued finitely additive measure* or a *charge* on A is a function $\mu : A \to G$ which satisfies $\mu(\emptyset) = 0$ and $\mu(A \cup B) = \mu(A) + \mu(B)$ whenever $A \cap B = \emptyset$. Suppose A and B are any fields of subsets of an arbitrary non-empty set X. Let $\alpha : A \to G$ and $\beta : B \to G$ be any G-valued charges which are consistent, i.e., agree on $A \cap B$. We consider the question of determining those abelian groups G for which any two such charges α and β admit a common extension $\gamma : \mathcal{P}(X) \to G$, where $\mathcal{P}(X)$ is the power set of X. We shall say that such groups G have the common extension property. There has been a series of papers on the problem of extension of charges beginning with Tarski's result [**HT**]), on extending a single charge $\alpha : A \to G$ to $\mathcal{P}(X)$ when G is the group of real numbers. This was extended in [**AR**], to the case of G algebraically compact and in [**CP**], to arbitrary G. The problem of common extensions, among other things, was considered in [RR] and [M] when G is the group of real numbers, in [BR] when both $\mathcal{A}$ and $\mathcal{B}$ are finitely generated, and more generally in [RS1] when only one of $\mathcal{A}$ and $\mathcal{B}$ is finitely generated. One of our main results answers the general question by showing that an abelian group G has the common extension property if and only if G is cotorsion. We accomplish this by reformulating the question in terms of Specker groups. In addition to new results, this algebraic reformulation leads to substantially simpler approaches to several of the known theorems. (See, for example, the Remark in §2 and Corollaries 3.2 and 3.3).

The question of characterizing the groups with the common extension property was posed to us by K.P.S. Bhaskara Rao and R.M. Shortt, who showed in [RS2] that an abelian group G with the common extension property also satisfies the so called Infinite Chinese Remainder Property (I.C.R.P.): i.e. Given any sequence of pair-wise relatively prime positive integers $\{n_i\}$ and any sequence $\{g_i\} \subseteq G$, the infinite system of congruences $x \equiv g_i \pmod{n_i G}$ always has a solution for x in G. We characterize the groups that have this property in §4. In §5 we introduce more general systems of congruences $x \equiv g_i \pmod{n_i G}$ where the n_i's are not necessarily relatively prime, but the g_i's satisfy the compatibility condition that $g_i \equiv g_j \pmod{d_{ij} G}$, where d_{ij} is the g.c.d. of n_i and n_j. We show that in an abelian group G the more general congruence has a solution if and only if G/G^1 is algebraically compact where $G^1 = \cap_{n=1}^{\infty} nG$. Likewise, (cf. §4) an abelian group G has the I.C.R.P. if and only if for every height sequence s, $G/G(s)$ is algebraically compact. The latter groups include all cotorsion groups and all semi-local abelian groups and are closed under direct products, homomorphic images and balanced subgroups.

The authors wish to thank Professors K.P.S. Bhaskara Rao and R.M. Shortt for introducing them to the common extension problem and for several pre-prints.

This work was done while the first author was a Van Vleck visiting Professor at Wesleyan University. He gratefully acknowledges the hospitality of the Department of Mathematics, Wesleyan University.

§2. Preliminaries.

All the groups that we consider here are additively written abelian groups. For general notation, terminology and basic results we refer to [F]. For the benefit of the reader we shall review some of these results. The symbols $\mathbb{Z}$ and $\mathbb{Q}$ denote respectively the additive groups of integers and rational numbers. An abelian group G is said to be divisible if $nG = G$ for all integers n. G is said to be reduced if $\{0\}$ is the only divisible subgroup of G. For any G, G^1 denotes $\cap_{n=1}^{\infty} nG$. If G is torsion-free, then G^1 is divisible. Given a prime p, the p-height of an element $x \in G$, written $h_p(x)$, is n if $x \in p^n G$ but $x \notin p^{n+1}G$. If $x \in \cap_{n=1}^{\infty} p^n G$, we say that x has infinite p-height and write $h_p(x) = \infty$. By a *height* (or a height sequence) s we mean a sequence $s = (\ldots, k_p, \ldots)$ indexed by the primes where each k_p is a non-negative integer or the symbol ∞. For such a sequence s, we denote by $G(s)$ the subgroup $\cap_p p^{k_p} G = \{x \in G : h_p(x) \geq k_p \text{ for all } p\}$. A subgroup H of an abelian group G is said to be pure in G if, for all positive integers n, $H \cap nG = nH$. An abelian group G is said to be *algebraically compact* if it is a direct summand of any group that contains G as a pure subgroup. For other equivalent properties, see [F]. G is said to be *cotorsion* if G is a direct summand of any containing group H for which H/G is torsion-free. It is equivalent to say $\text{Ext}(\mathbb{Q}, G) = 0$. If G is cotorsion and $G^1 = 0$, then G is algebraically compact. We shall use the fact that if A is a cotorsion subgroup of B and B is reduced, then B/A is reduced. Every abelian group G is embeddable as a subgroup of $G^* = \text{Ext}(\mathbb{Q}/\mathbb{Z}, G)$, a cotorsion group with G^*/G torsion-free divisible (see [F]).

Let X be a non-empty set. Let F be the group of all functions $f : X \to \mathbb{Z}$ which assume only finitely many distinct values in $\mathbb{Z}$. Clearly every non-zero $f \in F$ can be written uniquely in the form

$$f = k_1 I_{X_1} + \ldots + k_n I_{X_n} \tag{1}$$

where $k_i \in \mathbb{Z}$, I_{X_i} is the characteristic function of the set X_i and $X_1, \ldots, X_n$ are pairwise disjoint. A subgroup S of F is said to be Specker [S] if for each $f \in S$, the functions I_{X_i} in (1) are also in S for all $i = 1, \ldots, n$. A celebrated theorem of Nöbeling [N, F] says that F is a free abelian group and that any Specker subgroup S of F is a direct summand, i.e. we can write $F = S \oplus T$ for some T. Following [RS1], let us define, for any subfield $\mathcal{A}$ of $\mathcal{P}(X)$, the subgroup $\mathcal{S}(X, \mathcal{A}) = \{f \in F : f^{-1}(n) \in \mathcal{A}$ for each $n \in \mathbb{Z}\}$. It is clear that $\mathcal{S}(X, \mathcal{A})$ is a Specker subgroup of F. Conversely, if $A \subseteq F$ is Specker, then the subfield $\mathcal{A}$ generated by $\{Y \in \mathcal{P}(X) : I_Y \in A\}$ satisfies $\mathcal{S}(X, \mathcal{A}) = A$. Thus the correspondence $\mathcal{A} \to \mathcal{S}(X, \mathcal{A})$ takes subfields of $\mathcal{P}(X)$ to Specker subgroups of F and conversely. Moreover, for any abelian group G, G-valued charges $\alpha : \mathcal{A} \to G$ correspond to group homomorphisms $\alpha' : \mathcal{S}(X, \mathcal{A}) \to G$ given by $\alpha(Y) = \alpha'(I_Y)$ for all $Y \in \mathcal{A}$.

Remark: Observe that a charge $\alpha : \mathcal{A} \to G$ extends to a charge $\beta : \mathcal{P}(X) \to G$ exactly when the induced $\alpha' : \mathcal{S}(X, \mathcal{A}) \to G$ extends to a homomorphism $\beta' : F \to G$. Since the Specker subgroup $\mathcal{S}(X, \mathcal{A})$ is a direct summand of F, such an extension β' always exists. Thus Tarski's Theorem [HT] and its generalizations in [AR, CP] all follow from this observation.

§3 Common Extensions and Cotorsion Groups

We begin with the following Lemma which is implicit in [S] and proved in [N].

Lemma 3.1 [N] *If A and B are Specker subgroups of a Specker group F, then $A + B$ is a pure subgroup of F.*

Corollary 3.2 (Theorem 1 [BR]):   *If $\mathcal{A}$ and $\mathcal{B}$ are both finitely generated fields in $\mathcal{P}(X)$, then for any abelian group G, any consistent G-valued charges $\alpha : \mathcal{A} \to G$, $\beta : \mathcal{B} \to G$ have a common extension $\gamma : \mathcal{P}(X) \to G$.*

Proof: Let $A = \mathcal{S}(X, \mathcal{A})$, $B = \mathcal{S}(X, \mathcal{B})$ and $f : A \to G$, $g : B \to G$ be corresponding homomorphisms. These agree on $A \cap B$ so that they induce a homomorphism $h : A + B \to G$. By Lemma 3.1, $A + B$ is pure and finitely generated in the free group $F = \mathcal{S}(X, \mathcal{P}(X))$ and hence is a direct summand: $F = (A + B) \oplus C$. Clearly h extends to a homomorphism $F \to G$. ∎

Corollary 3.3:(Theorem 3.3 [RS1]):   *Corollary 3.2 holds if only one of the fields, say $\mathcal{B}$, is finitely generated.*

Proof: As before, let $A = \mathcal{S}(X, \mathcal{A})$, $B = \mathcal{S}(X, \mathcal{B})$ and $F = \mathcal{S}(X, \mathcal{P}(X))$. Since A is Specker, $F = A \oplus C$. Let $\pi : F \to G$ be the projection with kernel A. Clearly $A + B = A \oplus D$, where $D = \pi(B) \subseteq C$. By Lemma 3.1, $A + B$ and hence D is pure in F. Since D is finitely generated, it is a direct summand of the free group F and hence of $C : C = D \oplus E$. Then

$F = (A + B) \oplus E$ and so every homomorphism from $A + B$ to G extends to a homomorphism from F to G. ■

Our next result is based on an example of G. M. Bergman (cf. [RS1]) to whom the following lemma is due. Its proof is straightforward.

Lemma 3.4:   *Let R be the group of all bounded integer valued functions on $I\!N = \{0, 1, 2, \ldots\}$. Let*

$$S_1 = \left\{ f \in R : \ f(2i - 1) = f(2i) \quad \text{for all } i \geq 1 \right\}$$

and

$$S_2 = \left\{ f \in R : \ f(2i) = f(2i + 1) \quad \text{for all } i \geq 0 \right\}.$$

Then $S_1 + S_2 = \{ f \in R : \ \sum_{i=1}^{n} f(2i) - f(2i - 1) \text{ is bounded uniformly for all } n \}$.

Lemma 3.5:   *Let R, S_1, S_2 be as in Lemma 3.4. Then $R/(S_1 + S_2) \cong Q \oplus H$ for some group H.*

Proof: Let f be the characteristic function of the set of even integers and for each prime p and positive integer k, let $g_{p,k}$ be the characteristic function of the set of multiples of $2p^k$. Observe that

$$\sum_{i=1}^{n} g_{p,k}(2i) = \left[\frac{n}{p^k} \right]$$

where $[x]$ is the greatest integer function. Hence,

$$\sum_{i=1}^{n} [(f - p^k g_{p,k})(2i) - (f - p^k g_{p,k})(2i - 1)]$$

$$= \sum_{i=1}^{n} f(2i) - p^k \sum_{i=1}^{n} g_{p,k}(2i) = n - p^k \left[\frac{n}{p^k} \right] < p^k$$

for all n. Thus $f \notin S_1 + S_2$ while $f - p^k g_{p,k} \in S_1 + S_2$ for all p and k by Bergman's lemma. It follows that $(R/S_1 + S_2))^1$ is non-trivial and is divisible since $R/(S_1 + S_2)$ is torsion-free. Hence $R/(S_1 + S_2)$ has Q as a summand, as required. ■

Theorem 3.6:   *Let G be an abelian group. Then the following are equivalent:*

(1) G has the common extension property

(2) G is cotorsion.

Proof: (1) $\Rightarrow$ (2). Suppose G has the common extension property. Consider the Specker groups R, S_1 and S_2 of Lemma 3.5 so that $R/(S_1 + S_2) = Q \oplus H$ for some H. By hypothesis on G, every homomorphism from $S_1 + S_2$ to G extends to a homomorphism from R to G. Thus an application of the functor $\text{Hom}(\ , G)$ to the exact sequence

$$0 \to S_1 + S_2 \to R \to Q \oplus H \to 0$$

yields the exact sequence

$$\text{Hom}(R, G) \xrightarrow{i^*} \text{Hom}(S_1 + S_2, G) \to \text{Ext}(\mathbb{Q} \oplus H, G) \to \text{Ext}(R, G).$$

Since R is free $\text{Ext}(R, G) = 0$, and i^* is onto by hypothesis. Thus $\text{Ext}(\mathbb{Q} \oplus H, G) = 0$ which implies $\text{Ext}(\mathbb{Q}, G) = 0$. Hence G is cotorsion.

(2) $\Rightarrow$ (1). Let $\alpha : \mathcal{A} \to G$, $\beta : \mathcal{B} \to G$ be consistent G-valued charges from subfields $\mathcal{A}, \mathcal{B}$ of a field $\mathcal{F}$ of subsets of a non-empty set X. The corresponding consistent homomorphisms $\alpha' : \mathcal{S}(X, \mathcal{A}) \to G$ and $\beta' : \mathcal{S}(X, \mathcal{B}) \to G$ induce a homomorphism $f : \mathcal{S}(X, \mathcal{A}) + \mathcal{S}(X, \mathcal{B}) \to G$. Since, by Lemma 3.1, $\mathcal{S}(X, \mathcal{F})/(\mathcal{S}(X, \mathcal{A}) + \mathcal{S}(X, \mathcal{B}))$ is torsion-free and G is cotorsion, f extends to a homomorphism $g : \mathcal{S}(X, \mathcal{F}) \to G$. This proves (1). ∎

§4. The Infinite Chinese Remainder Theorem

Definition 4.1 (cf. [RS2])   *An abelian group G is said to have the Infinite Chinese Remainder Property (I.C.R.P.) if for every sequence of pairwise relatively prime positive integers $\{n_i\}$ and every sequence $\{g_i\} \subseteq G$, the system of congruences*

$$x \equiv g_i(\bmod\ n_i G), \quad i = 1, 2, \ldots \tag{*}$$

has a solution x in G.

Rao and Shortt [RS2] have shown that an abelian group G having the common extension property satisfies the I.C.R.P. We shall explore the groups with the I.C.R.P. in this section.

Observe that if G has the I.C.R.P. and $f : G \to H$ is an epimorphism, then H has the I.C.R.P., since $n_i H = f(n_i G)$ and f preserves congruences. Since in an abelian group a finite number of elements, whose orders are pair-wise relatively prime, generate a cyclic subgroup, it is clear that any finite system of congruences in (*) will be solvable in any abelian group G.

Let $G^1 = \cap_{m=1}^{\infty} mG$. Now for any positive integer $n, n(G/G^1) = nG/G^1$. It is then easy to see

Proposition 4.2:   *An abelian group G has the I.C.R.P. if and only if G/G^1 has the I.C.R.P.*

Also note that if $L = (\ldots, n_i, \ldots)$ is a sequence of integers, then there is a canonical homomorphism $\sigma = G \to \prod_{i=1}^{\infty} G/n_i G$ given by $\sigma(g) = (\ldots, g + n_i G, \ldots)$, and $\text{Ker } \sigma = L(G) = \cap_{i=1}^{\infty} n_j G$. Thus G has the I.C.R.P. if and only if for every sequence of pairwise relatively prime positive integers $\{n_i\}$, the canonical morphism $\sigma : G \to \prod_{i=1}^{\infty} G/n_i G$ is onto.

It is convenient to normalize things as follows. Let P be the set of all primes. For any integer $n > 0$, let $\text{supp}(n) = \{p \in P : p | n\}$. Suppose $L = \{n_i\}$ is a sequence of relatively prime positive integers with $\text{supp}(n_i) = X_i$. If $\cup X_i \neq P$, then partitioning $P \backslash (\cup X_i)$ by a collection of disjoint finite subsets, we get a larger sequence of relatively prime positive integers $\{m_j\}$ with $\text{supp}(m_j)$ forming a partition of P. Moreover the solvability of the congruence $x \equiv g_j \pmod{m_j G}$ for all $g_j \in G$ implies the solvability of the congruence $x \equiv g_i \pmod{n_i G}$ for all g_i. Thus we shall assume, without loss of generality, that for a given sequence $\{n_i\}$, $P = \cup \operatorname{supp}(n_i)$. Also note that if $n = p_1^{r_1} \ldots p_k^{r_k}$, then $nG = p_1^{r_1} G \cap \ldots \cap p_k^{r_k} G$. Thus, for a sequence $L = \{n_i\}$ of relatively prime integers, $L(G) = \cap_{i<\omega} n_i G = \cap_{p \in P} p^{r_p} G$ where the p^{r_p} are divisors of appropriate n_i's. Moreover, $G/L(G) \cong G/(\cap p^{r_p} G) = G/G(s)$, where s is the height sequence defined by the r_j.

In the following we say a height $s = (r_p)$ is *finite* if $r_p < \infty$ for all p.

Theorem 4.3:   *An abelian group G has the I.C.R.P. if and only if for all finite heights s, $G/G(s)$ is algebraically compact.*

Proof: Let $s = (\ldots, r_p, \ldots)$ be a finite height and $H = \prod_p (G/p^{r_p} G)$. Let $\sigma : G \to H$ be the canonical morphism given by $\sigma(g) = (\ldots, g + p^{r_p} G, \ldots)$ and let A be the image of σ. Observe that for each prime p, $A + \prod_{q \neq p} G/q^{r_q} G = H$ and that $pH \supset \prod_{q \neq p} G/q^{r_q} G$. This implies that H/A is a divisible group. But H/A must be reduced since H is reduced and $A \cong G/G(s)$ is algebraically compact. Thus $H/A = 0$ so that σ is onto. This proves that G has the I.C.R.P.

Conversely, if G has the I.C.R.P., then for any finite height $s = (\ldots, r_p, \ldots)$, $G/G(s) = \prod_p G/p^{r_p} G$ which is clearly algebraically compact. ∎

An abelian group G is called *semi-local* if $pG = G$ for almost all primes p.

Corollary 4.4:   *Suppose G is semi-local or cotorsion. Then G has the I.C.R.P.*

Proof: This follows from the observation that for any finite height s, $G/G(s)$ is algebraically compact. ∎

From Corollary 4.4 and Theorem 3.6 we get

Corollary 4.5: [RS2]   *If an abelian group has the common extension property, then it satisfies the I.C.R.P.*

If $G = \prod A_k$, then for any height s, $G(s) = \prod_k A_k(s)$ and $G/G(s) \cong \prod (A_k/A_k(s))$. Thus Theorem 4.3 yields:

Corollary 4.6:   *A direct product of groups with I.C.R.P. again has the I.C.R.P.*

Remark: Any balanced exact sequence $0 \to A \to B \to C \to 0$ and finite height s yield

$$
\begin{array}{ccccccccc}
 & & 0 & & 0 & & 0 & & \\
 & & \downarrow & & \downarrow & & \downarrow & & \\
0 & \to & A(s) & \to & B(s) & \to & C(s) & \to & 0 \\
 & & \downarrow & & \downarrow & & \downarrow & & \\
0 & \to & A & \to & B & \to & C & \to & 0 \\
 & & \downarrow & & \downarrow & & \downarrow & & \\
0 & \to & A/A(s) & \to & B/B(s) & \to & C/C(s) & \to & 0 \\
 & & \downarrow & & \downarrow & & \downarrow & & \\
 & & 0 & & 0 & & 0 & &
\end{array}
$$

with first two rows and all columns exact. By the 3×3 lemma the last row is also exact.

We use this in the proof of Proposition 4.7 below.

Proposition 4.7: *Suppose* $0 \to A \to B \to C \to 0$ *is balanced exact. Then B has the I.C.R.P. if and only if both A and C have the I.C.R.P.*

Proof: Let s be any finite height and consider the commutative diagram above. Since $\text{Hom}(\mathbb{Q}, C/C(s)) = 0$ the last row of the diagram yields the exact sequence

$$0 \to \text{Ext}(\mathbb{Q}, A/A(s)) \to \text{Ext}(\mathbb{Q}, B/B(s)) \to \text{Ext}(\mathbb{Q}, C/C(s)) \to 0.$$

Evidently, $B/B(s)$ is algebraically compact if and only if both $A/A(s)$ and $C/C(s)$ are algebraically compact. The result now follows from Theorem 4.3. ∎

§5. A General Congruence

As a generalization of the Infinite Chinese Remainder Property we consider the following systems of congruences. Suppose G is an abelian group, $\{g_i\} \subseteq G$ is an arbitrary sequence, and $\{n_i\}$ is a sequence of positive integers (not necessarily pair-wise relatively prime). We are interested in the extent to which the congruences

$$x \equiv g_i (\text{mod } n_i G) \tag{2}$$

$g_i \in G$, have a solution. First note that if the system of congruences (2) has a solution, then the g_i's must satisfy the necessary compatibility condition that for all $i, j, g_i - g_j \in n_i G + n_j G = d_{ij} G$, where d_{ij} is the greatest common divisor (g.c.d.) of n_i and n_j.

We shall begin by showing that any system (2) with finitely many n_i and compatible g_i is solvable in any abelian group. We shall use the following easily derivable remark in number theory.

Lemma: *For any positive integers* $a, b_1, \ldots, b_n$, $[(a, b_1), \ldots, (a, b_n)] = (a, [b_1, \ldots, b_n])$ *where $[c, d]$ and (c, d) denote respectively the least common multiple and greatest common divisor of c and d.*

Proposition 5.1: *Let G be an abelian group. Then for any positive integers $k_1, \ldots, k_m$ and $\{g_1, \ldots, g_m\} \subseteq G$, the finite system of congruences $x \equiv g_i \ (\text{mod } k_i G), 1 \leq i \leq m$, has a solution in G provided $g_i \equiv g_j \ (\text{mod } d_{ij} G)$ for all i, j where d_{ij} is $\gcd(k_i, k_j)$.*

Proof: Induction on m. Suppose $m = 2$. Write $d = \gcd(k_1, k_2)$, so that $d = k_1 s + k_2 t$ for some $s, t \in \mathbb{Z}$. By hypothesis, $g_2 = g_1 + du$ for some $u \in G$. Then $g = g_2 - k_2 t u = g_1 + k_1 s u$ satisfies $g \equiv g_i \ (\text{mod } k_i G)$ for $i = 1, 2$.

Suppose, by induction, that there exists $g' \in G$ such that $g' \equiv g_i \ (\text{mod } k_i G)$ for $i = 1, \ldots, m-1$. Write $g' = g_i + k_i u_i, u_i \in G, i = 1, \ldots, m-1$. Then $g' - g_m = g_i - g_m + k_i u_i = d_{im} u_{im} + k_i u_i \in d_{im} G$, since $d_{im} | k_i$. This holds for all $i = 1, \ldots, m-1$ and so $g' - g_m \in \cap_{i=1}^{} d_{im} G = dG$ where d is the *lcm* of the d_{im}. By the Lemma above, $d = \gcd(k_m, \ell)$ where ℓ is the lcm of $k_1, \ldots, k_{m-1}$. By the argument in case $m = 2$, there is $g \in G$ satisfying

$$g \equiv g' (\text{mod } \ell G)$$

$$g \equiv g_m (\text{mod } k_m G).$$

Clearly $g \equiv g_i \pmod{k_i G}$ for $i = 1, \ldots, m$. ∎

For convenience of expression we say an abelian group G *has the G.C.R.P.* (The Generalized Chinese Remainder Property) if the infinite systems of congruences (2) is solvable in G for all sequences $\{n_i\}$ of positive integers and all sequences $\{g_i\} \subseteq G$ whenever the compatibility conditions $g_i - g_j \in d_{ij}G$ hold for all i and j where $d_{ij} = \gcd(n_i, n_j)$.

First observe that for any positive integer $n, n(G/G^1) = nG/G^1$ where $G^1 = \cap\{mG : m = 1, 2, \ldots, \}$. Since passage to quotient groups preserves congruences mod nG, the proof of the following proposition is transparent.

Proposition 5.2: *An abelian group G has G.C.R.P. if and only if G/G^1 has the G.C.R.P.*

We are now ready to prove the main result of this section.

Theorem 5.3: *An abelian group G has the G.C.R.P. if and only if G/G^1 is algebraically compact.*

Proof: In view of Proposition 5.2, we may assume $G^1 = 0$. Suppose G satisfies the G.C.R.P. Now G is a (pure) subgroup of the cotorsion group $G^* = \mathrm{Ext}(\mathbb{Q}/\mathbb{Z}, G)$ with G^*/G torsion-free divisible. Since $G^1 = 0$, $(G^*)^1 = 0$ so that G^* is algebraically compact. Let $u \in G^*$. Since G^*/G is divisible, there exists a sequence $\{g_n : n = 1, 2, \ldots\} \subseteq G$ such that $u - g_n \in nG^*$ for all n. Now, for all m and n, $g_m - g_n \in G \cap d_{mn}G = d_{mn}G$, since G is pure in G^*, where $d_{mn} = \mathrm{g.c.d.}\ (m, n)$ as usual. By hypothesis, there is a $g \in G$ which is a solution of the system of congruences $x \equiv g_n \pmod{nG}$. Then $u - g \in \cap nG = 0$. Thus $u \in G$ and $G = G^*$ is algebraically compact.

Conversely suppose G is algebraically compact. For any sequence of positive integers $\{n_i\}$ and any sequence $\{g_i\} \subseteq G$, consider the congruences

$$x \equiv g_i \pmod{n_i G}$$

where for all i and j, $g_i - g_j \in d_{ij}G$, the d_{ij} being the g.c.d. of n_i and n_j. We can rewrite these congruences as a system of equations

$$x - n_i y_i = g_i \in G, \quad i = 1, 2, \ldots \tag{3}$$

By Proposition 5.1, every finite subsystem of (3) is solvable in G. Since G is algebraically compact, the entire system (3) is solvable in G. Thus G has the G.C.R.P. ∎

Remark: Clearly any cotorsion group G satisfies the G.C.R.P. since G/G^1 is algebraically compact. Conversely, if G/G^1 is algebraically compact and G is torsion or torsion-free, then G is necessarily cotorsion (indeed algebraically compact). However, it is easy to construct non-cotorsion mixed abelian groups G for which G/G^1 is algebraically compact.

References

[AR] V. Aversa and K.P.S. Bhaskara Rao, *Tarski's extension Theorem for group valued charges.* Proc. Amer. Math. Soc, **90**, 79-82 (1984).

[BR] A. Basile and K.P.S. Bhaskara Rao, *Common extension of group-valued charges.* (Preprint).

[CP] T. Carlson and K. Prikry, *Ranges of signed measures*, Period. Math. Hungaricae **13**, 151-155 (1982).

[F] L. Fuchs, "Infinite Abelian Groups", **I** & **II**. Academic Press. New York (1970, 1973).

[HT] A. Horn and A. Tarski, *Measures on Boolean Algebras*, Trans. Amer. Math. Soc. **64**, 467-497 (1948).

[M] D. Maharan, *Consistent extensions of linear functionals and of probability measures*. Sixth Symposium on Math, Stat. and Probability, **2**, 127-147 (1972).

[N] G. Nöbeling, *Verallgemeinerung eines Satzes von Herrn E. Specker*. Invent. Math. **6**, 41-55 (1968).

[RR] K.P.S. Bhaskara Rao and M. Bhaskara Rao, "Theory of Charges", Academic Press. New York (1983).

[RS1] K.P.S. Bhaskara Rao and R.M. Shortt, *Common extension of homomorphisms and group-valued charges*. (Preprint).

[RS2] K.S.P. Bhaskara Rao and R.M. Shortt, *Group-valued charges: common extensions and the Chinese Remainder Property*. (Preprint).

[S] E. Specker, *Additive Gruppen von Folgen ganzer Zahlen*. Portagaliae Math. **9**, 131-140 (1950).

Valuation Domains with Superdecomposable
Pure Injective Modules

Luigi Salce*

Dipartimento di Matematica Pura e Applicata
Università di Padova, Italy

INTRODUCTION

The first important extension of the theory of pure injective (i.e. algebraically compact) abelian groups, developed in the '50's by Kaplansky [K], Łoś [L], Maranda [Mar] and others, to modules over arbitrary associative rings with 1, was given by Warfield in [W1]. In section 6 of that paper, Warfield considered the special case of pure injective modules over a Prüfer ring R, and showed that such a module M decomposes as $M = E \oplus N$, where E is injective and the first Ulm submodule $N^* = \bigcap \{rN \mid 0 \neq r \in R\}$ of N is vanishing. If R is h-local (see [Mat]), then the injective summand E can be completely classified (see [W2]), and the summand N is a direct product $N = \prod N_{\mathfrak{m}}$ of pure injective modules over the localizations $R_{\mathfrak{m}}$ (where $\mathfrak{m}$ ranges over all maximal ideals of R), which are valuation domains. The last step in section 6 of [W1], in trying to determine the structure of general pure injective modules over valuation domains, is Theorem 5, which furnishes a complete classification in the torsionfree case.

The description of the structure of pure injective modules over a valuation domain R was improved by Fuchs and the author in the monograph [FS], where it was shown that such a module M decomposes in the following way:

$$M = E \oplus F \oplus T \oplus N$$

where E is injective, F is the pure injective hull of a direct sum of ideals of R (hence it is torsionfree), T is the pure injective hull of a direct sum of standard uniserial torsion modules (hence it is either torsion or mixed), and N is superdecomposable, i.e. it has no non zero indecomposable summands. Moreover, the first three summands are uniquely determined, up to isomorphism, by certain invariants of M, and also N is unique up to

* Lavoro eseguito con il contributo del Ministero dell'Università e della Ricerca Scientifica e Tecnologica

isomorphism. However, it is clear that not every valuation domain admits superdecomposable pure injective modules, as the structure of general pure injectives over a dicrete rank one valuation domain shows.

Eklof pointed out to the author that an explicit construction of a superdecomposable pure injective module over a fixed valuation domain cannot be found in [**FS**], even if their existence is taken for granted both in [**FS**] and in Fuchs' paper [**F**]. To make the subject more intriguing, one could mention the assertion, not supported by a proof, made by Ziegler [**Zi**] and Prest [**Pr**], that over general valuation domains no such modules exist.

The goal of this short note, originated from a question of Paul Eklof, is to characterize valuation domains which admit superdecomposable pure injective modules; from the proof one could easily deduce how to infer from [**FS**] the existence of superdecomposable pure injectives over the appropriate valuation domains. It turns out that these domains form the complementary class to the class of valuation domains investigated by Zanardo [**Za**] under the name "totally branched and discrete", and which are the local "generalized Dedekind" domains studied by Popescu [**Po**].

PRELIMINARIES

The main reference for all notions and basic facts concerning this note is the monograph [**FS**]. From now on, R will denote a valuation domain, P its maximal ideal and Q its field of quotients, that we always assume different from R. We shall use the notion of height of an element of an R-module M: given $a \in M$, define the *height ideal* of a in M as $H_M(a) = \bigcup \{r^{-1}R \mid r \in R, a \in rM\}$; obviously $R \leq H_M(a) \leq Q$. Setting $I = H_M(a)$ and $U = I/R$, define the *height* of a in M as

$$h_M(a) = \begin{cases} U & \text{if there exists a homomorphism } f \colon I \to M \text{ such that } f(1) = a \\ U^- & \text{otherwise.} \end{cases}$$

Accordingly, we say that U is a *non limit height*, and U^- is a *limit height*. A module is said to be *cohesive* if all elements have non limit heights. Pure injective R-modules are cohesive (see[**FS**, XI.4.3]).

We recall some other notions, that will be crucial in the proof of our main result. The *indicator* of an element a in M is the function

$$i_M(a) \colon \Gamma^+ \to \Sigma$$

where Γ^+ is the positive cone of the value group Γ of R and Σ is the set of the heights, $i_M(a)$ is defined as follows

$$i_M(a)(\gamma) = rh_M(ra)$$

where $r \in R$ satisfies $v(r) = \gamma$ ($v(r)$ is the value of r in Γ). The indicator $i_M(a)$ is *irregular* at the proper ideal $L > \text{Ann}(a)$ if, for all $r \in R \setminus L$, $s \in L$, the following inequality holds

$$rh_M(ra) < sh_M(sa);$$

the indicator $i_M(a)$ *increases on the right* at L if there are no $s_0 \in L$ such that $s_0 h_M(s_0 a) = \inf \{sh_M(sa) \mid s \in L\}$; the last inf is denoted by $i_M(a)^L$, and symmetrically we set: $i_M(a)_L = \sup\{rh_M(ra) \mid r \in R \setminus L\}$; finally, the indicator $i_M(a)$ *has a gap* at L if $i_M(a)_L < i_M(a)^L$.

A valuation domain is *discrete* if, for any two adjacent prime ideals $I > J$, R_I/JR_I is rank one discrete; it is *totally branched* if it has the property that each non zero prime ideal has an immediate successor (with respect to the inclusion relation), equivalently, $\mathrm{Spec}R$ is well ordered by the opposite inclusion. Totally branched discrete valuation domains have nice characterizations, given by Zanardo [**Za**] (see also [**FS**, XII.6]), in terms of their ideal structure and of their cohesive modules.

THE MAIN RESULT

In order to prove the promised characterization, we need two preliminary results.

LEMMA 1. *A pure injective module M over a valuation domain R is superdecomposable if and only if it does not contain non zero pure uniserial submodules.*

PROOF: If $0 \neq U$ is a pure uniserial submodule of M, then the pure injective hull $U^{\frown}$ of U is isomorphic to a summand of M, by [**FS**, XI.2.9]; $U^{\frown}$ is indecomposable, by [**FS**, XI.4.6] and [**FS**, XI.4.Exercise 2], hence M is not superdecomposable.

Conversely, if M is not superdecomposable, then it contains an indecomposable direct summand N, which is still pure injective; again by [**FS**, XI.4.6], N is isomorphic to the pure injective hull $U^{\frown}$ of a standard uniserial module U, which is pure in N, hence also in M. ∎

LEMMA 2. *Let R be a totally branched and discrete valuation domain. Then every cohesive non-zero R-module contains a non zero pure uniserial submodule.*

PROOF: This result is proved in [**Za**] (see also [**FS**, XIII.6.4, 3) $\implies$ 1)]) under the additional hypothesis that the cohesive module is torsion. Actually, a careful analysis of that proof shows that it is enough to assume that the cohesive module is not torsion-free. Since the claim is trivially true for torsion-free modules (even for arbitrary valuation domains), we are done. ∎

We can now characterize the valuation domains of the title.

THEOREM. *Let R be a valuation domain. There exist superdecomposable pure injective R-modules if and only if R is not totally branched and discrete.*

PROOF: Every pure injective module over a valuation domain is cohesive, i.e. all elements have non-limit height, by [**FS**, XI.4.3]. If R is totally branched and discrete, then all cohesive R-modules contain non zero pure uniserial submodules, by lemma 2, hence pure injective R-modules are not superdecomposable, in view of lemma 1.

Conversely, assume R not totally branched and discrete. We resume, for the sake of completeness, part of the proof in [**Za**] (see also [**FS**, XIII.6.4, 1) $\implies$ 2)]). There is a prime ideal J such that JR_J is not a principal ideal of R_J. Then it is possible to construct an R_J-module A, as in [**FS**, X.4], satisfying the following properties:

(i) A is cohesive;

(ii) A has no non zero pure uniserial submodules;

(iii) the indicator $i_A(a)$ of each non zero element $a \in A$ is constant up to an ideal $L > \mathrm{Ann}(a)$, and it increases at the right at L, where it has no gaps.

It is easy to see that properties (i) and (ii) are satisfied by A also as an R-module; moreover property (iii) also holds if A is viewed as an R-module, by [**FS**, VIII.1.6]. In order to conclude our proof, a new argument is needed. It is enough to prove, in view of lemma 1, that the pure injective hull $A^{\wedge}$ of A has no pure uniserial submodules. Let us assume, by way of contradiction, that U is a pure uniserial submodule of the R-module $A^{\wedge}$.

First we show that $U \cap A = 0$. If $0 \neq x \in U \cap A$, then $h_{A^{\wedge}}(x) = K/R = h_A(x)$ for some $R \leq K \leq Q$ (since $A^{\wedge}$ and A are cohesive). Then there exists a uniserial pure submodule of A isomorphic to K/I containing x, a contradiction. Therefore $U \cap A = 0$.

Now, since A is pure essential in $A^{\wedge}$, $A \oplus U/U$ is not pure in $A^{\wedge}/U$; thus there exist $a^{\wedge} \in A^{\wedge}$, $a \in A$, $u \in U$ such that $a^{\wedge} = u + a$ and

$$h(a^{\wedge}) = H/R > Y/R = h(u) = h(a)$$

(the heights are computed in $A^{\wedge}$, in U or in A, whenever possible). Since U is pure in $A^{\wedge}$, the indicator of u is constant up to $\mathrm{Ann}(u)$, hence also the indicator of a is constant up to $\mathrm{Ann}(u)$. If $\mathrm{Ann}(a) = \mathrm{Ann}(u)$ we are done, since then $Y/\mathrm{Ann}(u)$ is isomorphic to a pure uniserial submodule of A containing a. So let $\mathrm{Ann}(a) < \mathrm{Ann}(u)$; then for each $s \in \mathrm{Ann}(u) \setminus \mathrm{Ann}(a)$ we have

$$h(sa) = h(sa^{\wedge}) \geq s^{-1}H/R.$$

Hence we deduce

$$i_M(a)^{\mathrm{Ann}(u)} \geq H/R > Y/R = h(a) = i_M(a)_{\mathrm{Ann}(u)};$$

this shows that the indicator of a has a gap at $\mathrm{Ann}(u)$, contradicting property (iii) of A. Therefore, $A^{\wedge}$ has no non zero pure uniserial submodules and it is, consequently, superdecomposable. ∎

Remark. One could try to prove the above theorem by using the result in [**FS**, X.2.2], which would ensure that $A^{\wedge}$ has no non zero pure uniserial submodules, provided that A (had the same property and it) was smooth; but this is not the right way to prove the theorem, since the module A constructed in [**FS**, X.4] is not smooth in general (see [**FS**, X.Exercise 1]).

We would like to conclude recalling that the following problem is still unsolved: classify superdecomposable pure injective modules over valuation domains by suitable invariants.

REFERENCES

[F]   L. Fuchs, *On divisible modules over valuation domains*, J. Algebra **110** (1987), 498–506.

[FS] L. Fuchs and L. Salce, "Modules over Valuation Domains," Lecture Notes Pure Appl. Math. **97**, Marcel Dekker, 1985.

[K]   I. Kaplansky, "Infinite Abelian groups," Ann Arbor, 1954.

[L]   J. Łoś, *Abelian groups that are direct summands of every abelian group which contains them as pure subgroups*, Fund. Math. **44** (1957), 84–90.

[Mar] J.M. Maranda, *On pure subgroups of abelian groups*, Archiv der Math. **11** (1960), 1-13.

[Mat] E. Matlis, *Cotorsion Modules*, Mem. Amer. Math. Soc. **49** (1964).

[Po] N. Popescu, *On a class of Prüfer domains*, Rev. Roumaine Math. Pures Appl. **29** (1984), 777–786.

[Pr] M. Prest, "Model Theory and Modules," London Math. Soc. Lecture Notes Series No.130, Cambridge Univ. Press,1988.

[W1] R.B. Warfield Jr., *Purity and algebraic compactness for modules*, Pac. J. Math. **28** (1969), 699–719.

[W2] R.B. Warfield Jr., *Decompositions of injective modules*, Pac. J. Math. **31** (1969), 263–276.

[Za] P. Zanardo, *Valuation domains without pathological modules*, J. Alg. **96** (1985), 1–8.

[Zi] M. Ziegler, *Model theory of modules*, Ann. Pure Appl. Logic **26** (1984), 149–213.

Homological Dimensions of Completely Decomposable Groups

C. VINSONHALER[1]

W. WICKLESS

University of Connecticut, Storrs, CT 06269

This work is based on a simple construction that produces, for each positive integer n, a direct sum, G, of 2n+1 subgroups of the rationals, Q, which has flat (weak) dimension n over its endomorphism ring E (Theorem 5). The simplicity of the construction and the elegant result of Richman and Walker [RW] characterizing the completely decomposable groups (direct sums of subgroups of Q) which are flat over their endomorphism rings, prompted us to try to characterize the flat dimension of completely decomposable groups in terms of their typesets. The problem seems to be quite complicated, in general. Fortunately, some of the complications are avoided by restricting to finite rank and working with the divisible hull QG of G as a module over the quasi–endomorphism algebra QE $\simeq$ Q $\otimes_Z$E. In this setting, we are able to obtain a complete description of the flat dimension of QG in terms of the poset of types of summands of G (Theorem 11).

We begin with some known results about flatness and completely decomposable groups. For the moment, no finite rank restrictions are imposed.

Lemma 1. (see [R, Theorems 3.53, 3.54, 3.55]) A left module H over a ring E is flat if and only if either of the following hold:

   (a) For each right ideal J of E, the natural map J $\otimes_E$H $\to$ JH is an isomorphism.

   (b) Given an exact sequence $0 \to K \to F \to H \to 0$ with F flat, for each right ideal J
        of E, JF $\cap$ K = JK.

[1]Research supported, in part, by NSF grant DMS 9022730

If G is a completely decomposable group, then by the classic result of Baer, the set of types of rank one summands of G is an invariant of G. We will call this set of types the **critical typeset** of G. This terminology was introduced by Lady in [L]. The critical typeset is a partially ordered set under the usual ordering of types.

Proposition 2. [RW]   A completely decomposable group is flat as a module over its endomorphism ring if and only if any two elements in the critical typeset T which have an upper bound in T also have a lower bound in T.

Our next lemma allows us to apply Proposition 2 to fully invariant summands of the group G.

Lemma 3. If H is a fully invariant summand of an abelian group G, then H is $E(H)$–flat if and only if H is $E(G)$–flat.

Proof.   Denote $E = E(G)$ and let $e:G \to H$ be an idempotent projection onto H. Note that $Ee = eEe = E(H)$ since H is fully invariant. In particular,

$$(3.1) \qquad Ee \otimes_{eEe} H = eEe \otimes_{eEe} H \simeq H$$

as E–modules or as eEe–modules.

Assume H is eEe–flat, and let J be a right ideal of E. Then $J \otimes_E H \simeq J \otimes_E (Ee \otimes_{eEe} H) \simeq (J \otimes_E Ee) \otimes_{eEe} H \simeq Je \otimes_{eEe} H \simeq JeH = JH$, using (3.1) and the facts that Ee is E–flat and H is eEe–flat. Since all these isomorphisms are the natural ones, H is E–flat by Lemma 1(a).

Conversely, assume H is E–flat and let I be a right ideal of eEe. Then $J = IE$ is a right ideal of E with $Je = I$. Thus, $IH = JeH = JH \simeq J \otimes_E H \simeq J \otimes_E (Ee \otimes_{eEe} H) \simeq (J \otimes_E Ee) \otimes_{eEe} H \simeq Je \otimes_{eEe} H = I \otimes_{eEe} H$, and H is eEe–flat.

The next result illuminates the relationship between the E–module G and the QE–module QG. Recall that if τ is a type, $G(\tau) = \{x \in G \mid \text{type}(x) \geq \tau\}$, a fully invariant subgroup of G. If, in addition, the group G is completely decomposable, then $G(\tau)$ is a summand.

Lemma 4. Let G be a completely decomposable group with endomorphism ring E. Then,
    (a) G is flat over E if and only if QG is flat over QE.
    (b) For each type τ in the critical typeset of G, $G(\tau)$ is a flat E–module and $QG(\tau)$ is a flat QE–module.

Proof.   (a)   If M is a right QE–module, then $M \otimes_{QE} QG$ is naturally isomorphic to

$(M \otimes_E G) \otimes_Z Q$. Since Q is a flat Z–module, if G is a flat E–module, it is immediate that QG is QE–flat. Note that this direction does not use complete decomposability.

Conversely, suppose that QG is QE–flat. Using exactly the same arguments employed in [RW], one can show that any two critical types which have an upper bound in the critical typeset T of G must also have a lower bound in T. Thus, G is E–flat by Proposition 2. In the finite rank case, a relatively short argument using Lemma 1(b) can be given: Suppose QG is QE–flat and denote $F = \oplus QG(\tau)$, where the direct sum is over all minimal types τ in T. Form the exact sequence $0 \to K \to F \to QG \to 0$. The epimorphism $F \to QG$ is induced by the inclusion of each $QG(\tau)$ in QG; and K is the kernel of $F \to QG$. It is easy to check (see Proposition 7) that K is the sum of all submodules of the form $\{(x,-x) \in QG(\tau) \oplus QG(\tau') \mid x \in QG(\tau) \cap QG(\tau')\}$, where τ,τ' are distinct minimal types in T. With an eye to using Proposition 2, suppose τ and τ' are critical types which have an upper bound μ in T. If τ and τ' do not have a lower bound, then we may assume τ and τ' are minimal elements of the finite set T. Let A, A' and B be rank one summands of G of types τ, τ' and μ, respectively. Then there is an endomorphism λ of G which embeds A in B, A' in B and maps all other summands of G to 0. Then $0 \neq \lambda F \cap K$, while $\lambda K = 0$. By Lemma 1(b), QG cannot be flat as a QE–module. This contradiction shows that τ and τ' must have a lower bound; and G is E–flat by Proposition 2.

(b) That $G(\tau)$ is E–flat is a direct consequence of Lemma 3 and Proposition 2. Then $QG(\tau)$ is QE–flat by the argument in the first paragraph of the proof of part (a).

We now provide the construction promised in the introduction. The flat dimension of G as an E–module is denoted $fd(G)$, while $fd(QG)$ denotes the flat dimension of $_{QE}QG$.

Theorem 5. For every natural number n, there exists a completely decomposable group G of rank $2n+1$ such that $fd(G) = n$.

Proof. Let G be the direct sum of $2n+1$ subgroups of Q whose types form a poset represented by the following Hasse diagram:

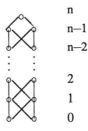

$$
\begin{array}{c}
n \\
n-1 \\
n-2 \\
\vdots \\
2 \\
1 \\
0
\end{array}
$$

The rows of the diagram are labelled by $0,1,...,n$ and the types in row i are labelled σ_i, τ_i

for $1 \leq i \leq n-1$. The top type is labelled σ_n. To show $\mathrm{fd}(G) = n$, we construct an E–flat resolution

$$0 \to F_n \to F_{n-1} \to \cdots \to F_0 \to G \to 0,$$

where $F_i = G(\sigma_i) \oplus G(\tau_i)$ for $0 \leq i \leq n-1$ and $F_n = G(\sigma_n)$. Note that if ρ is a critical type, then $G(\rho)$ is E–flat by Lemma 4(b). It follows that each F_i is a flat E–module. The epimorphism $\varepsilon : F_0 \to G$ is induced by the inclusion maps $G(\sigma_0) \to G$ and $G(\tau_0) \to G$. It is easy to check that the kernel of ε is,

$$K_0 = \{(x,-x) \mid x \in G(\sigma_0) \cap G(\tau_0)\} \subset G(\sigma_0) \oplus G(\tau_0)\}.$$

The map $\varepsilon_1 : F_1 \to K_0$ is given by $(x,y) \to (x+y, -x-y)$ for $(x,y) \in G(\sigma_1) \oplus G(\tau_1)$. Again it is easy to check that ε_1 is an epimorphism with kernel

$$K_1 = \{(x,-x) : x \in G(\sigma_1) \cap G(\tau_1)\} \subset G(\sigma_1) \oplus G(\tau_1).$$

The rest of the sequence is constructed in a similar fashion, with maps $\varepsilon_i : F_i \to K_{i-1} = \ker \varepsilon_{i-1}$; and

$$F_n \simeq K_{n-1} = \{(x,-x) : x \in G(\sigma_{n-1}) \cap G(\tau_{n-1}) = G(\sigma_n)\}.$$

The existence of this flat resolution proves that $\mathrm{fd}(_E G) \leq n$. To show equality holds, it suffices to prove that K_{n-2} is not flat [R, Theorem 9.13]. The E–module K_{n-2} is given by,

$$K_{n-2} = \{(x,-x) : x \in G(\sigma_{n-2}) \cap G(\tau_{n-2})\} \simeq G(\sigma_{n-1}) + G(\tau_{n-1}).$$

The latter group is a fully invariant summand of G which is a completely decomposable group of rank three, with critical typeset diagram $\overset{\circ}{\underset{\circ \quad \circ}{\diagup \diagdown}}$. This group is not a flat E–module by Proposition 2 and Lemma 3.

In [A], by employing a Corner type of construction, Angad–Gaur produced finite rank torsion–free groups of projective dimension n (over the endomorphism ring) for each natural number n. Theorem 5 allows us to obtain the same result with completely decomposable groups.

Corollary 6. For each natural number n, there is a completely decomposable group of rank $2n+1$ with projective dimension n as a module over its endomorphism ring.

Proof. If the subgroups of Q used in constructing the G in Theorem 5 are taken to be subrings of Q, then the resulting endomorphism ring $E = \mathrm{End}(G)$ is Noetherian and G is a finitely generated E–module. In this case, it is well known that $\mathrm{fd}(G) = \mathrm{pd}(G)$ (see [R, Corollary 3.58]).

The results in Theorem 5 and Corollary 6 motivated us to try and characterize the flat and projective dimensions of a completely decomposable group G over its endomorphism ring E in terms of the poset given by the critical typeset. This proved to be a difficult problem. Some indication of the difficulty involved is illustrated by the last example of this paper (Example 13). Nonetheless, we were able to obtain a complete characterization of $fd(_{QE}QG)$. The rest of the paper deals primarily with determining this dimension from the critical typeset of G. Some additional tools will be needed.

If $V = \oplus V_i$ is a direct sum of one dimensional Q–vector spaces V_i, the **support** of a nonzero element $x = \oplus x_i$ in V, denoted $spt(x)$, is $\{V_i \mid x_i \neq 0\}$. If S is a nonzero subspace of V, then $spt(S)$ is the union of the supports of the nonzero elements of S.

Proposition 7. Let $\varphi : \oplus V_i \to W$ be a Q–vector space homomorphism, with each V_i of dimension 1.

    (a) Any nonzero subspace of $\ker \varphi$ of minimal support in $V = \oplus V_i$ is one dimensional.

    (b) Two subspaces of $\ker \varphi$ with the same minimal support are equal.

    (c) $\ker \varphi = \Sigma C_j$, where the C_j are the one–dimensional subspaces of $\ker \varphi$ of minimal support.

Proof. (a) Let K be a subspace of $\ker \varphi$ of minimal support. That is, $spt(K)$ is minimal among $\{spt(L) \mid 0 \neq L$ is a subspace of $\ker \varphi\}$. Let $0 \neq x \in K$. By minimality, $spt(x) = spt(K)$. If $0 \neq y$ is another element of K, then for some nonzero rational r, $x - ry$ has smaller support than $spt(x) = spt(y) = spt(K)$. Hence, $x = ry$. This proves (a) and the proof of (b) is similar. For (c), let $0 \neq x \in \ker \varphi$. If $spt(Qx)$ is minimal, then $Qx = C_j$ for some C_j and there is nothing to prove. Otherwise, there exists a C_j with support properly contained in $spt(Qx)$. In this case we can choose $c \in C_j$ so that $spt(x - c)$ is properly contained in $spt(x)$. By induction, $x - c \in \Sigma C_j$, whence $x = (x - c) + c \in \Sigma C_j$.

We will use Proposition 7 to define a tier complex on a finite set of types. When the set of types is the critical typeset of a completely decomposable group G, then the length of the tier complex will determine the flat dimension of $_{QE}QG$.

Definition 8. The Tier Complex. Let T be a finite set of types. We will define a complex of Q–vector spaces

$$0 \to V_n \overset{\psi_n}{\to} V_{n-1} \to \cdots \to V_1 \overset{\psi_1}{\to} V_0 \overset{\psi_0}{\to} Q \to 0$$

based on the poset structure of T. Each $\psi_i : V_i \to V_{i-1}$ will be a vector space homomorphism with $im \, \psi_{i+1} \subseteq \ker \psi_i$.

For notational convenience, we define $V_{-1} = Q$ to be a (-1)–**cycle of** T. Denote $V_0 = V_0(T) = \oplus Q_\tau$, a direct sum of copies of Q, one for each minimal element $\tau \in T$. Let $\psi_0 : V_0 \to V_{-1} = Q$ be the map induced by the identity on each summand. If S is a subspace of V_0, then denote by **typeset(S)** the set of all types τ such that $Q_\tau \in spt(S)$. An upper bound for S is a type μ in T which is an upper bound (in T) for typeset(S). An upper bound for S is called **minimal** if μ is a minimal element of the set of upper bounds for S. A 0–cycle of T is a one dimensional subspace C of $ker \; \psi_0$ which has minimal support in V_0 and has an upper bound in T. It is easy to see that a subspace of $ker \; \psi_0$ of minimal support is of the form $Q(1,-1) \subseteq Q_\tau \oplus Q_{\tau'}$, where τ, τ' are distinct minimal elements of T. Thus, an upper bound for a 0–cycle C does not belong to typeset(C).

Let $k \geq 1$ and assume inductively that for $0 \leq i < k$, we have defined a complex of vector spaces and homomorphisms $\psi_i : V_i \to V_{i-1}$; and i–cycles which are one dimensional subspaces of V_i. More specifically, for $i < k$, each V_i is a direct sum $\oplus_C \oplus_\tau C_\tau$ where $C_\tau = C$ and the double sum is over all pairs (τ, C) such that C is an (i–1)–cycle and τ is a minimal upper bound for C with $\tau \notin$ typeset(C). The map ψ_i is induced by inclusion $C_\tau = C \subseteq V_{i-1}$. The i–cycles are the subspaces D of $ker \; \psi_i$ of minimal support which have a minimal upper bound in $T \setminus$ typeset(D). The terms "typeset", "upper bound" and "minimal upper bound" have the obvious meanings in this context. For example, typeset(D) = $\{\tau \mid C_\tau \in spt(D)\}$. Then,

(1) V_k is the vector space $\oplus_C \oplus_\tau C_\tau$, where $C_\tau = C$ is a (k–1)–cycle and $\tau \in T \setminus$ typeset(C) is a minimal upper bound for C;

(2) $\psi_k : V_k \to V_{k-1}$ is induced by the inclusion of each $C_\tau = C$ into V_{k-1};

(3) the k–cycles of T are the (one dimensional) subspaces D of $ker \; \psi_k$ of minimal support having a minimal upper bound in $T \setminus$ typeset(D).

The poset T is called **n–tiered** if n is the least non–negative integer for which $V_{n+1} = 0$. That is, T has (n–1)–cycles, but no n–cycles. It follows directly from the definitions, that every finite poset T of types is n–tiered for some $n \geq 0$. Indeed, this integer n will be no larger than the length of the longest chain in T. If G is a completely decomposable group of finite rank, we call G n–tiered if the critical typeset of G is n–tiered.

The complicated nature of these definitions calls for some examples.

Examples 9. (a) Suppose T is a finite set of types with a unique minimal element. Then T is 0–tiered: $V_0 = Q_\tau$, where τ is the unique minimal element, and $\psi_0 : V_0 \to V_{-1}$ is an isomorphism.

(b) Let $T = T_n$ be a set of $2n+1$ types with diagram as in the proof of Theorem 5.

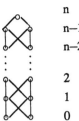

$$\begin{array}{c} n \\ n-1 \\ n-2 \\ \vdots \\ 2 \\ 1 \\ 0 \end{array}$$

This poset is n–tiered. The n "tiers" of T are indicated to the right of the diagram. There is one i–cycle for each $0 \leq i \leq n-1$, and $V_i = C_\tau \oplus C_{\tau'}$, where C is the (i–1)–cycle and τ and τ' are the elements in the i–th "tier". The last vector space is simply $V_n = C_\tau$, where C is the (n–1)–cycle and τ is the maximum type.

(c) Example 13 at the end of the paper describes a "complicated" 2–tiered poset.

Construction 10. Flat Resolution of QG. Let G be a finite rank completely decomposable group with endomorphism ring E. Suppose that G is n–tiered. We will construct a flat resolution of QG of the form

$$0 \to F_n \to F_{n-1} \to \cdots \to F_1 \to F_0 \to QG \to 0,$$

where each F_i is isomorphic to a direct sum of QE–modules $QG(\sigma)$, σ in the critical typeset of G. By Lemma 4, each such $QG(\sigma)$ (and therefore each F_i) is a flat QE–module. We first construct a QE–module complex based on the tier complex for the critical typeset of G given in Definition 8,

$$0 \to V_n \xrightarrow{\psi_n} \cdots \to V_1 \xrightarrow{\psi_1} V_0 \xrightarrow{\psi_0} Q \to 0.$$

Tensoring with QG (over Q) produces a complex of left QE–modules,

$$0 \to QG \otimes V_n \xrightarrow{1 \otimes \psi_n} \cdots \to QG \otimes V_1 \xrightarrow{1 \otimes \psi_1} QG \otimes V_0 \xrightarrow{1 \otimes \psi_0} QG \otimes Q \to 0.$$

For $0 \leq i \leq n$, we define a QE–submodule M_i of $QG \otimes V_i$ by $M_i = \oplus(QG(\tau) \otimes C_\tau)$, where the direct sum is over all C_τ such that C is an (i–1)–cycle and $\tau \notin$ typeset(C) is a minimal upper bound for C. We regard M_i as a submodule of $QG \otimes V_i = QG \otimes (\oplus C_\tau)$ in the obvious way. Note that M_i is a flat QE–module since each summand $QG(\tau) \otimes C_\tau \simeq QG(\tau)$ is flat.

Denote $\varphi_i = 1 \otimes \psi_i$. We show $\varphi_i(M_i) \subseteq M_{i-1}$ for $0 \leq i \leq n$. If $QG(\tau) \otimes C_\tau$ is one of the summands of M_i, then $\varphi_i(QG(\tau) \otimes C_\tau) = QG(\tau) \otimes C$ where C is a one–dimensional subspace of ker $\psi_{i-1} \subseteq V_{i-1}$. Moreover, if $D_\sigma \in$ spt(C), then $\tau > \sigma$ since τ is a (proper) minimal upper bound for C. Thus, $QG(\tau) \otimes C \subseteq \oplus(QG(\sigma) \otimes D_\sigma) \subseteq M_{i-1}$, where the sum is over all $D_\sigma \in$ spt(C). It now follows that the sequence

$$0 \to M_n \to \cdots \to M_1 \to M_0 \to QG \to 0$$

is a complex since

$$0 \to V_n \to \cdots \to V_1 \to V_0 \to Q \to 0$$

is a complex.

We use the complex of M_i's to construct a flat resolution of QG:

$$0 \to F_n \to F_{n-1} \to \cdots \to F_1 \to F_0 \to QG \to 0,$$

First, set $F_0 = M_0$. Then $F_0 \to QG$ is the homomorphism induced by $QG(\tau) \otimes Q_\tau \to QG(\tau) \otimes Q \to QG(\tau) \subseteq QG$.

The map $F_0 \to QG$ is an epimorphism since $QG = \sum QG(\tau)$, where the sum is over all minimal elements τ of the critical typeset. To define F_i for $i \geq 1$, we consider the exact sequences

$$0 \to \ker \psi_i \to V_i \to \text{im } \psi_i \to 0$$

$$0 \to \ker \varphi_i \to M_i \to \text{im } \varphi_i \to 0$$

Since we are working with complexes, $\text{im } \psi_{i+1} \subset \ker \psi_i$. More specifically, $\text{im } \psi_{i+1}$ is the subspace generated by the i–cycles in $\ker \psi_i$ and therefore is contained in the subspace W_i generated by all one dimensional subspaces U of $\ker \psi_i$ such that $\text{spt}(U)$ is minimal and typeset(U) has an upper bound in the critical typeset of G. Each such subspace U is either an i–cycle or typeset(U) contains a maximum element (which is the minimal upper bound). In the latter case we call U an **inessential cycle**. Suppose $\text{im } \psi_{i+1} \neq W_i$. Since W_i is generated by $\text{im } \psi_{i+1}$ and the inessential cycles, there is an inessential cycle D_1 contained in W_i with $D_1 \cap \text{im } \psi_{i+1} = 0$. Since D_1 is inessential, there exists $C_{\rho(1)} \in \text{spt}(D_1)$ with $\rho(1) \geq \tau$ for all $C_\tau \in \text{spt}(D_1)$. If $\text{im } \psi_{i+1} \oplus D_1 \neq W_i$, then there is an inessential cycle D_2 disjoint from $\text{im } \psi_{i+1} \oplus D_1$. Moreover, we can choose D_2 so that $C_{\rho(1)} \notin \text{spt}(D_2)$. To see this, note that if $C_{\rho(1)}$ is in $\text{spt}(D_2)$, then an upper bound for D_2 is also a bound for D_1, whence for $D_1 + D_2$. In this case we can choose a subspace of $D_1 + D_2$ of minimal support not containing $C_{\rho(1)}$. This subspace will be an inessential cycle with which we can replace D_2. Assume inductively that $D_1, ..., D_{h-1}$ have been constructed so that the sum $\text{im } \psi_{i+1} + D_1 + \cdots + D_{h-1}$ is direct but unequal to W_i. Then there is an inessential cycle D_h which forms a direct sum, $\text{im } \psi_{i+1} \oplus D_1 \oplus D_2 \oplus \cdots \oplus D_h$. Moreover, if necessary we can successively replace D_h by a subspace of $D_h + D_j$, $j = 1,2,...,h-1$, to guarantee that if $C_{\rho(j)} \in \text{spt}(D_j)$ with $\rho(j)$ a minimal upper bound for D_j, then $C_{\rho(j)} \notin \text{spt}(D_h)$ for $j < h$. Since W_i is finite dimensional, we will obtain, for some value of h,

$$W_i = \text{im } \psi_{i+1} \oplus D_1 \oplus D_2 \oplus \cdots \oplus D_h \subseteq \ker \psi_i.$$

We can regard $QG(\rho(j)) \otimes D_j$ as a submodule of $M_i = \oplus(QG(\tau) \otimes C_\tau) \subseteq QG \otimes V_i$, since $\rho(j)$ is an upper bound for D_j. Moreover, the sum $N_i = \sum QG(\rho(j)) \otimes D_j$ is direct since $C_{\rho(j)} \notin \text{spt}(D_k)$ for $k > j$. Additionally, since $C_{\rho(j)} \in \text{spt}(D_j)$, $M_i = F_i \oplus N_i$, where $F_i = \oplus\{QG(\tau) \otimes C_\tau \mid \tau \neq \rho(j), 1 \leq j \leq h\}$, a flat QE–module. We now have the modules F_i for our flat resolution.

Our next task is to show $\ker \varphi_i = \text{im } \varphi_{i+1} \oplus N_i$. Since the M_i's form a complex, im $\varphi_{i+1} \subseteq \ker \varphi_i$; and $N_i \subseteq \ker \varphi_i$ because each $D_j \subseteq \ker \psi_i$ and $\varphi_i = 1 \otimes \psi_i$. Let $x \in \ker \varphi_i \subseteq M_i$. There is an idempotent projection $e:G \to A$ of G onto a rank one summand A such that $0 \neq ex \in \ker \varphi_i$ (φ_i is a QE–map); and x is the sum of a finite number of such terms ex. Moreover, $C_\rho \in spt(ex)$ implies $\rho \leq type(A)$. Since A is a subgroup of Q, $A \otimes C_\rho$ is canonically isomorphic to C_ρ. There is then an induced isomorphism $\oplus(A \otimes C_\rho) \to \oplus C_\rho \subseteq V_i$, where the sum is over all $C_\rho \in spt(ex)$. It is easy to see that under this isomorphism, ex is mapped to element $\bar{x}$ in $\ker \psi_i \subseteq V_i$, since $\varphi_i = 1 \otimes \psi_i$. We can invoke Proposition 7(c) to say that $\bar{x}$ is a sum of elements of minimal support, with each such support contained in $spt(\bar{x})$, whence bounded above by $type(A)$. The one dimensional subspace generated by each of these elements of minimal support is therefore either an i–cycle or an inessential cycle. Consequently, $\bar{x}$ belongs to $W_i = im \, \psi_{i+1} \oplus D_1 \oplus \cdots \oplus D_h$. Returning to $\ker \varphi_i$ via the isomorphisms of $A \otimes C_\rho$ with C_ρ, it follows that $ex \in im \, \varphi_{i+1} + N_i$. We have shown that $\ker \varphi_i = im \, \varphi_{i+1} + N_i$. To show the sum is direct, we can take $x \in im \, \varphi_{i+1} \cap N_i$ and observe that the image $\bar{x}$ of ex in V_i belongs to $im \, \psi_{i+1} \cap (D_1 \oplus \cdots \oplus D_h) = 0$. We have established, for $1 \leq i \leq n$,

(*)
$$M_i = F_i \oplus N_i, \text{ and}$$
$$\ker \varphi_i = im \, \varphi_{i+1} \oplus N_i.$$

Recall that $M_0 = F_0$, so that if we take $N_0 = 0$, the above equations hold for $0 \leq i \leq n$. Also note that since G is n–tiered, $V_{n+1} = 0$ and $\ker \varphi_n = N_n$. The equations (*) therefore give a flat resolution,

$$0 \to F_n \to \cdots \to F_1 \to F_0 \to QG \to 0,$$

with maps $F_i \to F_{i-1}$ provided by (the restriction of) φ_i, followed by projection onto F_{i-1}.

We can now prove our main result.

Theorem 11. Let G be a finite rank completely decomposable group. Then G is n–tiered if and only if $fd_{QE(G)}QG = n$.

Proof. Suppose G is n–tiered. It follows immediately from the construction of our flat resolution for QG that $fd_{QE(G)}QG \leq n$. To prove the reverse inequality we show that, under the assumption G is n–tiered, the module $\ker \varphi_{n-2}$ of our resolution is not flat [R, Theorem 9.13]. Consider the exact sequence

$$0 \to \ker \varphi_{n-1} \to F_{n-1} \to \ker \varphi_{n-2} \to 0.$$

To prove that $\ker \varphi_{n-2}$ is not flat it is enough to find $\lambda \in QE(G)$ and $x \in F_{n-1}$ with $\lambda \ker \varphi_{n-1} = 0$ but with $0 \neq \lambda x \in \ker \varphi_{n-1}$. This is because if J is the principal right ideal of $QE(G)$ generated by λ we have $J \ker \varphi_{n-1} = 0$ but $0 \neq \lambda x \in \ker \varphi_{n-1} \cap JF_{n-1}$

(Lemma 1(b)). Since G is n–tiered, there exists at least one $(n-1)$–cycle D and a critical type $\tau \in$ typeset(D) with τ a least upper bound for D. Suppose some $\rho \in$ typeset(D) is an upper bound for the typeset of a different $(n-1)$–cycle D'. Note that $\rho < \tau$. Let τ', with $\tau' \le \rho < \tau$, be a minimal upper bound for D' and replace τ by τ' and D by D'. Since $\tau' < \tau$, after at most finitely many such replacements we arrive at a type τ and an $(n-1)$–cycle D with τ a minimal upper bound for D and with no $\rho \in$ typeset(D) an upper bound for any $(n-1)$–cycle. Let $\lambda \in E(G)$ satisfy $\lambda(H(\sigma)) = 0$ for each σ in the critical typeset of G with σ an upper bound for some $(n-1)$–cycle. Then $\lambda F_n = 0$ because $\lambda M_n = 0$ by definition of M_n. Since $\ker \varphi_{n-1} \simeq F_n$ as QE–modules, $\lambda \ker \varphi_{n-1} = 0$ as well. However, for $C_\rho \in$ spt(D), ρ is not an upper bound for any $(n-1)$–cycle. Thus, for all such ρ, λ can be chosen to satisfy the additional requirement $A \subseteq \lambda G(\rho)$, where A is a rank one summand of G of type τ. Then $0 \ne A \otimes D \subseteq \oplus\{A \otimes C_\rho \mid \rho \in$ spt(D)$\} \subseteq \lambda F_n \cap \ker \varphi_{n-1}$, while $\lambda \ker \varphi_{n-1} = 0$. We have shown that if G is n–tiered, then fd$(G) = n$.

For the converse, recall that any finite rank completely decomposable group G will be m–tiered for some positive integer m. If fd$(QG) = n$, then $m = n$ by what has already been shown. This completes the proof.

Remark. The theorem can be generalized slightly via Lemma 3 and Proposition 4: If H is a fully invariant summand of a completely decomposable group G with endomorphism ring $E = \text{End}(G)$, then fd$(_{QE}QH) = n$ if and only if H is n–tiered.

Corollary 12. If G is a finite rank n–tiered completely decomposable group, then fd$(_E G) \ge n$.

Proof. This follows from the theorem via the easy fact that fd$(_E G) \ge$ fd$(_{QE}QG)$. Indeed, any flat resolution of H can be tensored with Q to obtain a QE–flat resolution of QG.

The result in the Corollary cannot be sharpened in general. This is illustrated by the following (somewhat complicated) example.

Example 13. Let G be the direct sum of 14 subrings of Q, with 2 not a unit in any of them, whose types form a partially ordered set given by the following Hasse diagram.

μ

τ

σ

ρ

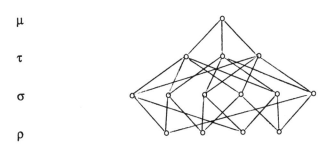

The types in each row are labelled by a Greek letter and numbered from left to right. For example, the types in row three are τ_1, τ_2, τ_3. We will show that G is 2–tiered while $fd(_EG) = 3$.

$$(*) \qquad 0 \to F_3 \to F_2 \to F_1 \to F_0 \to G \to 0,$$

using the methods of Construction 10. New notation, which we hope the reader will find self–explanatory, has been used for this example. Specifically,

$F_0 = G(\rho_1) \oplus G(\rho_2) \oplus G(\rho_3) \oplus G(\rho_4)$, with $\varphi_0 : F_0 \to G \to 0$ induced by inclusion;

$F_1 = G(\sigma_1) \cdot (13) \oplus G(\sigma_2) \cdot (12) \oplus G(\sigma_3) \cdot (23) \oplus G(\sigma_4) \cdot (24) \oplus G(\sigma_5) \cdot (34) \oplus G(\sigma_6) \cdot (14)$, with
$\qquad \varphi_1 : F_1 \to F_0$ induced by sending the element $g \cdot (ij)$ to $(g, -g)$ in $G(\rho_i) \oplus G(\rho_j)$;

$F_2 = G(\tau_1) \cdot (2134) \oplus G(\tau_2) \cdot (2143) \oplus G(\tau_3) \cdot (1423) \oplus G(\mu) \cdot (123)$, with $\varphi_2 : F_2 \to F_1$ induced
$\qquad$ by sending, for example, $g \cdot (2134)$ to $-g \cdot (12) + g \cdot (13) + g \cdot (34) - g \cdot (24)$ in F_1,
$\qquad$ and $g \cdot (123)$ to $g \cdot (12) + g \cdot (23) - g \cdot (13)$;

$F_3 = G(\mu)$, with $\varphi_3 : F_3 \to F_2$ induced by sending g to $g \cdot (2134) + g \cdot (2143) - g \cdot (1423) +$
$\qquad 2g \cdot (123)$.

By Proposition 4, each F_i is a flat E–module. The calculations needed to show the sequence $(*)$ is exact are omitted. The existence of the sequence $(*)$ shows that $fd(_EG) \leq 3$. To show equality, it suffices to show that the kernel, $\ker \varphi_1$, of $F_1 \to F_0$ is not flat [R, Theorem 9.13]. However, in this example, $E = End(G)$ is a left Noetherian ring and the F_i's are finitely generated (direct sums of cyclics). Thus, $\ker \varphi_1$ is flat if and only if it is projective [R, Corollary 3.58]. In this case, $0 \to F_3 \to F_2 \to \ker \varphi_1 \to 0$ is split exact. We complete the proof by showing that the map $\varphi_3 : F_3 \to F_2$ does not split. Suppose $\theta : F_2 \to F_3$ is an E–map which satisfies $\theta \varphi_3 = 1$ on F_3. Let $\beta \in E$ denote the map which embeds the summand of G of type τ_1 into the summand of G of type μ (as rings) and sends all other summands of G to 0. Then $\theta(G(\mu) \cdot (2134)) = \theta(\beta G(\tau_1) \cdot (2134)) = \beta \theta(G(\tau_1) \cdot (2134)) \subseteq \beta F_3 = \beta G(\mu) = 0$. Similarly, $\theta(G(\mu) \cdot (2143)) = \theta(G(\mu) \cdot (1423)) = 0$. Thus, for $g \in F_3 = G(\mu)$, $g = \theta \varphi_3(g) = \theta(2g \cdot (123) + g \cdot (2134) + g \cdot (2143) - g \cdot (1423)) = \theta(2g \cdot (123)) = 2\theta(g \cdot (123)) \in 2F_3$; that is, $F_3 \subseteq 2F_3$. But $2F_3 \neq F_3$ since we have assumed 2 is not a unit in any of our rings. This contradiction shows that no splitting map θ exists. Note that if the exact sequence $(*)$ were tensored with Q and regarded as a sequence of QE–modules, then the map $QF_3 \to QF_2$ would split. As a consequence, $fd(_{QE}QG) \leq 2$. It is not difficult to check that, in fact, $fd(_{QE}QG) = 2$, so that G is 2–tiered by Theorem 11. The 2–tiered assertion can also be verified directly from Definition 8.

References

[A]   H.W.K. Angad–Gaur, The homological dimension of a torsion–free abelian group of finite rank as a module over its ring of endomorphisms, Rend. Sem. Mat. Univ. Padova 57 (1977), 299–309.

[L]   E.L. Lady, Almost completely decomposable torsion free abelian groups, Proc. Amer. Math. Soc. 45 (1974), 41–47.

[RW]   F. Richman and E.A. Walker, Cyclic Ext, Rocky Mt. J. of Math. 11 (1981), 611–615.

[R]   J. Rotman, An Introduction to Homological Algebra, Academic Press, New York, 1979.

Index

\